T0210182

NATURE'S FLYERS

David E. Alexander

Nature's Flyers

Birds, Insects, and the Biomechanics of Flight

Foreword by Steven Vogel

The Johns Hopkins University Press
BALTIMORE AND LONDON

The Johns Hopkins University Press
2715 North Charles Street
Baltimore, Maryland 21218-4363
www.press.jhu.edu

Library of Congress Cataloging-in-Publication Data

Alexander, David (David E.)
    Nature's flyers: birds, insects, and the biomechanics of
    flight / David E. Alexander ; foreword by Steven Vogel.
       p.   cm.
    Includes bibliographical references and index.
    ISBN 0-8018-6756-8
1. Animal flight. 2. Flight. I. Title.
QP310.F5 A54 2002
573.7'98—dc21   2001000821

A catalog record for this book is available from
the British Library.

To Helen,
for her quiet confidence in my ability

and

To the Department of Entomology
at the University of Kansas,
for giving me such a wonderful
academic home for the past decade

CONTENTS

FOREWORD

*W*ho would have thought it possible? Flying—using thin air to sustain craft that weigh hundreds of tons miles above the ground, immune to gravity's siren attraction. What a preposterous notion, something akin to extrasensory perception or, closer to home, levitation. We are terrestrial creatures, tied to a two-dimensional world that is defined by the boundary between earth and sky. We can swim, if badly by normal animal standards, but we are incapable of flight, at least without extraordinary aerobic capability, ideal atmospheric conditions, and a huge and sophisticated prosthesis.

But other animals fly routinely, and they do so with no obvious heroics. At least four lineages have evolved powered flight, flight that includes the ability to take off and maintain altitude without the assistance of atmospheric motion. Three remain with us—birds, bats, and insects; one has disappeared—the great pterosaurian reptiles of the Mesozoic Era. A far wider range of organisms go in for gliding or other forms of controlled descent. Not just flying squirrels but their marsupial analogs do it, as well as a menagerie of lizards, snakes, big-footed frogs, flying fish, and squid. And a host of autogyrating seeds. We are looking at something that has appeared as an independent evolutionary innovation in dozens of lineages. Nature thus sends us an unmistakable message that flight must be within the realm of the possible.

And so we have dreamed of flight, and we've incorporated stories of flight in every cultural mythology, with angels, dragons, incubi, the winged horse Pegasus, the chariot of Helios, and Santa's reindeer. The primal inventor is Daedalus, who devised functional wings; the earthward plunge of Icarus symbolizes the hubris of human flight. And humans (of uncertain sanity) have attempted flight more often than seems credible, using bizarre equipment and with occasionally fatal results. Leonardo da Vinci made drawings of aircraft; perhaps his failure to publish spared a few lives. Yes, we now know how to fly, but we achieved practical flight a mere hundred years ago, when motors and aerodynamic savvy became, together, equal to the task. Few of us have not flown in airplanes, yet even fewer could explain the basis of flight.

Computers aside, what routine technology remains so mysterious to most contemporary humans?

All flying animals fly in much the same way. One or two pairs of wings move alternately upward and downward, or so it appears to the casual glance. Somehow this motion enables them to defy gravity. Do their wings work like paddles, pushing down with a hard stroke and then feathering (the word carries double meaning) for a gentle upstroke? They do not. Is the up-and-down motion critical for staying aloft? It is not. Watch a hovering hummingbird, noting how its wings go back and forth, not up and down, at least if gravity's downward force and the earth's surface define the frame of reference.

Lift must equal weight or we have a bone to pick with Sir Isaac Newton. And we must assume that the duck weighs as much when flying as it does on the butcher's poultry display. So how do wings—we must focus on wings, the inevitable feature of flyers—do the trick? Their game is at once subtle and wonderful, one whose physical basis was understood only after our flying machines first left the ground. So subtle is the way a wing generates lift that we have contrived a polite fiction to silence the doubts of students and of the general public who entrust their lives to the technology of flight. Still, the real story explains more than just lift; it rationalizes such things as why large birds often fly in V- or diagonal I-formations.

Nor does making sufficient lift to offset weight present the only problem without a self-evident solution. To get anywhere takes another force, thrust, which has to balance a craft's drag exactly if it is to go steadily onward. Getting thrust from wings? Maybe that's what the up-and-down movement somehow achieves. And must bird or airplane suffer drag? Nineteenth-century fluid mechanics explained the origin of drag no better than it did the origin of lift. Some shapes gave less drag than others, shapes commonly found among nature's swimmers and flyers, but explanation of the advantage of the rounded front and tapered, pointy rear again postdates practical flight by human technology.

And what about speed? Standing costs less energy than walking or running, so why does hovering have to be harder than flying forward? Toy helicopters can be easily contrived, and we have played with them for a century and a half. But full-size helicopters came along a generation after airplanes, and not for lack of attempts. For that matter they came years after fully successful flights by some unfamiliar rotary-winged nonhovering craft, the auto-gyros. At least some birds hover, but the difficulty evidently increases as they get larger: the larger the bird, the shorter the time over which it can sustain hovering. They are telling us something, but they give only the text and leave to us the deciphering of the subtext, the mechanisms beneath the phenomenon.

Cost of flight raises yet other peculiar questions. A rope holds something aloft indefinitely at zero cost—after all, it consumes no fuel. But free lift? We've yet to invent a gravity-insulator to put under our aircraft. Why does it cost so much to stay aloft? And why do craft with long and skinny wings stay up at lower cost than those with short, wide wings? Why must helicopters have long rotors rather than ordinary propellers tilted upward?

Although being small may make slow flight and hovering easier, it does something else that is both constraint and opportunity. We adjust airline schedules for prevailing winds, and we are buffeted by smaller-scale disturbances. But what are just minor atmospheric matters for the large and fast can dominate the flight of the small and slow. We build gliders that stay aloft for hours, taking advantage of atmospheric irregularities—if the air around you rises, then you can sink relative to that air without sinking relative to the ground. But birds exploit gliding in many more ways and for many more purposes. We recognize thermal soaring in rising vortices; declivity soaring next to hillsides; sea-anchor soaring by petrels with dangling, immersed legs; dynamic soaring by looping up and down in winds whose speeds vary with altitude. Large insects—butterflies and locusts—go in for gliding as well, but for them the speeds attainable are too low for the scheme to take them far. Still slower are the botanical gliders, most of whom just glide to slow their descents. Not that they would be damaged on impact—the slower the descent, the farther the wind can carry one from the release point. So progeny get launched.

Have we copied birds or bats or insects? Therein lies a story of sufficiently mixed evidence that (as with so much of history) one can support every view from an emphatic affirmative to the most damning negative. The idea of streamlining came from observations and measurements on animals—but dolphins and fish proved more enlightening than did birds. The idea that wings should be convex above and concave below came from birds, at least by 1890, perhaps much earlier—but it took years more before recognition dawned that the convexity at the top was what mattered, not the concavity at the bottom. One might as well put a plate across the bottom and store fuel within the wing. And we (the Wright brothers, specifically) learned how to use wing twisting rather than rudder movement to turn left or right—present-day ailerons preserve their avian-derived insight.

Birds, though, mislead as much as they led, perhaps more. We still do not build flapping-wing craft. And recognition of the advantage of separating propeller—rotating twisted airfoils—and wings proved prerequisite to successful aircraft. So, not one but two huge differences separate the two schemes. Flapping entails great mechanical complexity, not to mention a bunch of aerodynamic pitfalls. Animals manage, of course, and we are only now coming to understand some of their special and perhaps essential tricks. Still, one cannot escape the suspicion that if nature had done better at making wheels and axles, she'd have been no flapper and put up with nothing so awkward. Using wings for both lift and thrust? That's not too bad if you are a slow flyer, with lots of wing area relative to your weight; in short, if you are a bird or bat or insect rather than a craft that carries people. We now have engines big enough to pull it off, so to speak, and we make helicopters and Osprey tilt-wing craft—at the price of both complexity and inefficiency.

Worse, though, than diverting our attention to flapping flight, flying animals gave us no inkling of the problems of stability for a craft suspended in a turbulent atmosphere. Yes, a submarine faces precisely the same situation, but its neutral buoyancy renders the problem both less acute and more easily managed. Unfortunately for us, animals trade off stability for maneuverability—just watch a bird fly between the trunks or branches of trees or a hover fly stay poised in a sunbeam even as a breeze comes along. As aircraft-designer-turned-evolutionary-biologist John Maynard Smith pointed out, nature's contemporary flyers have been around long enough to have evolved nervous systems capable of dealing with substantial instability, and the gain in maneuverability holds great practical advantage.

Mostly, one must admit, we have looked to human technology for models in order to understand nature's technology rather than the other way around. I once described biomechanics as mainly the study of how nature does what engineers have shown to be possible. But not always, and not necessarily, and we are now in an era of major initiatives aimed at deriving technological insights, even specific devices, from studies of nature. Successes have sufficed to attract and excite investigators, even if they are as yet insufficient to find conspicuous niches in the marketplace.

Does nature show what is possible? Yes. But one hastens to declare the conditionality of the converse. What nature shows *must* be possible; but nature doesn't show all that *is* possible, some considerable contemporary mythology notwithstanding. That is a lesson we can draw from the first successful human-built flying machines—the French hot air and hydrogen balloons of the late eighteenth century. Nature's armamentarium of devices includes nothing remotely resembling either system for lighter-than-air flight. And, again, humans invented propellers, not nature (aside from a distant analog recently shown in bacteria). But we still must ask, if birds, bats, and bugs did not fly, would we have thought to try?

Natural flight fascinates quite beyond any prospect of imitation, in ways that even transcend its consummate grace and beauty. Flight taxes the animal engine. We learned much about metabolic processes from studies using the flight muscles of birds and insects. Insect flight muscle continues to provide some of our best material for studying the micromechanics of muscle in general. When investigating how organisms work, we have often derived key insights from focusing on nature's extreme cases, and flight provides a fine spectrum of such extremes—in engine, fuel-delivery system, strong and light materials, efficient construction, sophisticated control.

At the same time flight yields rich rewards for the animal that can manage to do it. The world becomes three-dimensional, not just a fractally bumpy surface. Travel becomes speedy, at least relative to body size. Birds quite commonly fly ten times as fast as racehorses run or the fastest marine animals swim, if we express speed as body lengths per second. A fruit fly, no particular speed demon, can go at not just ten but forty times the racehorse's speed. The flyer can seek food, nesting sites, and mates over a wide area. Mice in separate fields rarely trade genes with each other, but birds mix it up over long distances, and we use radar stations to track high-altitude movements of agriculturally threatening insects.

Yes, flight taxes the animal engine, extracting a high price relative to time aloft. But, again, flight is fast, even without adjusting for body size. At around forty miles per hour, a run-of-the-millpond duck can fly all day as fast as the winner of the Kentucky Derby can run for two minutes. And the duck does so with little if any underlying physiological superiority. So the cost relative to distance rather than time isn't at all bad. For any kind of travel, cost relative to weight and distance goes down as animal size increases. Migration by flight over globally significant distances proves practical even for small creatures. For a flyer—bird, bat, or insect—migration can pay off for creatures weighing a mere ounce. For walkers and runners, the minimum size that gets the creature out of the red ink rises to several pounds, and that only with favorable terrain.

Not that we don't follow similar rules. Flying is still flying, with the same problems, even if sometimes yielding to different solutions in human and natural technologies. Larger planes can go longer distances on less fuel relative to weight. And while faster may be more costly relative to time, it may afford economies relative to distance traveled or overall load moved. A large jet airplane holds fewer people or packages than a train, but for long distances it makes many more trips in the same time.

You can read excellent histories of human flight and learn nothing about how it actually happens. And (if you are like most of us) you will emerge only baffled further from an encounter with a textbook of aerodynamics. The best perspective on flight requires a look at every kind of active flyer, whether evolved or invented. We are fortunate that such great differences exist between what nature does and what we do—each can provide proper perspective on the other, and together they give a more richly textured view than either alone. But accessible accounts of flight in nature have been all too rare; even the technical accounts don't occupy much shelf space. Otto Lilienthal's prescient try—*Bird Flight as the Basis of Aviation*—dates from the nineteenth century, not the twentieth. The perspective that can be gained only from nature, though, is what you will find in the pages that follow.

Steven Vogel
Department of Biology
Duke University

I may be biased, but I think that animal flight is inherently fascinating. The sheer variety of flying animals is staggering, ranging from mosquitoes to eagles, hummingbirds to flying foxes. And yet they all rely on the same physical processes, the same basic mechanism of locomotion. Moreover, flying animals have had a major impact on nonflying organisms. Consider the evolutionary and ecological relationships between bees and flowers, or mosquitoes, malaria parasites, and humans; the world would literally be a very different place without flying animals. Finally, many of us simply envy the ability of so many animals to fly.

My goal with this book is to make available to a wide audience an account of our current scientific understanding of animal flight. I hope that people across the spectrum, from amateur birdwatchers and bug collectors to professional ornithologists and entomologists, will find at least some parts interesting. I have aimed this book primarily at nonscientists, but specialists may also find valuable information in it. Although some of this book is based on the science of aerodynamics, which in turn is based on some rather sophisticated mathematics, advanced mathematical skills are not needed to read this book. In most cases, I have been able to describe the important aerodynamic concepts qualitatively or graphically; the few equations I have presented use simple algebra.

New scientific knowledge usually grows out of careful observations and clever experiments. In the scientific enterprise, some researchers develop theories to explain how complex processes work, while others do experiments to see if those theories accurately describe real processes; some of the more brilliant scientists manage to do both. Although my focus is on the knowledge that we have gained from these endeavors, I have tried to include descriptions of a fair number of the studies themselves, to give the reader a better appreciation for the process of science. In some cases, these studies also show how our scientific knowledge undergoes change, which is sometimes dramatic.

As I wrote this book, something of a renaissance began in animal flight research, especially in the biomechanics of flight. Over the past couple of years, what had been a steady trickle of published studies in this area has become a veritable flood. Although this is an exciting time to be a researcher studying animal flight, I have found it a bit frustrating to review the current "state of the art" because that state changes so rapidly. I have made an effort to incorporate the latest findings where possible, but at some point I had to stop so that the book could be published. For instance, within a couple weeks after sending what I hoped was a final draft to reviewers, Charles Ellington published a paper on applications of insect flight research to designing tiny flying robots; Jeremy Rayner published a paper on flight muscle power and efficiency in birds; Ulla Norberg (and her colleagues) published a paper on soaring in flying foxes; and Robert Dudley's book on the biomechanics of insect flight came out. Thus, my account represents the state of the field at a point near the end of the twentieth century, but some details will undoubtedly become obsolete as new research is published in the next few years.

I had help from many people in completing this book. Several people read and commented on one or more chapters as I wrote them. I would like to thank Hugh Dingle, Joel Kingsolver, Sharon Swartz, and Vance Tucker for each reading a chapter, and Roy Beckemeyer, John McMasters, and Steven Vogel for reading two or more chapters. All made suggestions that significantly improved the book. Any remaining errors or inaccuracies are my own responsibility. The students in my Animal Flight seminar at the University of Kansas were the "beta testers" for several chapters, and one even contributed some illustrations for Chapter 5. The illustrations were a team effort, including drawings by Betsy Hart, Barbara Hayford, Olga Helmy, Jennifer Pramuk, Analia Pugener, and Sara Taliaferro; Sharon Hagen made prints of illustrations reprinted from other published sources. Jennifer, as my first illustrator, put up with a lot of changes and still managed to do about half the drawings for Chapters 2, 3, and 4. Sara came to the project late, drew approximately one-third of all the figures, modified several others, and generally helped to organize the artwork. To properly credit their work, I have given the initials of the illustrator who drew each original drawing in parentheses at the end of that figure's legend (with "B.Ht." for Betsy Hart and "B.Hd." for Barbara Hayford). I appreciate the efforts of all these talented people.

Steve Vogel was a great help at all stages, both directly in answering my many questions and indirectly by example. He was the first to suggest that I try writing for a nonscientific audience, and much later he cheerfully agreed to read the completed final manuscript. I also owe a great debt of gratitude to Natalie Dykstra, my editor. Her patient advice and continuing enthusiasm made it a pleasure for me to press on. I also thank Robert Harington, and later Ginger Berman, science editors at the Johns Hopkins University Press, for giving me a chance to try my hand at this kind of writing. The encouragement and support of my family, especially Helen Miller Alexander, were key to my decision to begin this project and motivated me to carry it to completion. Other friends and colleagues contributed in many small ways, probably without even realizing it, and I thank them collectively.

Note on citations: Biologists and physical scientists most often cite their sources using the "name, date" system, where the author's name and the year of publication are cited in the text, and an alphabetical list of references appears at the end of the book or article. This format tends to break up the flow of descriptions and may seem cumbersome and confusing to the nonscientist. Nevertheless, I feel strongly that a book of this type needs some mechanism to refer to original sources. I have therefore used numbers in square brackets in each chapter to refer to a list of references (titled Notes) that appears in an abbreviated format at the end of the book. Each source within a chapter is assigned a unique number, which therefore may be called out more than once in a chapter if the cite is repeated. Full bibliographic information is then given in a combined alphabetical listing of all references.

NATURE'S FLYERS

ONE

Introduction

*T*he study of flight begins with watching: a hawk or eagle soaring on motionless wings, a raucous, wheeling mass of seagulls, a line of gliding pelicans skimming effortlessly over ocean waves. Long before anyone tried to figure out how such creatures apparently defied—or outsmarted—gravity, or how airborne species fit into the larger web of life, people marveled at and envied flying animals. Poets tapped into these feelings with phrases like "flight of imagination," "soaring spirit," or "free as a bird." Even though the reality of animal flight is both more prosaic and more complex than these phrases suggest, it is no less fascinating.

Certainly, flight is a form of locomotion for animals, an efficient way to get from point A to point B. As John McMasters, an engineer interested in animal flight, puts it, humans fly commercially or recreationally, but animals fly professionally [1]. In other words, flying animals fly in order to make a living. Yet a scientist sees in animal flight a remarkable activity involving processes at many levels, from the physics of air, to anatomical and behavioral specializations, to new ways to interact with other organisms and the environment. Even from the animal's perspective, many remarkable processes and events can happen on the way from point A to point B. Bats catch and eat their prey in the air, male dragonflies patrol their territories from the air, many birds and insects have courtship rituals involving flight, and dragonflies and some birds even mate in the air. Moreover, different flying animals can use different types of flight—soaring, hovering, a couple of variations on flapping—to make the trip. As for our "flight as freedom" metaphor, a biologist would say that flight does, indeed, open up many new opportunities, but most animal flight is for a purpose, not simply for fun.

Indeed, the opportunities that flight opened up for an animal were key to the evolution of flight. Evolution, or change over time, can be caused by a variety of processes. The most potent of these is natural selection: Darwin's "survival of the fittest." Natural selection is not a directed process; it does not involve planning. Instead, any random genetic changes that improve an organism's ability to survive and reproduce are favored. Consider the advantages of being able to fly: it is a handy and effective means of escaping from predators, a way to reach resting, egg-laying, or nesting sites inaccessible to nonflyers, a way to reach otherwise inaccessible food, and a particularly efficient way to move long distances (Chapters 6 and 8). All these advantages do not necessarily accrue to animals in the early stages of evolving flight, but some of them surely benefited early flyers, and others must have helped fuel improvements in flight abilities in various species (Chapter 7).

Biologists categorize the uses of animal flight in a number of ways. The two categories to which I pay most attention in this book are local flight and migration. Local flight probably includes most flights made by animals. It includes flights to escape predators or catch prey and flights to find and use resources—food, territories, nest sites, mates. Migration occurs when members of a population move from one place to another to take advantage of some temporary condition, such as warm summers in Canada. Flying migrants include such familiar animals as robins, mallards, and geese, and such obscure ones as aphids and milkweed bugs. Although many songbirds, waterfowl, and butterflies migrate great distances, other flying animals, such as certain mountain-dwelling beetles and birds, only migrate a few dozen kilometers.

Cast of Characters

Many groups of animals (and even some plants) use flight to some degree. Most of these groups are gliders, meaning they cannot use powered flight (Chapter 3). Four groups of animals have evolved powered flight, using flapping to stay aloft at will. These four groups, in the order in which they evolved flight, are insects, pterosaurs, birds, and bats. Though these groups evolved powered flight independently, all use roughly the same flapping pattern (Chapter 4).

The wings of the three vertebrate groups—pterosaurs, birds, and bats—all evolved from a modified forelimb. Pterosaurs, extinct relatives of dinosaurs, had wings supported mostly by a single, enormously elongated fourth finger (Fig. 1.1A). Pterosaurs retained short first, second, and third fingers with claws on the front of their wings, and most of the wing surface was formed by a membrane stretched from the arm/hand skeleton to the side of the body. Since we only know of pterosaurs from fossils, we do not know much about the wing membrane. Until recently, paleontologists thought it was just a stretchy flap of skin, but some now believe it had a complex system of stiffeners that helped it to carry aerodynamic forces and maintain an aerodynamically efficient shape (Chapter 7).

Among living vertebrates, bats are the only mammals that have evolved powered flight. Bat wings are superficially similar to pterosaur wings in that they have greatly elongated fingers supporting a flexible membrane, but their anatomy is actually quite different. The arm skeleton (upper arm and forearm) supports a much larger proportion of the wing in bats than in pterosaurs, and the outer portion of a bat wing is supported by a fan of four elongated fingers (Fig. 1.1B). Only the first finger is separate from the wing membrane, retained as a short, clawed digit. The bat can adjust its four wing fingers independently, so it has a great deal of control over the shape of its wing, particularly the curvature of the wing membrane. This contrasts with the pterosaur, which only had direct control of the front edge of the wing, and thus would have had much less ability to adjust its wing's shape.

Birds are the remaining group of flying vertebrates. They have wings quite different from pterosaur or bat wings, because most of the surface of a bird's wing consists of feathers. The arm/hand skeleton extends barely half the length of the wing, and the fingers are reduced to little more than three short

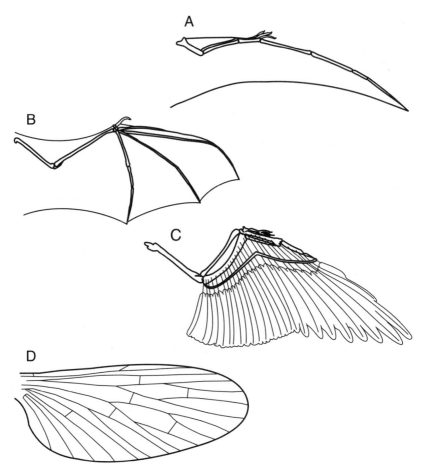

Figure 1.1. Structure of animal wings, showing the main skeletal supports.
A. Pterosaur wing. B. Bat wing. C. Bird wing. D. Insect wing. (S.T.)

spurs of bone (Fig. 1.1C). In pterosaurs and bats, the skeleton carries most of the flight loads out to the tip of the wing, and so is rather slender. In birds, most of the flight loads are carried by the feathers, which are anchored to the skeletal elements, so the bones of bird wings tend to be stouter, as well as shorter relative to the wing length.

Insects were the first animal group to evolve flight, and they are the only invertebrates to have done so. Insect wings are fundamentally different from vertebrate wings in several ways. Vertebrate wings are all modified legs (fore-limbs), while insect wings evolved separately from their legs. Insect legs are attached to the bottom of the thorax, and their wings are attached to the upper side of the thorax. Flying vertebrates have muscles as well as bones with joints out in the wing, so vertebrate flyers can directly control the shape and movements of their wings partly within the wing itself. In contrast, insect wings are built on a completely different plan. The wing is made of an extremely thin, nonliving, transparent membrane supported by a branching arrangement of stiff rods called veins (Fig. 1.1D). The veins contain living tissue and carry blood vessels and air passages. There are no joints or muscles out in the wing. Instead, the veins are anchored to a complex set of tiny skeletal structures, *axillary sclerites,* that make up the wing hinge or articulation. By adjusting the position of the axillary sclerites, the insect can push or pull on different veins, which can, in turn, adjust the shape of the wing. The anatomy of insect wings thus seems as if it should limit how much an insect can change its wing shape. True, insects cannot flex their wings—bend them to shorten their span (length)—as birds do routinely while flapping, and all vertebrates do to fold their wings when not in use. However, by adjusting the articulation, insects can do a surprising amount of wing twisting and fore-aft bending. Moreover, a number of insects have evolved modifications to get around this limitation. Beetles and earwigs use some parts of veins as springs and others as hinges, and they can perform startlingly complex folding and unfolding. This allows them to stow their wings under rigid wing covers much smaller than the wings themselves.

Despite these anatomical differences, all animal wings share certain properties. They must be rigid enough to carry the animal's weight without being unnecessarily heavy. They must have the proper shape for producing lift to keep the animal aloft while minimizing drag. The animal must be able to flap them with the appropriate motions (similar in all flying animals) to produce the thrust that drives it through the air. Finally, in all vertebrates and most insects, the animal must be able to fold or stow its wings in some way so that it can move around easily on foot.

Animal Aerodynamics

Any comprehensive discussion of flight depends on aerodynamics, which is the physics of air movement. This book is not intended to turn the reader into an aerodynamicist, although I hope it would provide the basis for an intelligent conversation with one. Although aerodynamics can appear forbiddingly complex when viewed in rigorous mathematical detail, many of the key concepts are straightforward enough for anyone to understand. In addition, some of these concepts lead to surprising and even comical consequences.

In the past few years, animal aerodynamics has attracted increasing research activity. This increase has been driven by advances in both experimental and theoretical methods. New viewing methods, along with clever mechanical wing flappers, gave researchers a look at airflow patterns on an entirely new and smaller scale [2, 3, 4, 5]. At the same time, ever more complex processes could be analyzed by the advances in both engineering and biological aerodynamics [6, 7, 8]. Airplane wings and animal wings both move through the same air and obey the same general physical principles, but airflow patterns around a flapping bird or fly wing have some significant differences from flow patterns around the wings of a Piper Cub or an airliner. These differences require an extra layer of specialized theory on top of conventional or engineering-type aerodynamics. As a result, biological aerodynamics has become a separate subdiscipline in its own right.

Flight Plan

My goal for this book is to open up the world of animal flight research to anyone interested in learning more about how animals fly. I aim to guide the reader along the easier paths through the often-intimidating thicket of research on animal flight, but I shall also try to maintain a sense of wonder at the many marvelous aspects of flight in the natural world.

Modern research on animal flight encompasses an enormous range of topics. In order to bring some structure to what could easily become a confusing sea of information, I have grouped my topics into those that are mainly physical, or physiological, and those that act on more ecological or behavioral levels. The former extend from aerodynamics to the power requirements of flapping, and the latter range from flight evolution to the effects of flight on nonflying organisms. The final topic, which is outside these two groups, is a look at how our understanding of animal flight has (and has not) contributed to aviation technology.

The topics in this book are often technical, as they must be to describe our current knowledge of the physics and biology of flight. Although some readers may feel that such descriptions detract from the poetic beauty of flight, for me the opposite is true. When an animal's flight appears effortless, it is actually demonstrating its complete mastery of aerial locomotion; as I show in this book, flight is anything but effortless. The more I learn about flight, the more amazement and wonder I feel that animals (and a few plants) have evolved such a complex and specialized ability.

How Wings Work

Animals overcame some major challenges in evolving the ability to fly. Flight is a physical process, and understanding this process requires understanding some basic Newtonian physics. You can only appreciate the constraints on the biological aspects of flight if you understand the underlying physical principles. How is gliding "powered by gravity"? Why do animals need to flap their wings? Why is hovering limited to small animals, and why do large birds spend so much time soaring? The answers to these questions depend on the physical and aerodynamic properties of wings.

In most everyday situations, liquids and gases obey the same principles, and they have similar flow patterns under similar conditions. To a physicist, the term *fluid* includes both gases and liquids, and their flowing behavior does not really diverge until speeds approach the supersonic. In addition, the forces and flow patterns are the same, regardless of whether an object moves through a stationary fluid or whether the object stands still and the fluid flows over it. This property is helpful when trying to understand flight, because it is usually easier to visualize air moving over a stationary wing than trying to imagine what happens as a wing slices through a given blob of air.*

Forces and Coefficients

From a physical point of view, flight is all about forces. Weight is a force that a flyer needs to overcome; lift is a force that opposes weight. Any object moving through air experiences the force of drag, slowing it down. A successful flyer, then, must have some way to produce a force opposed to drag, and this force is called thrust. These two pairs of forces—weight and lift, drag and thrust—must be roughly balanced in order for an animal or machine to fly (Fig. 2.1).

What Is a Force?

Force means something very particular in physics. To a scientist, force is defined by Newton's Second Law: force = mass × acceleration. A force can be any push or pull that could cause a change in the motion of some object: starting it moving from rest, speeding it up or slowing it down, or even changing its direction. If there is no *change* in the object's motion, either there are no forces acting on it or the forces exactly balance each other. In our everyday experience, however, when a person stops pushing on an object, it slows down. It decelerates because we live in a world full of friction, which provides the slowing force. When you push something at a constant speed, your push exactly balances the force due to friction, so there is no *net* force on the object. In the vacuum of outer space where there is no friction, when you stop pushing on an object, the object keeps moving.

A force is a *vector,* which is shorthand for saying that it has both a size (or magnitude) and a direction. Forces in the same direction can be added together, forces in the opposite direction subtract from each other, and forces at intermediate angles can be combined graphically (or with trigonometry)

---

* There are even names for these two viewpoints: watching fluid move over a stationary object is called "Lagrangian," after the French mathematician Joseph-Louis de Lagrange (1736–1813), whereas watching an object move through a stationary fluid is called "Eulerian," after the great Swiss mathematician and physicist Leonhard Euler (1707–1783).

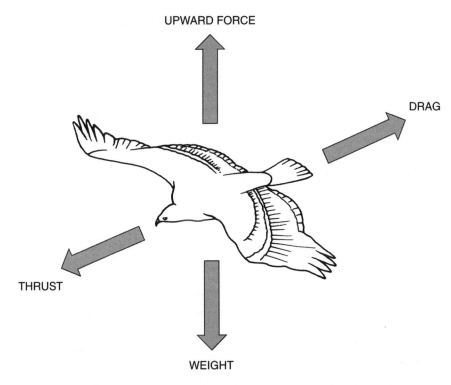

Figure 2.1. Forces associated with flight. The arrows show the four forces that must be balanced in order to fly steadily. Contrary to popular belief, "lift" is not necessarily exactly upward. The upward force is that portion of the lift force that is exactly vertical and balances weight. (J.P.)

to yield a single, equivalent *resultant* force (Fig. 2.2). Weight is a force. A 5-pound bag of flour has a weight with a size (5 lb or 108 N)* and a direction, which is down or vertical. Tug-of-war teams each produce a force, which in this case is horizontal. If the teams are well matched, there may be little or no movement at first, because the forces are of equal magnitude and opposite direction, canceling each other.

---

* The correct Système International (metric) unit of force is the newton, N. Unfortunately, pounds (as used in the United States) and kilograms (as used everywhere else) do not measure the same thing. Pounds measure weight, which is a *force*, but kilograms measure *mass*. Weight is how hard you push down on the ground, whereas mass is a measure of how much there is of you. I may weigh 620 N on Earth and 103 N on the moon, but my mass is 63 kg in both places (and anywhere else in the universe).

Aerodynamic forces rarely act on a single point; they are typically distributed over a surface as a *pressure,* which is defined as force per unit area. Oddly enough, pressure is not a vector—it has a direction in the sense that it must act at right angles to any surface, but it does not have an independent direction of its own. A pressure has only a magnitude. Pressure is usually spread out over a surface, and it may actually vary from place to place on the surface. Even so, it is often useful to treat the pressure as if it were a single force acting at a single point on the surface. If you know the average pressure, then the magnitude of the force is simply the average pressure times the area of

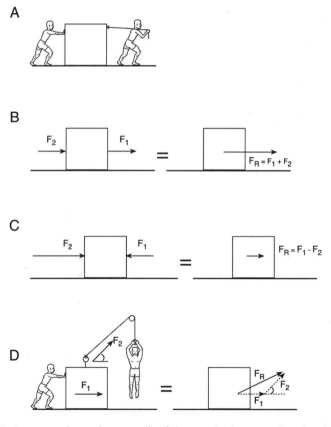

Figure 2.2. Forces can be pushes or pulls. If they are in the same direction (A), they can be added together (B). If they are in opposite directions (C), they subtract, and if they are at angles to each other, they can be combined graphically (D). F_1 and F_2 are single forces, and F_R is the *resultant* force, which represents a combination of forces. (J.P.)

the surface it is acting upon. The direction of this force is a bit trickier to de-termine, but in the simplest case, it is the average direction at right angles to the surface.

The Forces of Flight

The most crucial forces for flight are vertical. Gravity produces weight, which is a force in the downward direction. Under the proper conditions, wings produce a force in the upward direction that counteracts weight (Fig. 2.1). The wings produce a force called *lift,* and the upward force is the verti-cal component of lift. The direction of the lift is always perpendicular to the movement of the wing: if the wing movement is exactly horizontal, the lift will be directly upward and the upward force will be the same as the lift. If the wing's path is tilted, however, then the lift is no longer vertical, and the upward force is just a portion of the lift. Flapping wings hardly ever move ex-actly horizontally, so the upward force is only a fraction of the lift produced by flapping wings (see Chapter 4). If the upward force is greater than weight, the object will rise; if they are equal, the object will fly at a constant altitude; and if upward force is less than weight, the object will descend. Unbalanced upward or downward forces cause upward or downward *accelerations.* In other words, in a steady climb or descent, the forces are actually balanced. An upward force can be produced in various ways. For example, the thrust of the engines of a Space Shuttle produces an upward force during liftoff. The most efficient way to produce an upward force, however, is with an airfoil or wing, and bird wings, helicopter rotors, and airplane wings all use the same general mechanism to produce this force.

Drag

The importance of vertical forces for flight may be obvious, but what is less obvious is the almost equal importance of horizontal forces. Wings only pro-duce lift when moving forward through the air, and this forward movement involves two new forces: thrust and drag. Any object moving through air or water experiences a slowing force called *drag,* and any force in the opposite direction from drag is *thrust* (Fig. 2.1). At low speeds, drag is caused largely by viscosity, the "thickness" or "goopiness" of a fluid, or the tendency of a fluid to resist being deformed. Honey, for instance, has a high viscosity compared to water, which has a lower viscosity. Viscosity is a form of friction: it is caused by the tendency of adjacent regions in a fluid to *resist sliding past each other,* or as an engineer would say, the resistance of fluid to *shear.* When fluid flows over a solid object, there is an infinitesimally thin layer of fluid in contact with the surface of the object that does not move; this defines the *no-slip condition.* Between the freely flowing fluid and the object's surface, there is a zone or layer where lots of shearing occurs, and this shearing is resisted by viscosity. The layer of intense shear, where the speed changes from zero to the free stream speed, is often called the *boundary layer.* The object feels the resistance to shear in this layer as a force on the object tending to pull the object along in the direction of flow. This same process occurs when an object moves through fluid. Imagine friction causing the object to pull some of the fluid along, and—like any other friction—slowing the object. This viscous or friction slowing force is one form of drag. *Drag* has an appealingly intuitive meaning: formally, it is any force opposing movement through a fluid. Thus, drag is always in the opposite direction from the movement.

Another important type of drag is inertial or pressure drag.* Under certain conditions, an object moving through a fluid produces a turbulent wake (Fig. 2.3). The pressure in the wake is lower than the pressure on the front of the object. Thus, the wake actually exerts a small suction on the back of the object. This suction acts as a force directed backward from the object, and this force is called the pressure drag. Both pressure and viscous drag always act on any object that moves through a fluid, but they are usually quite different in magnitude. In practice, we usually measure the total drag: the sum of the pressure drag, viscous drag, and any other drag. Wings, for example, experience a third type of drag, induced drag, which comes from the same process that generates lift.

Imagine that you have been given a device that allows you to measure the drag on any object moving through air. (It turns out to be surprisingly difficult to measure drag on objects with irregular shapes under natural conditions, but assume we have figured out how to get around this.) You would

---

* The reason that this type of drag is called "inertial" is beyond the scope of this book. Suffice it to say that, thanks to viscosity, a tiny blob of fluid starting at the front of a blunt object may not have enough inertia to go all the way around to the back. For complete and lucid descriptions of the sources of drag, see *Life's Devices* (1988) or *Life in Moving Fluids* (1994) by Steven Vogel [1, 12].

A

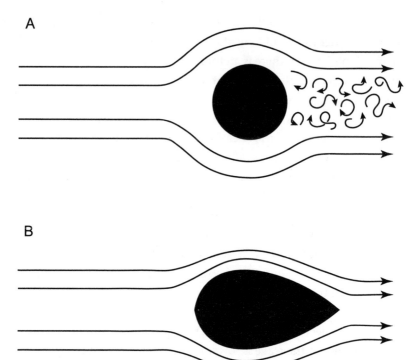

B

Figure 2.3. Flow patterns over blunt and streamlined objects. When a fluid flows over a blunt object such as a sphere, or flat plate perpendicular to the flow, the flow tends to separate from the back of the object, which produces a wide, turbulent, low-pressure wake (A). When fluid flows over a streamlined shape, the flow tends to follow the tapered back part and produce little or no wake (B). The amount of pressure drag is a direct reflection of the size of the turbulent wake. (S.T.)

probably not be surprised to find that the drag on a Formula One race car was lower than on a semitrailer truck, or that a sparrow had lower drag than an eagle. But what if you could scale the sparrow up to the size of the eagle, or the truck down to the size of the race car? If the drag on both birds or both vehicles were now equal, we could confidently say that the differences in drag between the normal-sized objects were solely due to size. However, more than just size is involved in the amount of drag on an object. In our example, we have scaled them up or down so that differences in size alone do not affect the amount of drag. Any remaining differences in drag must therefore be due to shape differences, or as a biologist would say, differences in morphology. Engineers developed a measurement of the relative "dragginess" of differently shaped objects *independent of size* called the *drag coefficient,* C_D. The drag coefficient is defined by the equation*

Equation 2.1 $$C_D = \frac{D}{\frac{1}{2}\rho v^2 S}$$

where D equals the magnitude of the drag force, ρ is the density of the fluid, v is the speed of movement, and S is some reference area.† The "$\frac{1}{2}\rho v^2 S$" term defines a sort of reference force based on the speed of the fluid, which comes from Bernoulli's equation. The beauty of the drag coefficient is that it is dimensionless: the numerator and denominator are both forces, so the units cancel. Thus, the drag coefficient allows us to compare the drag characteristics of bodies of any shape and size, and it does not matter what system of units—metric or traditional U.S.—we use to calculate it, as long as any one calculation uses just one system.

At least four different reference areas can be used for the reference area, S. The most commonly used area is the frontal area (Fig. 2.4), the maximum area of the object projected onto a surface at right angles to the direction of movement. In other words, the frontal area is the area you would see if you stood directly in front of the object as it came toward you. For simple, streamlined shapes (bodies of trout, hulls of submarines), the total surface area or *wetted* area is often used. This, however, becomes highly impractical to measure on complex shapes like swimming crabs or flying insects. For lighter-than-air ships such as blimps, $V^{2/3}$ (volume raised to the $\frac{2}{3}$ power) is used, and in his book *Life in Moving Fluids,* Steven Vogel [1] makes a good case for it being the most biologically relevant area. Finally, for winged objects, the wing area is normally used. Some of my own research demonstrates how choosing

* The more commonly seen form of this equation, $D = \frac{1}{2}\rho v^2 S C_D$, is not in any way a definition of *drag.* This so-called drag equation is simply a convenient shorthand for showing that the behavior of drag under various conditions is largely determined by the drag coefficient under those conditions.

† The "S" comes from surface area, and the "v" comes from velocity; strictly speaking, "velocity" is a vector and includes a direction, while "speed" is the magnitude of this vector (and is what most people mean when they say "velocity" in an attempt to sound more technical).

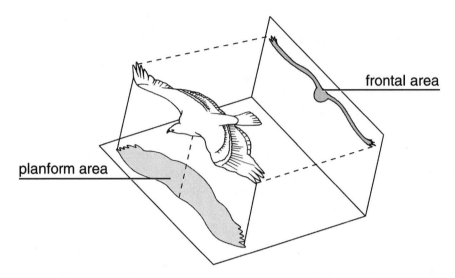

Figure 2.4. The *frontal area* is the object's area projected onto a vertical surface perpendicular to the direction of movement. The *planform area* is the area of a wing or
lifting surface that is projected onto a horizontal plane. By convention, the planform
area treats the wings as if they were connected through the body, but does not include the front or back part of the body. In contrast, the frontal area normally includes the body. (J.P.)

different reference areas can dramatically change the conclusions drawn
from comparisons of drag coefficients [2, 3]. I compared the drag of two different species of small marine isopod crustaceans, and found that one species
had higher drag coefficients when I based the coefficients on the frontal area,
but the second species had higher drag coefficients when I based them on the
wetted area. Thus, knowing which reference area was used is crucially important if such comparisons are to be valid. Indeed, Vogel rightly argues that
studies reporting values for drag coefficients but without specifying which
reference area was used are worthless. Fortunately, for our purposes no choice
is necessary, because by convention, the reference area for wings is the *planform* area. This is the area of the wing projected onto a horizontal surface, or
the area you would see looking down on the wing from above (Fig. 2.4).

So far, I have described the drag coefficient as if it were a constant for a
given shape. This, however, is not true, because the same body can have different drag coefficients under different conditions. A quantity that helps predict drag coefficients exists, and it is called the Reynolds number, Re, after the

British hydrodynamicist Osborne Reynolds (1841–1912). The Reynolds number is defined as

Equation 2.2 $\mathrm{Re} = \dfrac{\rho v l}{\mu}$

where l is a characteristic length (usually the length of the object along the axis of movement—for example, the body length of a fish, or the width of a wing from front to back) and μ is the dynamic viscosity.* In addition to predicting changes in the drag coefficient, the Reynolds number is also a measure of the relative importance of viscosity versus inertia. In other words, the Reynolds number determines whether viscous drag or pressure drag should be dominant. At Reynolds numbers above 1.0, pressure drag is more important; at Reynolds numbers between zero and 1.0, viscous drag is more important. For example, a flying duck has a Reynolds number of over 10,000, so pressure drag is important and viscous drag is negligible; a microscopic swimming protozoan would have a Reynolds number of 0.1 or less, so viscous drag is dominant. In most cases, Reynolds numbers are rough approximations, and their importance is not so much their exact values but their order of magnitude—differences of less than a factor of ten are insignificant. Size and speed tend to affect the Reynolds number together: small creatures move slowly, and have low Reynolds numbers, while big animals move fast and have high Reynolds numbers. For most flying animals, pressure drag is of much more importance than viscous drag, although for the smaller flying insects, viscous drag and the effects of viscosity become significant.

Producing Lift: The "Basic" Model

Bernoulli's Equation

The basis for the basic model of lift production is Bernoulli's equation. The model is sometimes even inaccurately called the "Bernoulli model"; this is

---

* The dynamic viscosity should not be confused with the kinematic viscosity, v, which is dynamic viscosity divided by density (μ/ρ). The dynamic viscosity is a measure of what we think of intuitively as viscosity, whereas v is heuristically useful in many equations, but not easy to describe in everyday terms. Even biologists confuse the two, sometimes in print.

rather unfair because the Bernoulli equation is much more general than just a description of wing properties. Given the frequency with which it is blithely applied, Bernoulli's equation is subject to some rather startling assumptions—including the assumption that the fluid has no viscosity. Fortunately, the equation is rather robust to minor violations of its assumptions, and for airflows outside boundary layers (where viscosity has most of its effect) it is generally quite accurate. This equation can be expressed in a variety of forms. A useful one for our purpose is

Equation 2.3 $P + \frac{1}{2}\rho v^2 = \text{constant}$

where *P* is the *static pressure*. The static pressure is what a pressure gauge measures, as when a bicyclist measures the pressure in a bicycle tire. The "$\frac{1}{2}\rho v^2$" term is sometimes called the *impact pressure*, because it is the pressure that flowing air molecules would exert if they were brought to a sudden halt. (It is related to the kinetic energy, for those who remember their physics.) The key concepts revealed by the equation are: first, the pressure is proportional to the *square* of the speed; second, pressure and speed vary inversely, so that if the pressure goes up, the square of the speed must go down by an equal amount to maintain the constant in the equation. This result is counterintuitive and merits a deeper look.

If a wind can blow objects about that do not move in still air, how can higher wind speeds lead to lower pressures? In fact, two different pressures are involved. Imagine a big steel plate with a small hole in it for a pressure gauge. If air flows over the plate parallel to its surface, the pressure gauge will measure the *static pressure*.* If the wind now blows directly at the plate (in other words, perpendicular to its surface), the pressure gauge will measure the *impact pressure*, which does increase as the speed goes up. A Venturi tube shows the effect of air speed on the static pressure quite explicitly (Fig. 2.5). As the flow speeds up to pass through the constriction, the pressure drops. This effect is easy to demonstrate: hold a narrow strip of paper (perhaps 1 inch wide by 6 inches long, or 2 × 15 cm) up to your chin and blow over it along the length of the strip. The air over the top is moving, so there is lower pressure on top than on the bottom: the paper rises as you blow.

---

* Often called the *hydrostatic pressure*, but that gets a bit confusing when referring to air!

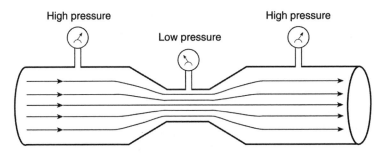

Figure 2.5. A Venturi tube. Fluid speeds up as it passes through the constriction, which lowers the pressure. The contourlike curves are *streamlines,* which represent the paths followed by tiny packets of fluids. They can be "read" like a contour map: the closer they are together, the faster the fluid is moving. (A.P.)

The same effect applies to wings. A *wing* is the whole three-dimensional structure such as you find attached to a bird or an airplane, in contrast to an *airfoil,* the cross-sectional shape of the wing, as if it were sliced from front to back and you looked at the cut end (Fig. 2.6).* Figure 2.7 shows a common type of airfoil. Note that it is more convex on top than on the bottom, making it a *cambered* airfoil—the degree of upward convexity is the *camber* (Fig. 2.6). Now imagine air flowing past it, which is, remember, equivalent to the airfoil moving through the air. If two little blobs of air approach the front or *leading edge* of the airfoil together so that one goes over the top while the other goes under the bottom, the one going over the top will go a greater distance. Assuming that packets that start together at the leading edge must end together at the end, or *trailing edge,* the packet going over the top will have to go faster to cover a longer distance. This is where Bernoulli's equation comes in—the difference in pressures between the top and bottom of the airfoil will be proportional to the differences in the *squares* of the speeds. Simply put, a small difference in speeds can make a large difference in pressures. Because the high speed is on top, the pressure will be lower on top than on the bottom. The low pressure on top pulls up, the high pressure on the bottom pushes up, and summing the pressures together over the wing's surface gives a force called *lift.* (Note that *lift* is defined as being *perpendicular to the direc-*

* Many authors (including some experts) are rather casual about this distinction and use wing and airfoil interchangeably, but such usage is confusing and ambiguous.

tion of movement. Because the movement in this case is horizontal, lift is upward. If the wing moved at a different angle, lift would be tilted away from the vertical.)

The basic model is adequate to describe lift production qualitatively for this type of airfoil moving horizontally. Why, then, do so many people (especially pilots) argue about its validity? First, note that Fig. 2.7 shows that this model requires the airfoil to be more convex on top than on the bottom, or cambered. Some airplanes, and flying animals under some conditions, use uncambered, or symmetrical, airfoils in their wings. Also, as anyone who has been to an air show can attest, airplanes can fly upside down. Bats must be able to fly upside down to reach their perches on ceilings. Finally, why should blobs of air that start out together at the front be required to reach the rear of the airfoil together? As a matter of fact, I have been told by several aerody-

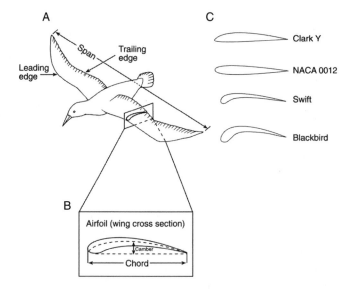

Figure 2.6. Wing and airfoil terminology and examples. The whole wing is shown in (A), and the cross section of the wing, or airfoil, is shown in (B). Note that the length of the chord changes along the span of the wing (except for rectangular wings). Most airfoils are *cambered,* which means more convex on top than on the bottom; the degree of upward convexity is the camber. Some airfoils from real wings are shown in (C). The upper two are from airplanes [9] and the lower two are from birds [5]. (Airplane airfoils are named by their designers as part of a series of similar airfoils, not by the airplanes in which they are used.) Note that the lower airplane airfoil (second from the top, NACA 0012) is *symmetrical,* i.e., not cambered. (J.P.)

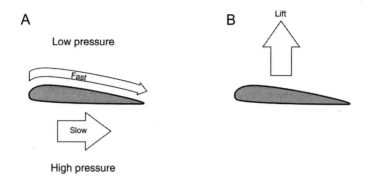

Figure 2.7. Airflow over an airfoil (A) produces pressure differences which sum to yield a net upward force, lift (B). (A.P.)

namicists (and seen computer simulations showing) that the air over the top of a wing arrives at the trailing edge appreciably *before* the air underneath, when wings with well-designed airfoils are tested in wind tunnels.

I want to emphasize that there is nothing incorrect about Bernoulli's equation. With accurate measurements of the air speed over the top and bottom of a wing (outside the boundary layer, of course), Bernoulli's equation predicts the pressure differences, and hence lift. The problem is the assumption that little packets of air that separate going over the leading edge must rejoin at the trailing edge. The mechanism is far more subtle than the explanation given by the basic model.

Producing Lift: The "Engineering" Model

Lift from Rotating Cylinders

How *do* symmetrical airfoils generate lift? How *do* wings work upside down? To answer these questions, we need to look first at a device that produces lift like a wing, but does not look very much like a wing. Imagine an airplane with a rather peculiar modification: this airplane has cylinders sticking out from its side where wings should be (Fig. 2.8A). Furthermore, these cylinders can be rotated by a motor in the airplane separately from the propeller. The cylinders rotate so that as this airplane taxis down the runway, the top of each cylinder is moving in the same direction as the wind flowing over it (Fig. 2.8B). Thanks to the no-slip condition, air will be pulled along faster

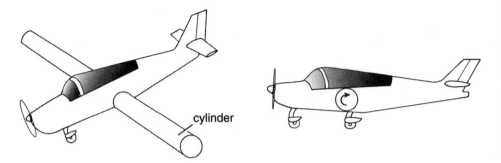

cylinder

Figure 2.8. A. Airplane with cylinders in place of wings. B. Direction of rotation of cylinders. (J.P.)

over the top, and dragged to a lower speed under the bottom. Just as with the basic model, faster air movement over the top and slower air movement underneath will lead to a pressure difference between the top and bottom of the cylinders. With a low pressure on top and higher pressure below, the net result is an upward force—lift. These rotating cylinders are called Flettner rotors,* and to the best of my knowledge, they have never been used on full-sized airplanes. However, a number of flying model airplanes, including some relatively large remote-controlled ones, have flown successfully with Flettner rotors instead of wings. More surprisingly, Flettner rotors are found in nature: several different types of trees use them as seed dispersal mechanisms (Chapter 3).

We do not normally encounter lift-producing cylinders in our daily lives, but rotating spheres that produce lift are common in sports. For example, in baseball, a pitcher throws a "curve ball" by giving the ball a sideways spin so that the ball produces lift to one side. This lift pulls the ball along a curved path. "Hooks" and "slices" in golf are produced the same way: when the golfer hits the ball in a way that gives it a sideways spin, the ball experiences lift directed to the side, and the ball veers off a straight path. Lift produced

---

* Named after Anton Flettner, who built ships that used Flettner rotors as sails. It is easy to see why they were commercial failures, because they combined the worst features of sailing ships (dependence on wind, inability to sail directly upwind) and steamships (need to carry fuel to power engines).

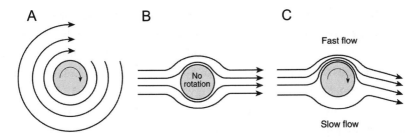

Figure 2.9. Airflow patterns around cylinders (end view). A. Rotating cylinder in still air, forming bound vortex. B. Stationary cylinder in flowing air, with symmetrical flow over the top and bottom. C. Air flowing over a rotating cylinder, going faster over the top than under the bottom. (A.P.)

by rotation is called the *Magnus effect,* and it clearly works as well on spheres as on cylinders.

The Bound Vortex

What does a Flettner rotor or the Magnus effect have to do with airfoils? Even if they are not particularly practical, Flettner rotors graphically illustrate an important phenomenon also associated with airfoils: the *bound vortex.** Looking at the end of a rotating cylinder in still air, we would see that the cylinder pulls air around itself because of viscosity and the no-slip condition. This mass of revolving air is called the bound vortex (Fig. 2.9A). If we stop the rotation of the cylinder and allow a wind to flow over it, the wind flows symmetrically over the top and the bottom (Fig. 2.9B). If we combine the rotating cylinder with flowing air, we get the situation shown in Fig. 2.9C. An aerodynamicist would say that we have "superposed" a bound vortex with a symmetrical flow, which brings about the lift-producing pattern of faster flow on top and slower flow beneath. Thus, we have combined the effects of a bound vortex and a symmetrical, horizontal flow.

The Bound Vortex on an Airfoil

A wing is thus a device that produces a bound vortex, not by rotating, but because of its shape: specifically, its airfoil-shaped cross section. What aspect

---

* A *vortex* (plural, *vortices*) is a flow pattern where the fluid follows a circular path, as in a whirlpool or tornado.

of this shape brings about the bound vortex? The airfoil must possess a sharp trailing edge. The camber (upward convexity) is also involved, but a change in orientation can also substitute for camber. An accurate technical description of the basis of the bound vortex involves rather daunting mathematics, but some key features of this flow pattern are easy to understand. First, the camber of the airfoil causes the flow to behave as it does in a Venturi tube, squeezing the top streamlines together. Put another way, a given mass of air has to squeeze over the top of the airfoil, so the air speeds up. This causes a reduction in air pressure on top. Now the sharp trailing edge comes into play: air naturally tends to flow from high pressure to low pressure—in this case the air would have to flow from the bottom, around the trailing edge, to the top. With a sharp trailing edge, however, the air's momentum prevents the air coming off the bottom of the trailing edge from reversing direction, turning the sharp corner, heading back up and going over the top. Thus, the combination of camber and a sharp trailing edge forces the air to flow around the airfoil in a pattern mathematically equivalent to a bound vortex combined with a symmetrical free stream flow. The bound vortex (described as a *net vortex* [4]) around an airfoil is real, but it can only exist when combined with a breeze blowing across the wing (or when the wing moves through the air). In other words, you can never see the bound vortex alone on a real wing, even though it is not difficult to separate the bound vortex from the free stream flow mathematically (Fig. 2.10). Thus no little blobs of fluid actually circumnavigate the airfoil.

The magnitude of lift produced by an airfoil is proportional to both the speed of the movement through the air and the strength of the bound vortex (its speed of rotation, or *circulation*). Since the bound vortex itself is cre-

A   B   C

Figure 2.10. This diagram shows how a bound vortex (A) and symmetrical flow (B) can be mathematically combined ("superposed") to give the actual flow pattern around an airfoil (C). Keep in mind that (A) is a mathematical component of (C) and can never occur alone; (C) occurs on a real wing during lift production and (B) only occurs at the angle of attack where no lift is produced. (A.P.)

ated by the airflow, the circulation is also proportional to the speed of the airfoil's movement. The lift on an airfoil, then, is proportional to the *square* of the speed—once for the airflow itself, and once for the strength of the bound vortex.

Lift Coefficient

Just as a dimensionless coefficient for drag is useful, a dimensionless coefficient for lift is also useful in many situations. The form of the defining equation is almost identical:

Equation 2.4 $C_L = \dfrac{L}{\frac{1}{2}\rho v^2 S}$

where L is the magnitude of the lift. Remember, this allows us to compare lift production from one airfoil at different Reynolds numbers or from different airfoils at the same Reynolds number. For our purposes, the lift and drag coefficients may be most useful when used to define the lift-to-drag ratio.

Angle of Attack

When an airfoil, in a horizontal airflow or moving horizontally, is tilted leading-edge-up, the amount of lift it produces increases (Fig. 2.11). The angle between the airfoil's chord and the direction of movement is called the airfoil's angle of attack, α, and lift increases as the angle of attack increases. As the angle of attack increases, it is just like narrowing the channel in a Venturi tube: the flow speed increases over the top of the wing, an effect illustrated by the streamlines being squeezed together (e.g., Figs. 2.7 and 2.10C). This effect increases the difference between the air speeds over the top and under the bottom of the wing, another way of saying that the circulation

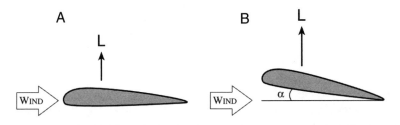

Figure 2.11. Angle of attack, α. A. No angle of attack, moderate lift. B. As the angle of attack, α, increases, lift also increases. (A.P.)

(strength of the bound vortex) increases as angle of attack increases. Increasing the circulation causes lift to increase, so *increasing angle of attack causes an increase in lift.* This effect of angle of attack on lift production allows a symmetrical airfoil to produce lift, and also allows any airfoil to produce lift when upside down.

Symmetrical and Cambered Wings, and Inverted Flight

At an angle of attack of zero degrees, a wing with a symmetrical airfoil produces no lift. As its angle of attack increases, the wing starts to generate a bound vortex and produce lift (Fig. 2.12). This type of wing also flies upside down. Because the top and bottom are the same, if the wing has a negative angle of attack (i.e., tilts the leading edge down), it will produce lift directed downward (a seemingly contradictory concept). By flipping this wing upside down, the negative angle of attack becomes positive, and lift is once again in its comforting upward direction. A cambered wing will also produce "negative," or downward, lift if the angle of attack becomes sufficiently negative,

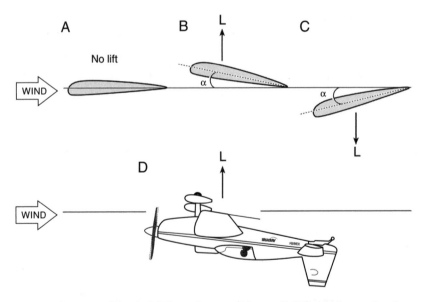

Figure 2.12. A. A cambered airfoil produces no lift at a slight negative angle of attack (a symmetrical airfoil produces no lift at zero angle of attack). B. With a positive α, lift is upward. C. With a significant negative α, lift is downward. D. When inverted, a negative α becomes positive, and inverted flight is possible (α exaggerated). (A.P. and S.T.)

Figure 2.13. Progressively increasing angle of attack (A, B) leads eventually to stall (C). (Note visible downwash in B.) (S.T.)

so most wings can be made to fly upside down with appropriately adjusted angles of attack. Why use cambered wings at all? They must have some advantage, because they are more common than symmetrical airfoils on airplanes and nearly universal among flying animals. The answer is simple: a cambered airfoil (1) produces lift at zero angle of attack; (2) produces more lift than a symmetrical airfoil at any given angle of attack; (3) produces less drag than a symmetrical airfoil for a given amount of lift. Given that very few animals (or airplanes, for that matter) routinely fly upside down, the benefits of a cambered airfoil clearly outweigh its less-than-stellar inverted performance. On the other hand, when animals flap their wings, the wings move at a negative angle of attack on the upstroke: this causes them to produce negative lift! (See Chapter 4.) Thus, the concepts needed to understand inverted flight have relevance for flying animals as well as airplanes in air shows.

Stall

Increasing the angle of attack to get more lift sounds beneficial, but it has a very specific limit. When a wing's angle of attack exceeds a certain critical value, *stall* occurs: the airflow over the upper surface stops flowing smoothly along the surface and peels away (Fig. 2.13). This leaves a large, turbulent wake, which causes the wing to suffer a large loss of lift and a great increase in drag. (Note that this sort of "stall" has nothing to do with engines: paper gliders, birds, and sailplanes can all experience it, and none of them has an engine.) The critical angle depends partly on the airfoil shape and partly on the Reynolds number. The critical angle for a large bird is typically around 20°, for a grasshopper, about 30°, and for a fruit fly, about 50°.

Because most flying animals or machines stop flying and start falling when their wings stall, stalling is usually considered to be detrimental. During landing, however, a well-timed stall can be very useful. In fact, just before birds touch down on a perch or on the ground, they often sharply increase the angle of attack of their wings. Timing is crucial: the idea is to use the increased drag of the stall to slow quickly just when low enough that the loss of lift does not matter. If the bird stalls too early, it may stop before arriving at its perch and fall; too late, and the bird may be going too fast to grasp the perch or bounce back in the air if landing on the ground. I am not merely speculating that birds use stalls this way. Photographs and slow-motion movies often show that the feathers on the upper surfaces of birds' wings lift up and flutter just before touchdown, clearly indicating the turbulence of a stall [5].

Downwash

Another component of the lift-producing process is *downwash*. When air flows over a lift-producing wing and the top and bottom flows join at the trailing edge, they are tilted slightly downward (Fig. 2.13B). The angle through which the flow is tilted after passing over the wing is the *downwash angle*. This angle is typically less than half the angle of attack, and partly dependent on the wing shape (mainly the aspect ratio, or ratio of span to chord). Technically, the *downwash* itself is the slight downward velocity added to the airflow as it passes over the wing that tilts the flow slightly downward. The downwash is actually an expression of Newton's Third Law: For every action there is an equal and opposite reaction. The process that produces an upward force on the wing also produces a downward force on the air flowing across the wing, which deflects it downward. Another way to think of it is that the air pushes (or, more accurately, pulls) up on the wing to produce lift, so as an equal and opposite reaction, the wing must push down on the air, causing downwash. But remember that downwash is *not* a separate process from the others described above for wings; it is simply another aspect of the same process.

Vortices on a Wing

Aerodynamic theory says that a vortex cannot begin or end in the air. What this means for a wing is that the bound vortex cannot simply end at the wing tip. How can this be? To see what happens at the tip, we first need to see what happens when the wing just begins to generate lift. As the wing

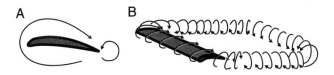

Figure 2.14. A. Bound and starting vortices on an airfoil at the instant it begins to produce lift. B. Schematic view of the three-dimensional vortex system of a wing. (S.T.)

starts moving, it produces a *starting vortex,* which rolls off the trailing edge (Fig. 2.14A) and turns in the opposite direction to the bound vortex. As the wing moves away from the starting vortex, the bound vortex stays attached to it by a *trailing* or *tip* vortex off the end of each wing tip (Fig. 2.14B). The vortices form a system much like a distorted smoke ring. The starting vortex stays at the point in the air where the wing started to move, and viscosity eventually causes it to dissipate. The tip vortices stream off each end of the wing and trail along behind the wing as long as the wing produces lift. In theory, the trailing vortices should stream all the way back to the starting vortex and stay connected. In reality, the trailing vortices are produced and stream back from the wing as long as it produces lift, but viscosity causes the starting vortex and the distant parts of the trailing vortices to die out (rapidly for small wings, more slowly for large ones). One reason that geese and pelicans fly in neat V-formations is that by maintaining a very precise spacing, birds can get a slight upward boost from the trailing vortex of the bird in front: the outer side of the trailing vortex of the bird in front has an upward component. Of course, the leader at the point of the V gets no such benefit, which is undoubtedly why birds exchange the lead position relatively often in such flocks. Trailing vortices can thus be of benefit, as in geese formations. But they can also be harmful, as when small airplanes are thrown out of control by the trailing vortices of very large ones. The vortex system of a fixed wing (i.e., not flapping, as when gliding) is quite simple, in practice consisting of the bound vortex on the wing and a trailing vortex off each wing tip that gradually dissipates far behind the wing. The vortex system of a set of flapping wings is quite different, as we shall see when we look at how flapping works, with ring-shaped or complex ladderlike vortices forming in the wing's wake (Chapter 4).

Limits on Wing Performance

Induced Drag and Wing Tips

Wing stalling is one major limit on wing performance: lift can only be increased by increasing the angle of attack up to the critical angle. Another important constraint on a wing is the *induced drag*. Induced drag is produced by the same process that produces lift and is in addition to the pressure and viscous drag that a wing experiences whether or not it produces lift. According to airfoil theory, if we could somehow make an infinitely long wing (i.e., a wing with infinite span), such a wing would have no induced drag even when producing lift. Engineers come quite close to this condition by putting wing models in wind tunnels where the wing runs all the way from one wall of the tunnel to the other. What does a real, finite wing have that our theoretical infinite one does not? The infinite wing has no tips, so there are no tip vortices, and tip vortices are part of the cause of induced drag. A real wing cannot generate lift without tip (trailing) vortices, and the wing "feels" the vortices as extra drag. The exact relationship between the tip vortex and the induced drag is rather complex, but it has an intuitively simple basis: the wing exerts a force (actually an array or distribution of forces) on the air that causes the tip vortex, so the air exerts an "equal and opposite" reaction on the wing, which is the induced drag. It is worth emphasizing that when a wing moves through the air at its zero-lift angle of attack (for example, 0° for a symmetrical wing, or some negative angle for a cambered wing), it produces no lift, has no tip vortices, and therefore has no induced drag. It *does* still have viscous and pressure drag, and it has these two types of drag whether or not lift is being produced.

Lift-to-Drag Ratio

A crucial characteristic of a given airfoil (and a wing with that airfoil) is the *lift-to-drag ratio,* commonly given the symbol L/D or C_L/C_D. (It is left as an exercise for the reader to convince him- or herself that L/D = C_L/C_D; see equations 2.1 and 2.4.) The drag in this case is the total drag, which is the sum of viscous, pressure, and induced drags. The lift-to-drag ratio influences wing operation in several ways. First, the drag of a wing represents the "cost" of producing lift, hence of flying. Remember, the lift requirement is fixed: the wing must produce enough upward force to balance the weight, so the weight sets the lift requirement. Since drag must be balanced by thrust, the higher the lift-to-drag ratio, the less thrust you need to achieve that required lift. In almost all cases, having a high lift-to-drag ratio is beneficial.

The lift-to-drag ratio, however, is not constant. Both lift and drag change as the angle of attack changes, but not in direct proportion; the relationship between lift and drag is thus rather complicated. To illustrate this complex relationship, Gustave Eiffel, designer of the Eiffel Tower, developed the *polar diagram.* It compares the lift and drag coefficients: lift coefficient is plotted on a graph against drag coefficient, and the values for angle of attack are indicated on the curve. An example is shown in Fig. 2.15. Some features of this diagram are worth noting. First, the curves intersect the horizontal (drag coefficient) axis above zero, which means that the lift can go to zero but there

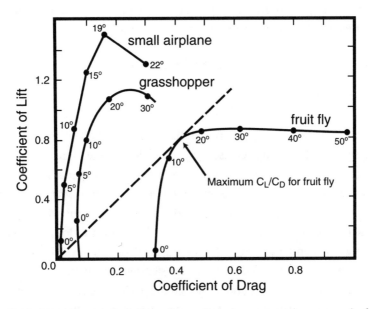

Figure 2.15. Polar diagram for wings of a small airplane, a grasshopper, and a fruit fly. Angles of attack are shown on each curve. The upper end of each curve represents the point where stall occurs. A straight line tangent to the curve that passes through the origin touches the curve at the angle of attack where the maximum C_L/C_D occurs (e.g., dashed line, 18° for fruit fly). Note that as size (and hence Reynolds number) decreases, the maximum C_L decreases, but stall is also delayed to higher angles of attack. Modified from S. Vogel 1967 [10] by permission of the Company of Biologists Ltd. (S.T.)

always remains some drag. In other words, lift and *induced* drag can disappear, but we can never eliminate viscous and pressure drag. Second, at the upper right end, the curve usually has a peak and then starts to fall before abruptly ending. The peak shows the maximum lift coefficient, and the curve ends where the angle of attack reaches the critical angle, or in other words, when stalling occurs.

Oddly enough, the angle of attack of maximum lift coefficient is *not* the most useful angle of attack. The angle of attack that gives the *maximum* C_L/C_D (i.e., the highest lift-to-drag ratio) is normally the most useful angle of attack, and it occurs at a much lower angle of attack than maximum lift coefficient. (The angle of attack at which maximum C_L/C_D occurs can easily be obtained from the polar diagram, see Fig. 2.15.) The angle that gives the maximum lift coefficient, with its disproportionately higher drag coefficient, may be tolerable during takeoff and positively useful during landing. Maximizing C_L/C_D (i.e., lift-to-drag ratio), however, is much more important than maximizing lift coefficient alone when traveling a long distance.

Shapes of Wings and Why It Matters

Aspect Ratio and Tip Effects

Several elements of wing shape can have important effects on lift production. Remember, a small increase in speed yields a large increase in lift. The planform area of the wing also helps determine the amount of lift produced: the more area of wing, the more area to generate lift. The *aspect ratio*—the ratio of the wingspan to the average chord*—is another feature that has a central role because of its effect on tip vortices. The aspect ratio is a measure of how long and slender a wing is. Now imagine this: you have two wings with the same planform area. One is long and slender, thus having a high aspect ratio, and the other is short and broad, giving it a low aspect ratio. Which will have the stronger tip vortex? The wing with low aspect ratio (short and broad) will effectively have more "tip" for its area, and so will have a stronger

---

* Technically, the aspect ratio is defined as the square of the span divided by the planform area, which is identical to span divided by chord for a rectangular wing.

tip vortex. Thus the low-aspect-ratio wing will have more induced drag than the high-aspect-ratio wing. In other words, all else being equal, long, narrow wings will have lower induced drag and higher lift-to-drag ratio than short, stubby wings. If efficient travel over long distances is of primary importance, it pays to have high-aspect-ratio wings; the albatross is a wonderful example. Why don't *all* flying animals have high-aspect-ratio wings? Imagine flying through a dense forest with long wings. Long, slender wings are awkward in such cluttered habitats—in addition to simply getting in the way, they tend to reduce quickness in turns (Chapter 5)—and they also tend to be fragile or heavy. In nature, just as in engineering, all designs are compromises: maneuverability and structural concerns often must be traded off against high aspect ratios.

Airfoil Shape

The details of the shape of the airfoil of a wing can have significant effects on the performance of a wing, especially at high Reynolds numbers. Thickness affects total drag (thicker airfoils usually being draggier), which affects speed. Within limits, increasing the camber increases lift at any given angle of attack, at the expense of higher drag. Thickness and camber both affect stall characteristics. Such features as bluntness and droopiness of the leading edge, location of maximum thickness, sharpness of taper toward the trailing edge, curvature or flatness of the lower surface, can all influence the lift coefficient, the drag coefficient, and stall behavior. Furthermore, these effects may also change at different Reynolds numbers. Indeed, engineers spend considerable effort refining airfoils for particular purposes. Conversely, as size decreases, the Reynolds number also becomes lower, and at low Reynolds numbers the shape of the airfoil becomes much less important. Boundary layers get thicker as Reynolds number decreases, and the thick boundary layers apparently hide or smooth over details of the airfoil shape.

While the cross section of a bird wing looks reasonably similar to an airplane-type airfoil, the cross section of an insect wing looks distinctly different (Fig. 2.16). With its sharp peaks and valleys, the insect wing cross section hardly looks like an airfoil at all! However, researchers have shown that at the relatively low Reynolds numbers experienced by insects, the air flows smoothly over their wings as if the peaks and valleys were filled in to form a

A

B

Figure 2.16. Corrugated or pleated cross section of a dragonfly wing (A) and a hover fly wing (B). On the cross sections, circles represent wing veins and straight lines represent membrane between veins. Drawings of cross sections are based on author's original data (A) and data from C. J. Rees 1975 [11]. (J.P.)

smooth upper and lower surface [6, 7]. This indifference to airfoil shape at low Reynolds numbers might seem like a great advantage, but it comes at a hefty price: the increased influence of viscosity makes it harder to form the bound vortex, so values of the lift coefficient and lift-to-drag ratio are lower in general (compare the height of the curves on Fig. 2.15). Indeed, for the smallest flying insects, the lift-to-drag ratio may be less than 2. This contrasts with lift-to-drag ratio values for a typical bird of around 10, and for good gliders (for example, large vultures) of over 15. High-performance sailplanes, which operate at much higher Reynolds numbers than birds and have wings with much simpler functional requirements, can have values of lift-to-drag ratio exceeding 40 (Chapter 3).

Planform

The shape of a wing's planform also affects its performance. Lift, remember, is the net result of a pressure distribution acting over the entire upper and lower surface of the wing.* Early in this century, airplane designers determined that there is one particular type of pressure distribution that should theoretically have the lowest induced drag for a given amount of lift and a given planform area—the *elliptical* pressure distribution. For many years, airplane designers believed that the only way to get an elliptical lift distribution was to have wings with elliptical planforms.† This led to many graceful, elegant airplane designs, of which the Spitfire (of World War II fame) is probably most famous. Elliptical airplane wings are expensive and time-consuming to build compared to wings that are straight-edged; moreover, very few flying animals have elliptical wings. Aerodynamicists eventually discovered that wings could have elliptical or near-elliptical pressure distributions without having elliptical planforms, so it is entirely possible that many flying animals have more efficient pressure distributions than originally supposed. For example, C. W. Burkett [8] demonstrated that wings with pointed, swept-back tips have higher values of lift-to-drag ratio than blunt-tipped wings with the same area. Indeed, falcons, swifts, and swallows evolved exactly such scimitar-shaped wings long before there were engineers to study them.

Wings: Too Good to Be True?

This is a good place to pause, take a big step back from the details, and ponder the almost miraculous nature of what a wing does. For the price of pushing hard enough to overcome drag, the wing generates a much larger force at right angles to the push. For a wing with an unspectacular lift-to-drag ratio of 10, pushing on it with a horizontal force of 10 lb (216 N), gives 100 lb (2160 N) of lift. In physics, as in economics, "there is no such thing as a free lunch," but it seems to me that getting lift from a wing is an awfully good bargain.

---

  * This pressure distribution is essentially a huge array of small forces, each force acting on its own little square centimeter of surface area, all summing together to have the same net effect as our original single lift force. These little forces are not all the same over the whole wing: they are higher toward the center and taper to nothing right at the tip.

  † An elliptical planform is one where the leading edge and the trailing edge each form the smooth curve of half an ellipse. The front and back need not form the *same* ellipse.

Gliding and Soaring

Gliding is the simplest form of animal flight. Flying animals typically flap their wings to produce thrust (a forward force to overcome drag), but if they stop flapping and keep their wings stretched out, they can glide. So we can define *gliding* as using wings to produce lift, but not actively producing any thrust. Gliding is the aerial equivalent of coasting on a bicycle: just like a coasting cyclist, a glider will decelerate unless it gets a boost from gravity by going "downhill." While some flying animals rarely or never glide, most birds, bats, and large insects glide occasionally, particularly when descending to land. Some birds glide most of the time that they are aloft: if you have seen hawks or vultures circling high in the sky on motionless wings, you have seen them using a special form of gliding called *soaring*. Soaring allows birds to stay aloft for long periods without flapping. The most specialized soaring birds travel great distances without flapping. Storks migrate between Europe and Africa entirely by soaring, and eagles, vultures, and hawks all travel long distances mainly by soaring. Albatrosses may soar over the ocean for days at a time, and shearwaters migrate from England to Brazil mainly by soaring. Some animals can glide, but cannot use powered flight, because they cannot flap their wings. Flying squirrels, along with two other families of mammals, two families of fish, and several tropical lizard and frog species (among other animals), can all glide, but cannot maintain flight under their own power. Many plants even fit into this category, because they have evolved gliding seeds. The simplicity of gliding, compared to flapping, undoubtedly explains why gliding has evolved in such diverse organisms.

How Does Gliding Work?

To maintain level flight, an animal needs an upward force to oppose weight, and a forward force (thrust) to oppose drag. Since gliding occurs with no active thrust production, a glider traveling at a constant speed must always descend relative to the air it moves through. Gliding is thus "powered" by gravity.

When an object is in steady motion, all forces must balance. A wing moving horizontally through the air, operating at some angle of attack near its maximum lift-to-drag ratio (L/D_{max}), will have lift and drag vectors that look something like those of Fig. 3.1A: the motion is horizontal, so lift is vertical and drag is horizontal. In order to maintain this steady, horizontal flight, some source of thrust is needed to overcome the drag, or the animal will decelerate until it stalls. Remember, adding the lift and drag vectors gives the *resultant* force, R. The resultant force will be tilted backward by the angle labeled θ. Figure 3.1A also shows that θ is determined by the lift-to-drag ratio: for example, if lift increases relative to drag, θ gets smaller, and conversely, if drag increases relative to lift, θ gets larger. The relationship between the lift-to-drag ratio and θ is a crucial factor in gliding,* and the lift-to-drag ratio is probably the most important "figure of merit" for a glider. A wing's aspect ratio (wingspan divided by the chord) has a strong influence on its lift-to-drag ratio. Long, narrow—high-aspect-ratio—wings tend to have higher lift-to-drag ratios than short, broad—low-aspect-ratio—wings. So the more effective gliders tend to have higher aspect ratios.

---

* For those interested in the mathematics, the exact relationship can be obtained with trigonometry: tangent (θ) = D/L, or cotangent (θ) = L/D.

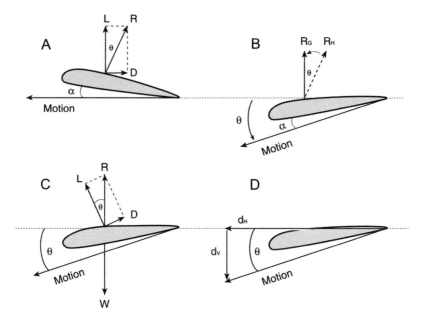

Figure 3.1. Airfoil section of a wing in horizontal motion and steady gliding, to show relationships among forces, angles, and speeds. A. Steady horizontal motion at angle of attack α, showing lift (L), drag (D), and the vector sum of lift and drag, called the resultant force (R). Angle θ is the angle between lift and R. This wing is moving horizontally, so it must be under power, not gliding. B. Tilting the path of movement downward by angle θ makes R exactly vertical, which is the orientation in a steady glide. R_H is the resultant force direction in horizontal flight, and R_G is the resultant force direction when gliding. C. Forces on a wing in a steady glide. The aerodynamic forces sum to give R, which is exactly equal and opposite to the weight (W). D. The distance moved horizontally, d_H, divided by the distance moved vertically, d_V, in a given time is the glide ratio. The trigonometric relationship between these distances and the glide angle θ is $\text{cotangent}(\theta) = d_H/d_V$. This ratio turns out to be the same as the lift-drag ratio, L/D. Dividing d_V by the time it took to descend that distance gives the sinking speed. (A.P.)

The Glide Angle

The relationship between the lift-to-drag ratio and θ clarifies how the mechanism behind gliding works. The principle is very simple: by tilting the direction of motion slightly downward, the direction of the resultant force, R, becomes vertical. How much does the direction need to be tilted down? By exactly the angle θ, because that will make the resultant force exactly vertical (see Fig. 3.1B). When the wing moves forward and down with an angle of θ, the resultant force is vertical, exactly opposite the weight of the glider (Fig. 3.1C). The angle θ now becomes the *glide angle,* and because the lift-to-drag ratio determines the glide angle (θ), the lift-to-drag ratio thus determines the angle by which the path of the glider must be tilted down.

The geometry of gliding produces both an obvious and a more subtle effect. The obvious effect is that the lift-to-drag ratio sets the *glide ratio,* which is the ratio of the distance traveled forward to the distance descended in a given time. Furthermore, since distance divided by time equals speed, we can easily obtain the sinking speed from the glide ratio and the flight speed. The equality of the lift-to-drag ratio and glide ratio is why the lift-to-drag ratio is so important for a glider: if its lift-to-drag ratio is high, a gliding animal may move many meters forward for every meter of altitude it descends, while if its lift-to-drag ratio is low, it moves forward a shorter distance for every meter of altitude lost. Table 3.1 gives some representative values of the lift-to-drag ratio for a variety of flying animals and machines.

The more subtle effect involves the relationships among the resultant force, weight, glide ratio and flight speed. Recall from Chapter 2 that for any given wing, the lift-to-drag ratio is controlled by the angle of attack, α. Because the glide ratio and the lift-to-drag ratio are identical, the angle of attack also controls the glide ratio, and hence, the glide angle. What would happen if a glider became heavier without changing anything else? Think of a vulture before and after a meal. The weight would be larger than the resultant force (R), so the forces do not balance. If the angle of attack stays constant, the glider will accelerate to a new, higher flight speed (which increases both lift and drag) until the resultant force is again as large as the total weight. The glide angle, however, *does not change.* The speed increases, but the path the glider follows does not change once it is back in equilibrium. Thus, if a glider becomes heavier but the angle of attack stays constant, it will come down to earth in less time, but it will cover the same distance as it did before adding the weight. Weight per unit area of wing is called *wing loading,* so this effect can be restated: all else being equal, a glider with a high wing loading will glide faster, but follow the same path, as a glider with a low wing loading. The heavier glider will cover the same horizontal distance, but because of its higher speed, it will have both a higher horizontal *and vertical* speed.

Some of my students once tried to demonstrate this effect by adding weights to a large toy glider. The loaded glider did fly faster than when un-

TABLE 3.1 Aspect Ratios and Lift-to-Drag Ratios of
Some Biological and Human-Made Wings

Flyer	Aspect ratio	Lift-to-drag ratio
Animals		
Fruit fly[a,b]	5.5	1.8
Bumblebee[c,b]	6.7	2.5
Crane fly[d,b]	6.9	3.7
Sparrow[e]	5.3	4.0
Swift[e]	11	10
American kestrel[e] ("sparrow hawk")	7.5	n/a
Falcon[e]	8.5	10
Red-tailed hawk[e]	7.1	10
Turkey vulture[e,f,b]	7.0	15.5
Great blue heron[e]	7.4	9
Brown pelican[e]	11.3	n/a
Stork[e]	7.8	10
Black-browed albatross[e]	13.1	n/a
Bald eagle[e]	6.6	n/a
Wandering albatross[e]	15.0	19
Airplanes		
Hang glider[e]	7	8
Standard sailplane[e]	21	40
Small 4-seat airplane[g] (Cessna 172)	7.3	10
Jumbo jet (Boeing 747)[e]	7	15
Competition sailplane[e]	38	>50

Sources: a. S. Vogel 1967 [34]. *b.* S. Vogel 1994 [8]. *c.* R. Dudley and C. P. Ellington 1990a and b [35, 36]. *d.* W. Nachtigall 1977 [37]. *e.* H. Tennekes 1996 [9]. *f.* C. J. Pennycuick 1971 [38]. *g.* J. Roskam 1979 [39].

n/a = not available

loaded, but they found it impossible to launch the glider with a consistent starting speed. Thus variations in the time to achieve equilibrium gliding conditions—all forces balanced and a steady flight speed—ended up being so large that they masked any differences in flight speed and distance. I encourage any motivated and energetic readers to improve on our experiment, perhaps using a catapult or some other launcher, and recording the *equilibrium* glide angle with a video camera. Note that the added weight must be placed at the center of mass ("center of gravity") of the glider; also, the tail surfaces must not be adjusted between glides, or the angle of attack will not be the same.

Size and Lift-to-Drag Ratio

The values for lift-to-drag ratios in Table 3.1 suggest that the lift-to-drag ratio may be related to size: over the size range spanned by flying animals, small animals tend to have low lift-to-drag ratios, and lift-to-drag ratios increase with size. Not surprisingly, this is due to changes in Reynolds number. At low Reynolds numbers, the greater contribution of viscosity inhibits formation of the bound vortex and reduces its strength, and so reduces lift production. At the same time, the viscous component of drag increases. Both of these effects cause lift-to-drag ratios to decrease as the Reynolds number decreases. At the small end of the scale, with lift-to-drag ratios (and glide ratios) of less than 2, gliding is only slightly better than falling, and becomes very impractical for small flyers like insects. At the other end of the spectrum, a high-performance sailplane operates at a Reynolds number at least 10 times higher than a hawk or vulture and so can have a significantly higher lift-to-drag ratio. Not all of this difference is due to differences in Reynolds number, however; at least some of the difference comes from the more complex operating requirements of the vulture wing. The sailplane wing can be manufactured to high tolerances to optimize its lift-to-drag ratio, whereas the vulture wing must be able to flap, resist impact and collision damage, and fold into an extremely small bundle for walking on the ground and perching in trees—all this in addition to gliding. As with any design, adding versatility reduces specialized abilities. Even if they operated at the same Reynolds number, a bird wing would be unlikely to achieve the lift-to-drag ratio of a sailplane.

The foregoing description of forces, movements, angles, and ratios should not be allowed to obscure the central concept of gliding: gliding is powered by gravity, so a glider must *always descend* relative to the air through which it flies. The big advantage of gliding, therefore, is that it does not require any thrust-producing mechanism. At the same time, the price of this type of flight is that a glider cannot stay level or climb continuously *relative to the surrounding air*. Yet there *are* ways for gliders to climb relative to the ground.

Parachuting versus Gliding

Parachuting refers to the use of structures that produce lots of drag in order to reduce the speed of falling. Dandelion and milkweed seeds commonly use parachuting to take advantage of wind currents for dispersal. The gliding gecko, a lizard from Southeast Asia, and the South American "flying" frog use parachuting for a different reason: to avoid injuries when leaping from trees to avoid predators. Parachuting relies on drag to slow descent, but true gliding requires lift production. In practical terms, it is sometimes difficult to distinguish between the two; in spite of their names, the flying frog and the gliding gecko are actually parachuters. To make a useful operational distinction between gliding and parachuting, biologists often use the glide angle: if the glide angle is less than 45°, the lift-to-drag ratio is greater than 1, so anything with a glide angle less than 45° is considered gliding. If the glide angle is greater than 45°, the lift-to-drag ratio is less than 1, and this is considered parachuting. The glide angle criterion is rather arbitrary, and it is perfectly possible for an animal to be producing lift while gliding with a glide angle of more than 45°, but such an animal would be going down faster than forward, and thus clearly not covering significant horizontal distances while aloft. Based on the glide-angle criterion, the gliding gecko and the South American flying frog, as well as at least four closely related tropical species of "flying" snakes, are all parachuters rather than true gliders, whereas several Southeast Asian flying frogs are true gliders [1]. The criterion therefore makes a useful practical distinction.

Many arboreal (tree-dwelling) animals may have some parachuting ability even without obvious structural specializations. For example, the common North American anole ("American chameleon" lizard) has no obvious anatomical specializations for parachuting, but the biologist James Oliver noticed that these lizards occasionally leap to the ground from great heights with no ill effects. He found that anoles always assume a characteristic spread-eagled posture when falling, a posture that is stable, that produces some horizontal travel in addition to the vertical fall, and that ends in a relatively soft landing. Other species of lizards of similar size did not assume the "parachuting" posture, so they fell faster and more vertically, and landed harder than the anoles [2].

Unpowered Flight in Animals

Capable gliders can be found among fish, amphibians, reptiles, and mammals. These animals are structurally and ecologically diverse, but they all use the same basic aerodynamic principles to achieve flight without power.

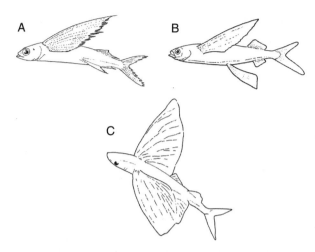

Figure 3.2. Flying fish. A. A two-winged flying fish. B. A four-winged flying fish. C. A two-winged flying fish gliding in air. (A) and (B) are from preserved specimens, (C) is from a photo of a fish in flight. A and B redrawn from J. Davenport 1992 [40]; C redrawn from H. E. Edgerton and C. M. Breder 1941 [3]. (A.P.)

Flying Fish

A whole family of species of marine fish, Exocoetidae, are highly adapted to flying in air (Fig. 3.2). They have greatly enlarged pectoral fins, which they spread out and use as wings in air. For several decades in the late nineteenth and early twentieth centuries, biologists argued about whether flying fish achieved true flapping flight or were simply gliders. After almost a century of inconclusive observations, Harold Edgerton* joined forces with a prominent marine biologist, C. M. Breder, in the early 1940s to study the question [3]. They made stroboscopic photographs of flying fish at night, which clearly showed that the flying fish are gliders, not flappers. Even so, recent work by John Davenport and other biologists shows that the flight of these fish is still quite remarkable. For instance, attaining high speed under water is much harder than in air, so when flying fish leap out of the water into air, they are not moving fast enough to glide any great distance. To accelerate to a sus-

---

* Edgerton was the photographic genius who invented the strobe and took pictures of bullets in flight and drops forming splashes in milk.

tainable flight speed, they dip the enlarged lower part of the tail back into the water and beat it furiously (50 to 70 strokes per second). After "taxiing" this way for about 10 meters, the fish nearly doubles its speed, and it rises up into the air for the free gliding portion of its flight, trading some airspeed for altitude in its initial climb. After a glide of about 50 meters, the fish may dive back into the water. Often, however, it dips its tail into the water for another bout of taxiing acceleration while still airborne, and then rises up for another glide. This process may be repeated several times before the fish submerges, and observers on ships have reported flight distances of over 400 meters [4].

Why do flying fish fly? Some biologists have argued that gliding may be an energy-conserving mechanism for long-distance travel, given that the fish is presumably not working very hard while gliding. If the benefit of the free gliding portion outweighs the energy cost of the taxiing portion of a flight, flight might reduce the energy needed for long-distance journeys. Most evidence, however, supports the idea that flight is mainly an effective escape from large, fast predatory fish. Large fish tend to swim faster than small fish, and for a marine fish, flying fish are average to small. Gliding in air thus may be a way to both escape from the immediate reach of a predator and flee at a much higher speed than would be possible by swimming [4].

Reptiles

Turning now to land animals, lizards in the Southeast Asian genus *Draco* have evolved very unusual wings. These lizards have a set of ribs that can be spread out to the side, supporting a fold of skin. This arrangement forms a rather low aspect-ratio wing with an oval, almost circular, planform (Fig. 3.3). At first glance, these structures would not seem to be particularly effective wings. Using measurements from a diagram in Edwin Colbert's 1967 paper on *Draco*, I calculated an aspect ratio of 1.7 for a large lizard. Other data in his paper imply aspect ratios for other lizards of 2.0 to 2.3 [5]. With such a low aspect ratio, these lizards would seem to be poor gliders. And yet people who have observed their flight behavior describe just the opposite. How, then, can we account for reports of 50-meter glides with a loss of only 2 or 3 meters of altitude? Note that this represents lift-to-drag ratios of 17 to 25, and glide angles of 2½° to 3½°. Several factors could be at work. For example, the lizards may take advantage of rising air currents to flatten and extend their glides, just as some soaring animals do. Second, they probably generate a significant amount of lift

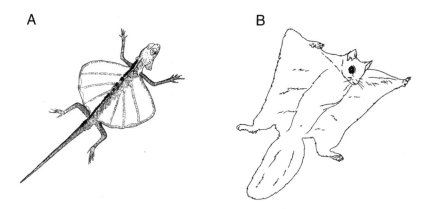

Figure 3.3. Other gliding animals: a gliding lizard in the genus *Draco* (A) and a North American flying squirrel (B). (A.P. and O.H.)

on the rest of the body, in addition to their wing membranes, which could increase the lift-producing area 20 percent or more. Finally, to achieve the flat glide paths reported by several biologists, these lizards must actually *avoid* steady-state gliding. They normally start their flights with a steep dive of a couple of meters to build up airspeed, and then level out. In the near-level phase of the glide, they must gradually and continuously increase their angle of attack to produce enough lift as they decelerate, trading speed for altitude (actually, for reduced loss of altitude). These lizards are usually gliding very slowly at the end of a flight, according to Colbert and others [5]. They need only a short, brief climb at the end of the flight to reduce speed for landing. Colbert also contrasts these flat glides toward a specific landing site with what he describes as "flight of animals not toward a target," in which glides were "relatively steep, on the order of 30°." In other words, when attempting to glide toward a target, the lizards are able to judge quite accurately the amount of deceleration acceptable in order to arrive at the target just before running out of speed and stalling. When not headed toward a specific target, however, the lizards have no need for finely tuning their glide, and were almost certainly in steady-state glides. A glide angle of 30° yields a lift-to-drag ratio of 1.7, which seems very reasonable for the structural arrangement of these lizards.*

* The equality of the value of the lift-to-drag ratio with the aspect ratio of one specimen is sheer coincidence and has no aerodynamic significance.

Draco's remarkable precision-landing ability hints at an equally remarkable maneuvering ability. Several eyewitnesses have commented on the lizards' apparently effortless avoidance of obstacles, as well as their highly aerobatic chases during territorial disputes. One observer even describes a lizard performing a complete barrel roll. These lizards clearly have a well-developed ability to maneuver in three dimensions. Most remarkable of all, this maneuvering seems to be accomplished primarily by distorting the wings. The legs and tails play little or no role in changing direction, according to films of *Draco* in flight made by James McGuire and Robert Dudley of the University of Texas. The lizards must have very precise control over the shape of their wing membrane to achieve such maneuverability.

Mammals

Flying squirrels have a large fold of skin along their flanks. This wing membrane, or *patagium*, retracts inconspicuously when the squirrels are clambering about on tree limbs. When gliding, the front and hind legs are held directly out to the side, and the wing membrane is stretched between the foreleg and hindleg (Fig. 3.3). Thus, unlike *Draco*, a flying squirrel's wings are supported entirely by its legs. Generally, flying squirrels do not have unusually long legs compared with squirrels that do not glide. An exhaustive study of the body proportions of dozens of species of flying and nonflying squirrels turned up only one difference: medium-sized and larger gliders tended to have forelegs about 20 percent longer than the nongliders of the same size. Gliding and nongliding squirrels showed no differences in hindleg lengths, and among small squirrels, there were no differences in foreleg lengths [6]. Several biologists have observed the gliding performance of flying squirrels, and the consensus is that flying squirrels have glide angles of about 18° to 26°, or lift-to-drag ratios of 2 or 3 [6]. The gliding behavior of *Draco* and flying squirrels shows both similarities and differences. Both often begin a flight with a short, steep drop before flattening out the glide. Unlike *Draco*, the squirrels apparently enter a steady glide, rather than decelerating; but, like *Draco*, flying squirrels are highly maneuverable. In her fascinating book on flying squirrels, Nancy Wells-Gosling describes their excellent ability to avoid obstacles. She also recounts observations of flights of dozens of meters with one or more 90° turns [7].

Despite their vastly different appearances and ancestry, it is no coincidence that *Draco* and flying squirrels have some important physical similarities. For example, they both have wing loadings of 40 to 50 N/m², although the lizards have body masses ranging from 1 to 6 grams and the squirrels range from 20 to over 250 grams [5, 6]. To a large extent, wing loading controls speed, so gliding lizards and flying squirrels must fly at very similar speeds. In animals that use powered flight, wing loading increases with body size. For example, wrens have wing loadings of about 25 N/m² [8] compared with 30 N/m² for a dove [9], 66 N/m² for a stork [10], and 100 N/m² for a condor [8]. Likewise, the aspect ratio of the lizards is about 1.7 to 2.2, and for the squirrels, it ranges from 1.2 to 2.2. These similarities indicate a shared combination of physical and environmental constraints acting on these otherwise very dissimilar animals.

But what about the low aspect ratios of these terrestrial gliders? More specifically, given that the lift-to-drag ratio (and hence, glide angle) should be higher for animals with wings of *high* aspect ratio, is there a benefit to these animals in having such stubby flight surfaces? Or are they just poorly adapted because they are at an early stage of evolving flight ability? For an arboreal animal, wings that are easily stowed while climbing are clearly advantageous, and this may limit the size of the wings. An aerodynamic factor may be involved, however. In the early 1930s, when aerodynamicists were still exploring low-speed flight and working toward a standard configuration for airplanes, C. H. Zimmerman discovered a peculiar property of wings with aspect ratios between 1 and 3: they could operate at extremely high angles of attack, which gives them surprisingly high maximum C_L values. As a bonus, these wings are also more stable [11]. Of course, these very low aspect-ratio wings also had very high drag, so they did not have especially beneficial lift-to-drag ratios. For lizards and squirrels, this type of wing would allow flight at unexpectedly low speeds, as well as sharp decelerations and brief, steep climbs at the end of a flight for softer landings. The precise soft-landing ability of *Draco* is just as well-developed in flying squirrels [7]. The stabilizing effect may have been an important advantage for an animal early in the evolution of gliding, because it reduces the complexity of the required control mechanisms (i.e., nervous system and behavior; Chapter 5). Finally, a low-aspect-ratio wing can double as a parachute for emergency vertical descents, while high-aspect-ratio wings make terrible parachutes. Thus, the similarities between *Draco* and flying squirrels represent adaptations to similar conditions: a wing small enough not to interfere with tree-climbing ability, large enough to give useful glide angles, and with a low aspect ratio to improve landing ability and stability.

Both types of animal fly mostly in thickly forested regions where high winds may be rare, so they can use low wing loading to advantage. Far from being an early step on the way to becoming bats or pterosaurs, these animals are actually well adapted to their lifestyles and habitats.

Flight in Plants: Gliding Seeds

Although plants have not yet evolved powered flight, a large number do make use of gliding or parachuting. Inasmuch as all familiar plants are firmly rooted to the ground, the idea of flying plants may seem highly unlikely. Many, however, have evolved gliding mechanisms to aid in seed dispersal. Dispersing one's offspring widely has many advantages—for example, getting out of the parent plant's shade or colonizing new or recently disturbed areas—and gliding can be a much more effective means of dispersal than just falling.

Parachuting

Many plants use parachuting for dispersal. Although this is not strictly speaking gliding, the similarities and the differences between the two are instructive. Very light-weight plant seeds are suspended from fluffy plumes to enable them to parachute, giving the whole "craft" a very high drag coefficient. Familiar examples include the seeds of dandelions, milkweeds, and sycamore trees. Their light weight and high drag gives the seed a very low sinking speed. A fairly strong breeze is usually needed to dislodge the seed from its parent plant, and this moves the seed horizontally as it falls. Thus, the slower the fall, the farther the seed will be carried by the wind. Contrary to popular belief, a horizontal wind does not in any way slow the seed's sinking speed: the seed will take the same time to fall a meter vertically with wind or without wind. With wind, the seed simply moves horizontally as well as vertically. In still air, a plumed seed would fall right under its parent plant, but because the seed needs a wind to detach it, the chances are good that such a seed will get blown some distance from its parent.

Gliding

Seeds that use true gliding for dispersal require some sort of wing or lifting surface. Winged seeds are called *samaras*. (Many samaras are actually "fruits" in the botanical sense.) For our purposes, *samara* refers to the whole structure, consisting of the aerodynamic surface or wing, and the payload or seed containing the plant embryo. The two general types of samaras are *simple gliders* and *autogyrating gliders*. The simple glider is, paradoxically, fairly rare, with only one species known to have highly developed gliding seeds, and a handful of others (like elms) with limited gliding abilities. The only really stable simple glider is the "Javanese flying cucumber," the seed of an Indonesian rain forest tree, which is not a cucumber at all. The tree is called *Alsomitra macrocarpa* but is better known by its obsolete genus name, *Zanonia*. Autogyrating samaras are, by contrast, quite common, and examples include the seeds of maple, pine, ash, and tulip poplar trees.

Simple gliders: Simple gliders are rare among samaras, because such gliders must be very stable (passive stability, Chapter 5). Wings with flat or typical airfoil cross sections are unstable in pitch. When the leading edge of this type of wing tilts up, the aerodynamic forces shift forward, causing the wing to pitch up even further. Eventually, the wing either stalls or tumbles out of control. Flying animals have evolved various active mechanisms to overcome this inconvenient action, and it is the main reason why airplanes have tail surfaces. As in airplanes, some birds use their tails to counteract this pitching tendency. Most flying animals, however, use active wing movements, both to control wing orientation and to keep the aerodynamic forces and center of gravity in appropriate locations relative to each other. In contrast, no samara has yet evolved a tail. Moreover, samaras lack muscles and nervous systems, so they cannot make any active movements to control their flight path. From this perspective, the evolution of *any* simple gliders among samaras is surprising.

The samara of *Alsomitra*, the flying cucumber, uses some of the same stabilizing techniques that human engineers use to stabilize flying-wing aircraft. Most of our knowledge of this samara comes from work of Akira Azuma and his colleague Yoshinori Okuno of the University of Tokyo [12]. The flying cucumber has what an engineer would call a "flying-wing" configuration. The seed sits in the center, with the thin wing membrane extending out to each side. Viewed from above, the wing's planform is somewhat reminiscent of the size and shape of a cucumber (see Fig. 3.4), but the resemblance is distant at best.

These samaras have swept-back wings with dihedral and reflexed trailing edges. *Dihedral* describes wings whose tips are raised above the root or middle region, a geometry that provides a stabilizing effect in roll (rotation around a longitudinal axis: raising or lowering a wing tip) and yaw (rotation around

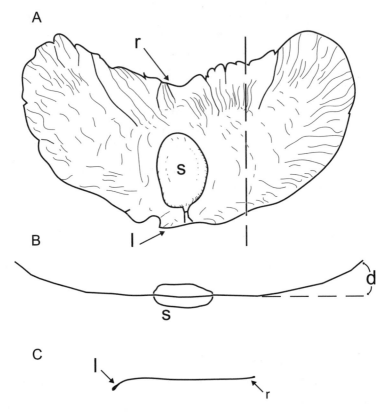

Figure 3.4. The samara of *Alsomitra macrocarpa,* a simple glider. A. Top view. B. Front view. C. Wing cross section at dashed line shown in A. s: seed; d: dihedral; l: leading edge; r: reflexed trailing edge. Redrawn from data in A. Azuma and Y. Okuno 1987 [12]. (S.T.)

a vertical axis: turning the nose left or right). Many gliders and airplanes have wings with dihedral. Wings with *sweepback,* or *swept* wings, have the wing angled back from root to tip. A *reflexed* trailing edge is one where the trailing edge is bent up slightly. Sweepback and reflexed trailing edges are both mechanisms for reducing the pitch-up tendency of a tailless craft [13]. Although most "flying-wing" (tailless) airplanes have either sweepback or reflexed trailing edges, the wings of the flying cucumbers have both of these features (Fig. 3.4). (Most jet airplanes also have swept wings to reduce drag at speeds near the speed of sound, but this effect is irrelevant for samaras gliding at less than 1 percent of the speed of sound.)

Why have both swept wings and reflexed trailing edges? Most flying-wing aircraft have one or the other, but not both. In addition to their stabilizing effect, reflexed trailing edges have both a benefit and a cost. The benefit is quick recovery from stalls, but the cost is an increase in drag. The flying cucumbers get part of their stability from reflexed trailing edges and part from sweepback. The portion of their stability from reflexed trailing edges comes with a compromise: a slightly lower lift-to-drag ratio, and thus a reduced total glide distance, in return for quicker stall recovery. As a craft with no active control mechanisms, the flying cucumber benefits from being able to recover quickly from upsets caused by gusts or collisions with tree branches. These advantages clearly outweigh the reduction in horizontal travel. As a further compromise, the trailing edge is apparently not quite reflexed enough to provide all the needed stability, so sweepback serves simultaneously to increase stability.

Autogyrating samaras: The vast majority of samaras are the autogyrating type, such as the familiar maple seed "helicopters" that fascinate small children. These samaras have a wing with the seed located at one tip. The center of gravity of the whole structure is located at or near this seed mass, which puts it out near a wing tip. When released into the air, a maple seed starts to drop with the blunt seed at one end downward, but the samara quickly begins rotating about the seed. The geometry of an autogyrating samara is deceptively complex (Fig. 3.5). The glide path is actually a helix or corkscrew shape,* with the center of rotation at the center of gravity of the samara. Because of the rotation, the tip (at the opposite end from the seed) moves fastest and thus produces the most lift. With lots of lift on the tip and none on the seed end, the tip of the samara tends to tilt up. At a certain rotation rate, the outward centrifugal force of the rotation balances the upward and inward tilt due to lift, and the samara adopts an equilibrium *coning angle* (Fig. 3.5) [14]. Autogyration is inherently stable. For a given combination of weight and wing shape, the samara inevitably assumes the one combination of angle and rotation rate in which all forces balance [15, 16]. Furthermore, rotation itself produces stability because any rotating body resists being tilted. For example, balancing on a bicycle is easier when it is moving than when it is

---

* A corkscrew is shaped like a helix, not a spiral. A spiral is a flat, two-dimensional shape, like the side view of a roll of tape. A "spiral" staircase is helical, not spiral.

Rotational
axis

Rotational
axis

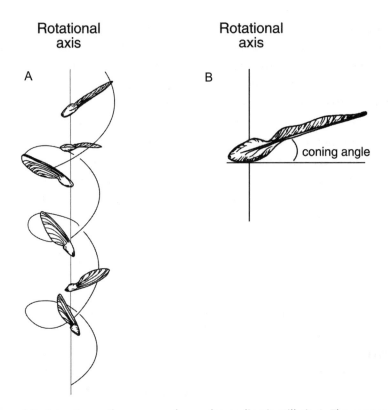

Figure 3.5. An autogyrating samara, shown descending in still air. A. The samara descends in a tight helix, with the axis of rotation at the center of mass and the wing tip following a corkscrew-shaped path. B. A snapshot of the samara at the instant it is parallel to the plane of view. The wing tip is raised above the horizontal by an angle called the coning angle. The coning angle is determined by a balance between lift and centrifugal force. (J.P.)

standing still, spinning tops are more stable when they spin faster, and spinning gyroscopes are able to retain their orientation in space.

Autogyration is thus a special kind of gliding, where the samara glides in a tight helix. There is one problem with gliding in a helix: in still air, an autogyrating samara glides almost straight down, although considerably more slowly than an acorn falls. Why go to the trouble to evolve a structure that glides straight down, albeit in a corkscrew pattern? Aerodynamically, an autogyrating samara is a glider, but in practice it operates like a parachute. Lift from the wing is used to slow the samara's fall. The only way that a maple samara can travel away from the parent tree is if the samara is released into a wind. Then, just as with a parachuting dandelion seed, the slow descent of the samara gives the wind plenty of time to blow the seed away from its launching site. However, the advantage of an autogyrating wing, as opposed to a bit of fluff, is that a much larger payload of seed can be carried. Extra weight can even be an advantage for an autogyro—increasing weight increases the glide speed, which increases the rotation rate, which, in turn, increases stability, which is a trade-off against a faster vertical descent rate. Moreover, because of the relationship between lift and the square of the speed, the descent rate does not increase as fast as weight increases: a four-fold increase in weight only doubles the descent rate. A disadvantage of autogyrating samaras is that they must be dropped from a great height, because they have to fall some distance to establish their equilibrium rotation and slow down. So, although autogyrating samaras may be quite useful for sizable trees, they cannot be used effectively by small herbaceous plants.

Three types of autogyrating samaras: Based on their structural complexity, rotating samaras can be placed in three general categories. Rolling or *autorotating* samaras have the simplest shapes, followed by the types with uncambered airfoils. The samaras with cambered airfoils have the most complex shapes.

Autorotating samaras—those of ash and tulip poplars, for example—are the most symmetrical. In cross section, these samaras are thickest in the middle, taper toward both edges, and have approximately the same shape on the top and bottom (Fig. 3.6). Thus, there is no way to distinguish between a front and back edge or a top and bottom surface. This cross section does not look like an airfoil, so how does the samara glide? When an ash samara falls, it first begins

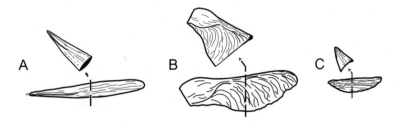

Figure 3.6. Three types of rotating samaras, showing the whole samara and a cross section through the wing. A. An ash seed, which is an autogyrating, autorotating (rolling) samara. B. A maple samara, which has a symmetrical wing. C. A pine samara, which has a cambered wing. (S.T.)

to turn about its long axis, somewhat like a child rolling down a hill. The ash samara, however, rolls the other way, lower surface forward and upper surface back. Then it begins to rotate in a tight helix, just as a maple samara does. These seeds are actually Flettner rotors, and Flettner rotors do not have to be cylinders. In fact, for a Flettner rotor to begin spinning by itself, a long flat plate is better than a cylinder. By painting one side of an ash samara black and one side white, C. W. McCutchen [17] showed that they did, indeed, spin about their long axis (*autorotate*)* while simultaneously autogyrating about a vertical axis. This effect is easy to demonstrate with weighted strips of index cards. Any narrow flat plate will start to spin this way: take an index card, or better yet, half an index card cut lengthwise, and drop it. Most of the time it will start rotating and glide off laterally. Taper it or weight one end slightly, and voilà, it autogyrates as well as autorotates. Because the weight of the seed is concentrated near one end of the samara, it autogyrates like a maple seed, but it produces lift as a Flettner rotor rather than with an airfoil. Samaras such as those of ash and tulip poplar samaras are the only common examples of Flettner rotors among either natural or human-designed devices.

The nonrolling samaras come in two forms: those with uncambered wings—uncambered airfoils—and those with cambered wings. The uncambered samaras, such as those of maples, have well-defined, thickened leading edges and thin trailing edges forming a crude, but recognizable, airfoil (Fig. 3.6), with no camber. Since they are not cambered, they have no defined upper or lower surface. As a result, they can autogyrate with either surface up. Imagine putting a mark on one side of a maple samara's wing. When thrown

* The terms *autorotation* and *autogyration* are not used consistently by researchers. *Autogyrating* is used by Steven Vogel [8, 41] and Carol Augspurger [42] to refer to helically gliding samaras. These samaras produce lift in the same way as an autogyro, which is a craft that looks like a helicopter but has an unpowered rotor and a separate propeller for thrust. They reserve *autorotation* to refer to the rolling, Flettner-rotor rotation of ash and poplar samaras. These are excellent, unambiguous terms that distinguish between samaras that rotate about their long axis while gliding and those that do not. Unfortunately, these terms are used irregularly. Because *autorotation* also refers to an unpowered descent of a helicopter (an emergency procedure where the helicopter functions as an autogyro), some researchers have used *autorotation* in the same sense that Vogel and Augspurger use *autogyration* [e.g., 14, 43, 44]. Autorotating autogyros like ash samaras have also been called "rolling" samaras, as opposed to "nonrolling" samaras like maples [45].

into the air, the samara is just as likely to autogyrate with the marked side up as with the marked side down.

The cambered autogyros, like those of hornbeam and pine trees, have the most complex shapes (Fig. 3.6). They have defined leading and trailing edges, as well being both cambered and curved along the wingspan. These curvatures define an obvious top and bottom, because such samaras always autogyrate with their convex sides up. No matter how often I fling a pine seed into the air, it will always rotate with the convex side upward. Finally, although the cambered autogyros have the most complex shapes among autogyrating samaras, the simple glider, *Alsomitra,* clearly has the most complex shape of all: these samaras have camber, sweepback, dihedral, and a reflexed trailing edge.

Functional Differences among Samaras

All the autogyrating samaras require a reasonably brisk wind to detach them and blow them away from their parent tree. Autogyration, with all its inherent stability, provides a parachuting effect without the necessity for a plume the size of a soccer ball. Samaras show a trade-off between simplicity of design and aerodynamic effectiveness. Autorotating autogyros like ash samaras require the longest drop to establish equilibrium rotations (1 or 2 meters). Being symmetrical, if they start falling with the wing vertically above the seed, they often fail to rotate at all, and drop like stones. The nonautorotating samaras with uncambered airfoils, such as maple seeds, generally require a shorter drop to start spinning (about half a meter or less for a medium-sized maple samara). Moreover, they only occasionally fall without spinning. The cambered autogyrating samaras start rotating with only a few centimeters of initial drop, and forcing a normally shaped hornbeam samara to drop without rotating is extremely difficult. Some trees have clearly evolved simpler but less aerodynamically effective samaras, while natural selection has caused other trees to evolve more structurally (and developmentally) complex samaras with better aerodynamic performance.

Wind, Tree Heights, and Gliding

Although wind is important for dispersing autogyrating samaras, it is less important, and in some ways even detrimental, for a simple glider like *Alsomitra.* Strong winds are usually gusty, which challenges the stability of these flying wings. Furthermore, the flying cucumbers only glide at 1 or 2 meters per second, so even a fairly mild wind could cancel or negate the samara's progress. The Indo-Malaysian rain forests where *Alsomitra* are found have unusually tall canopies, 45 to 60 meters [18]. In contrast, canopy heights in Panamanian, Amazonian, and African rain forests are much lower (30 to 40 meters). The unusually tall trees of Southeast Asian rain forests may reduce or block most winds, as well as providing unusually high launching sites for the flying cucumbers. Flying lemurs, *Draco*, flying frogs, and parachuting snakes and lizards are also found in these rain forests. Robert Dudley and Phil DeVries suggest that the abundance of gliding animals in Indo-Malaysian forests might be related to these tall canopies, and it seems reasonable that some of the same factors might have favored the evolution of the simple gliding samaras of *Alsomitra*.

Soaring: Staying Aloft by Gliding

Gliding is gravity-powered flight where the movement of the glider always has a downward tilt. But many birds are capable of *ascending* without flapping their wings, and this is called *soaring*. Soaring uses energy in the atmosphere, such as rising air currents, to increase time aloft while gliding. Vultures, albatrosses, eagles, storks, and many hawks are masters of soaring. Several extinct flying reptiles (pterosaurs) had dimensions and wing shapes that suggest that they, too, were soarers [19]. Because lift-to-drag ratios are low at low Reynolds numbers, insects are generally poor gliders. Only the largest insects have much hope of soaring, and then only under special conditions. Conversely, airplanes can soar quite well if specifically designed to do so—such craft are called sailplanes (Chapter 2).

How does a bird go about soaring? The usual trick is to find air that is rising as fast or faster than the gliding bird's sinking speed. For example, a turkey vulture (North American "buzzard"),* might glide at about 13 meters

---

* The birds called buzzards in Europe are broad-winged buteo hawks, closely related to the North American red-tailed hawk. The early European settlers in North America saw vultures soaring and mistook them for the hawks with which they were familiar. Thus, *buzzard* has come to mean *vulture* in North America. Indeed, in everyday use, *buzzard* has come to have a disparaging connotation in North America, which must be perplexing to European birdwatchers.

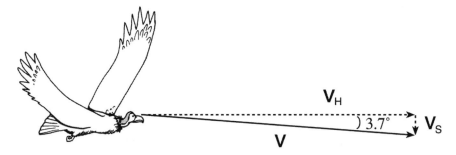

Figure 3.7. A vulture gliding at a speed of 13 m/s (v) along a path tilted 3.7° below the horizontal. Sinking speed (v_s) is 0.8 m/s, and horizontal speed (i.e., ground speed in still air, v_H) is 12.9 m/s. (J.P.)

per second, with a glide angle of 3.7° (lift-to-drag ratio = 15.5), giving her a sinking speed of about 0.8 meters per second. If the vulture can find a place where the air is rising at 0.8 meters per second, she will be able to maintain a constant altitude, and if she finds air rising faster than that, she will be able to climb (Fig. 3.7). An updraft of 1 meter per second (2 mph or 3 km/hr) is not at all dramatic, and a variety of common conditions can produce much stronger updrafts. This explains how vultures can fly for hours without ever flapping their wings.

Sources of Rising Air

Two common processes produce updrafts or rising air. When heated air rises, it is called a *thermal,* and when wind blows up a hill or over a large obstacle, it is called *ridge lift* or *slope lift.* Thermals occur when the sun heats some parts of the ground more than others. For example, a freshly plowed field may heat up faster than an adjacent meadow. The warm ground heats the air above it, and the air starts to rise. As the warm air rises, cool air from the surrounding terrain is pulled in over the warm ground to replace it, and this new air is, in turn, heated until it rises. Thermals may be continuous "chimneys" of rising air, or a series of discrete, doughnut-shaped bubbles ("ring thermals") formed at intervals by the warmed ground (Fig. 3.8). If they could be made visible, ring thermals would look like giant, rising smoke rings. Some airplane pilots and biologists disagree about the exact form of the continuous thermal chimneys. Pilots have traditionally interpreted thermals as large, tall columns of rising air, usually with a cumulus cloud marking the top of the column.

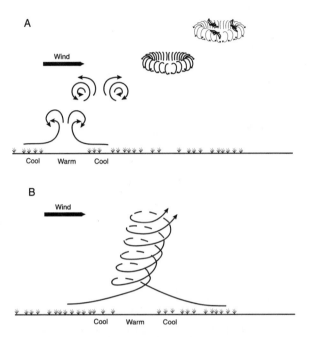

Figure 3.8. Ring and column thermals. A. As air is warmed over a warm patch of ground, it starts to rise, but curls over and pinches off. As one ring rises and blows downwind, a new one forms over the same spot on the ground. B. Columnar thermals form a more continuous "chimney" of rising air, often with a significant rotating component. Small columnar thermals with intense rotation form "dust devils." (B.Hd.)

Careful observers of animal flight find only small, localized thermal chimneys, which usually take the form of "dust devils." In a 1975 review of flight mechanics, Colin Pennycuick, a sailplane pilot and prolific researcher on bird flight, discounts thermal chimneys and recognizes only ring thermals as sources of large-scale, long-lasting updrafts [20]. In any case, thermals can rise two or three kilometers above the ground. Also, they tend to increase in size and intensity as they rise, sometimes reaching over 1000 meters in diameter [21]. Thermals are usually capped by a cloud, because the upper limit of a thermal is set by the altitude where the temperature is low enough to condense water vapor in the thermal, which cools the air and forms a cloud.

As long as the upward speed of the thermal is greater than the sinking speed of a glider, the glider will ascend in the thermal. Of course, the glider will quickly fly out of the thermal if it flies in a straight line, so it must circle to stay in the rising air. (A glider should stay on the inside of the ring, because the air on the outer edge of the ring is actually rolling downward; see Fig. 3.8.) Imagine a vulture ascending to 1500 meters above the ground by circling in a ring thermal. From this height, she will be able to fly out of the thermal and glide for about 30 minutes (traveling over 23 km) before she runs out of altitude and needs to either start flapping, find another thermal, or land. Many soaring birds use just this pattern: climbing up in a thermal, gliding a long distance, then finding another thermal in which to soar. This type of flight is an efficient way to cover long distances at a low energy cost, making it a handy way to migrate or search for food.

Slope soaring is useful when wind blows up a slope. The wind's upward component can be calculated in the same manner that the sinking speed of a glider is calculated. If the upward component of this wind is greater than or equal to the sinking speed of a glider, the glider will be able to maintain altitude. Such ridge lift has a characteristic that is both an advantage and disadvantage: ridge lift is usually predictably tied to a particular slope, so it is easy to find. But it is only usable in that fixed, local area, which usually limits foraging distances. Our vulture, for example, could use ridge lift to climb up above the top of a hill, but if she flies out over a nearby plain, she loses her source of ridge lift. The predictability of ridge lift has some interesting consequences. For example, there is a hill in Minnesota near Lake Superior known as Hawk Ridge. Because of the orientation of the hills relative to the lake, the winds reliably produce substantial ridge lift at particular times, including the migratory period for many species of hawks. At that time, these hawks use the ridge and its updrafts as a staging area to get a great boost in altitude before setting off on the next leg of their migration. At peak times, large numbers of birds of many species (falcons, buteos, eagles, vultures, etc.) can be seen cruising along the ridge, gradually working up to great altitudes.

Wind blowing over any large object can produce some ridge lift. A clump of trees, a sand dune, or a building can be a source of rising air for a skilled and determined (or desperate) soarer. I once watched a turkey vulture work the lift over a series of small woodlots and wooded borders surrounding some small agricultural fields. There may have been some slight thermal activity, but because of the time of day, most of the lift was certainly from wind blowing over the edges of the wooded areas. I observed the vulture for several minutes. It never got higher than three or four meters above the treetops, or lower than three or four meters above the fields, and it spent most of its time a meter or two above the trees at the edge of the fields. During the whole time I watched, it never once flapped its wings. This virtuoso performance continued as the big bird soared low over the trees and out of my field of view. I have also seen smaller birds like grackles and starlings take advantage of the fact that the houses in my neighborhood are all set back about the same distance from the street. When the wind is perpendicular to the street, a bird can glide down the row of houses, getting just enough of a boost over each roof to glide across the gap to the next house, effortlessly cruising the length of the block.

Soaring without Updrafts: Dynamic Soaring

A third type of soaring takes advantage of changes in wind speed with height, rather than updrafts. This is called *dynamic soaring* and is used primarily by seabirds over the open ocean. Thermals do not form over most of the ocean. Likewise, if the surface is calm or the waves are small, there are no slopes to produce ridge lift. Instead, dynamic soaring takes advantage of the boundary layer. Just as a boundary layer forms over an object moving through air, a boundary layer forms over the ocean surface when wind blows over it continuously. In this layer, the wind speed gradually changes from the full, "free-stream" wind speed well above the surface to almost still air very close to the surface. This decrease in wind speed near the surface is called a *vertical velocity gradient*. A boundary layer over the ocean may be many meters thick, depending on the free stream wind speed. In dynamic soaring, the bird climbs by gliding into the wind, starting near the surface. As the bird climbs through the boundary layer, it encounters increasingly faster wind, so that its airspeed increases enough to compensate for the drag that would slow the bird down in a steady wind (Fig. 3.9). When the bird reaches the top of the boundary layer, it turns around and glides downwind. On the downwind leg, the bird sinks back into the boundary layer while using the tailwind to build up momentum for the next climb. Just above the water, the bird wheels about to face into the wind, and then climbs up through the boundary layer once again. If the bird tried this outside of the boundary layer, where the wind speed is the same at any height, it would quickly lose airspeed and stall. But because of the increasing wind speed as the bird climbs through the boundary layer, the lift on its wings stays high even as the bird slows down relative to the ocean surface. Open ocean seabirds like albatrosses may travel thousands of kilometers this way, never needing to flap their wings [22].

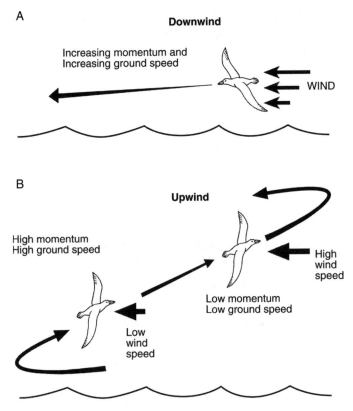

Figure 3.9. Dynamic soaring. A. Bird glides downwind in a shallow descent, building up momentum with a tailwind. B. Bird turns into the wind, climbing steeply, at first using momentum gained in downwind glide. As bird loses momentum, the faster wind speeds at greater heights maintain lift until the bird turns back downwind. (J.P.)

Designed for Soaring

The goal of soaring may differ for different animals. In some cases, such as migration or returning to the nest after foraging, the goal might be to cover the greatest distance with the least loss of altitude. In other cases, such as in searching for food or mates, the goal may be to maximize flight time. Whether a soaring animal's goal is to maximize time aloft or travel as far as possible, wings with high lift-to-drag ratios are always beneficial. Even when not operating at their best lift-to-drag ratio, such wings will allow a glider both to stay up longer and glide farther than wings with a lower maximum lift-to-drag ratio at the same speed. One of the most important factors affecting a wing's lift-to-drag ratio is its aspect ratio: long, narrow wings, with high aspect ratios, also tend to have high lift-to-drag ratios. Thus the best wings for soaring ought to be those with the highest aspect ratios (Table 3.1). Albatrosses surely possess wings with the highest aspect ratio in nature, 13 to 15 [9], and they are, indeed, the quintessential dynamic soarers. Having a wingspan of over 3 meters (10 feet) is acceptable if, like the albatross, you spend 95 percent of your time over the open ocean and the rest nesting in open, treeless areas of remote islands. But what if your habitat includes trees and shrubs, hills and ravines, rocks and other solid objects—in short, a typical terrestrial habitat? Long, narrow wings like the albatross's would be quite impractical in most terrestrial habitats. Indeed, a captive albatross released on the deck of a sailing ship is effectively trapped, because it has no space to spread its wings or for the long takeoff run needed to build up flight speed.

The wings of most birds that soar over land are much broader than albatross wings: for example, vultures, buteo hawks, and eagles all have aspect ratios of roughly 6 or 7 [9, 21]. In other words, a vulture may have a similar total wing area to an albatross, but the vulture's wings are much shorter and broader. We might expect vultures to have a much lower lift-to-drag ratio, but their specialized wing tips give them higher lift-to-drag ratios than their aspect ratios suggest. Recall from Chapter 2 that induced drag comes largely from tip vortices. Many land birds, including all the terrestrial gliders, fly with the large feathers at the tip conspicuously spread apart (Fig. 3.10). These feathers, being the largest, are called the *primary* feathers, so birds that fly with such spread feathers are said to have *separated primaries* (sometimes also called *slotted wing tips*). A simple way to think about the action of separated primaries is to consider each individual primary feather as a small, high-aspect-ratio wing. Thus, instead of having one broad wing tip, the bird has several small, high-aspect-ratio tips, which reduces the overall induced drag. Lower drag, in turn, leads to higher lift-to-drag ratios—glide angle measurements on living vultures and condors give them lift-to-drag ratios around 14–16, not much lower than the value of 18 for albatrosses [23, 24].

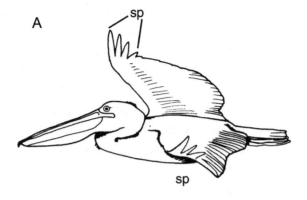

Figure 3.10. Gliding birds showing separated primary feathers (sp). A. Side view of a gliding pelican. B. Front view of a condor from slightly below. Redrawn from G. Rüppell 1977 [10]. (J.P.)

Speed and Wing Loading

High lift-to-drag ratios are always useful for soaring, but some other physical characteristics also affect the type of soaring for which a bird is best suited. Speed, which is closely tied to wing loading (body weight divided by wing area), determines whether a large bird can effectively use thermals. Most thermals are relatively narrow, especially near the ground, so a bird attempting to use the thermal needs to be able to turn tightly to stay in the rising air. Remember that the air rises fastest on the inside of a ring thermal. Just as an automobile is more likely to skid out of a turn when going around a curve too fast, it takes more aerodynamic force to hold a bird in a tight turn at a high speed than at a lower speed. Because a bird with low wing loading can fly slower than a bird with the same weight but higher wing loading, thermal-soaring birds tend to have lower wing loadings than other birds of similar size. For example, golden eagles weigh roughly the same amount as some of the smaller albatross species, but the eagles have only two-thirds of the wing loading of the albatrosses. Similarly, turkey vultures have similar body weights to great blue herons (which do not soar in thermals), but vultures only have three-quarters of the wing loading of herons [9].

In contrast, low wing loading can actually be detrimental for slope and dynamic soaring. Some hawks, such as the American kestrel (or sparrow hawk), are particularly adept at slope soaring, and they tend to have higher wing loading than their thermal soaring cousins of similar size. In both slope and dynamic soaring, a bird must to be able to progress into strong winds, so fast flight is useful. Sailplane pilots call this "wind penetration." Some competition sailplanes even have ballast tanks of water so that wing loading can be adjusted to suit conditions. Without wind penetration, a slope soarer gets blown backward, and a dynamic soarer does not have enough momentum to climb up through the boundary layer. Thus, for slope and dynamic soaring, higher wing loading is an asset. As already noted, albatrosses have much higher wing loadings than any terrestrial soaring bird (140 N/m² for a wandering albatross, compared with 60 N/m² for a bald eagle [9]). What about all the thermal-soaring birds that also make use of ridge lift? Birds can do something that no ordinary airplane does—they can adjust their wing area. If a hawk needs to glide faster, it simply bends the joints in its wings—which correspond to our wrists and elbows—to shorten the wing and reduce its area (Fig. 3.11) [25, 26, 27].

Sinking Speed

In some cases, maximizing the lift-to-drag ratio is not the best tactic for a soaring animal. When an eagle searches for a mate, its goal may be to stay aloft for as long as possible, rather than covering the greatest distance with the least loss of altitude. In this case, a low sinking speed is desirable. Because of the complex relationship between lift, drag, and speed, the flight speed that gives the minimum sinking speed—in other words, the longest time aloft—is somewhat slower than the speed that gives the maximum lift-to-drag ratio. Even though the glide angle is a bit steeper and the bird travels less distance horizontally, the slower flight speed means that the bird takes longer to come down. Interest-

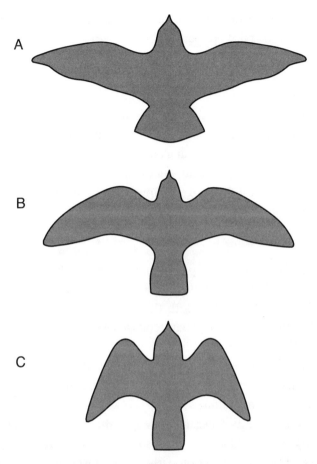

Figure 3.11. Top view silhouette of a gliding falcon. A. At low gliding speeds, the bird's wings and tail are fully extended to maximize planform area. As speed increases (B and C), the bird needs less wing area to maintain constant lift, so it flexes its wings to reduce their area. Redrawn from V. A. Tucker and G. C. Parrott 1970 [25] by permission of the Company of Biologists Ltd. (S.T.)

ingly, samaras of *Alsomitra* glide at speeds right between the speed of maximum lift-to-drag ratio and the speed for maximum flight time [12].

Who's Who among Soarers: The Effect of Size

For soaring to be at all practical, a flying animal needs a reasonably high lift-to-drag ratio, perhaps 10 or more. Because the Reynolds number decreases as animals get smaller, lift-to-drag ratio also goes down for small animals. Body size thus puts a constraint on soaring ability.

Birds

Most passerine birds (i.e., familiar songbirds) are too small to be effective at soaring: they do not have high enough lift-to-drag ratios to soar in thermals, and their wing loadings are too low for extensive slope soaring. Most soaring birds are large, and familiar examples include many hawks (particularly buteos), vultures, eagles, gulls, pelicans, and storks. Indeed, the two living animal species with the longest wingspans, wandering albatrosses and Andean condors, are both skilled soarers.

Bats

Bats are the other living group of flying vertebrates, but do any bats soar? The vast majority of bats are in the nocturnal, insect-eating group known as "microbats" (Microchiroptera). Both their small size and nocturnal habits work against soaring. As a group, bats tend to have lower wing loadings than birds [28], and just as for small birds, most bats are too small to have really high lift-to-drag ratios. Furthermore, nocturnal conditions do not favor soaring. In most places, thermals do not form at night, and wind speeds usually drop as well, which limits slope soaring.

As their name suggests, "megabats" (Megachiroptera), or flying foxes, are much larger than microbats, and at least some are day-flying. Flying foxes are at the lower end of the size range of soaring birds: the largest flying foxes have masses of about 1.5 kg, comparable to a medium-sized hawk or small vulture. Scientists do not know why there are no bats as big as eagles or condors. One biologist [1], however, has suggested that mammals are not as good at thermoregulating as birds, and also have inherently heavier skeletons. If true, bats would have an increasingly difficult time dumping the excess heat produced in flight as they get bigger, as well as having to carry a heavier skeleton. These factors would thus constrain the body size of bats more than that of birds. Also, the structure of bat wings is completely different from that of birds (see Chapter 7). Since a bat cannot take advantage of separated primaries, its effective aspect ratio is even lower. Soaring must be very rare in microbats, although some microbats might slope soar in canyons. Given the nocturnal habits of these bats, such soaring is difficult at best to confirm or refute. On the other hand, megabats in the genus *Pteropus* do soar. Several species occasionally slope soar, and the Samoan species *Pteropus samoensis* (the largest known bat, with a wingspan of over two meters) commonly uses thermals for soaring over the rain forests it inhabits [29]. *Pteropus samoensis* is restricted to virgin rain forests. Because of its large body size and solitary habits, this species requires large areas of rain forest to maintain a successful population. Sadly, this fascinating and unusual bat species is on the verge of extinction due to habitat destruction (Chapter 10).

Pterosaurs

Pterosaurs were flying reptiles related to dinosaurs [30], and like the dinosaurs, they are extinct (Chapter 7). Most pterosaurs were probably capable soarers, and pterosaurs tended to have higher aspect ratios and lower wing loadings than bats or most birds according to measurements from all known pterosaur fossils [19]. Moreover, the largest pterosaurs were much larger than any living bird that can fly. *Pteranodon* had a wingspan of 7 meters, and incomplete skeletons of *Quetzalcoatlus* suggest that it had a wingspan of up to 13 meters; paleontologists have estimated that the body masses for these animals were almost ten times higher than those of any flying bird alive today [31, 32]. The larger pterosaurs were almost certainly limited by weight and muscle physiology to soaring, and many smaller species would also have been good at soaring based on their high aspect ratios and low wing loading. The best fossils form in sediments on seabeds, so most pterosaur fossils are from marine or coastal habitats. These pterosaurs show striking similarities in wing design and body size to modern seabirds such as skimmers, which are dynamic soarers, and frigate birds, which are thermal soarers.

Insects

By and large, insects are prevented by small size from soaring—their size restricts them to low lift-to-drag ratios. Insects do soar, however, in some specialized situations. The best-known cases are migrating butterflies. Butterflies usually have very low wing loading (poor penetration) and low lift-to-drag ratios (steep glide angles), so we might not expect them to be especially effective at soaring. But monarch butterflies migrate thousands of kilometers, and, surprisingly, they make great use of soaring along the way. As with other butterflies, monarchs have low flight speeds. Even with low lift-to-drag ratios, their low flight speeds give them low sinking speeds. Thus, monarchs only need a mild updraft (i.e., a weak thermal) to stay aloft. A monarch's poor penetration means that it cannot fight the wind. Instead, monarchs only soar when the wind is blowing in the direction in which they are traveling: in other words, monarchs only soar with tailwinds. The butterfly just allows itself to be carried with the wind. On days when there are no tailwinds, monarchs may use flapping flight or simply may not fly at all.

Insects are perhaps the last flyers we would expect to be able to slope soar, but large dragonflies do occasionally take advantage of ridge lift under certain conditions. If any insects could slope soar, dragonflies would be the best candidates, because of their high-aspect-ratio wings and high wing loading (for insects). In fact, dragonflies occasionally slope soar in mild breezes blowing over sand dunes and buildings near seashores [33]. If the wind is weak enough for the dragonflies to penetrate, and the slope is steep enough to give a large vertical component to the wind's direction, dragonflies are able to slope soar as they patrol for prey or mates.

Gliding and Soaring in the Repertoire of Flying Animals

Powered flight requires a lot of energy, while gliding takes roughly the same amount of energy as sitting still. (Of course, the glider may have expended a great deal of energy getting high enough to begin its glide.) A bird may glide just to take a brief rest from flapping or because its desired path is downward anyway, just as a rider coasts downhill on a bicycle. Most flying vertebrates and some of the larger flying insects probably make use of occasional glides in this way. For those animals such as *Draco* and flying squirrels—which glide but which are not capable of powered flight—gliding may be an energetically efficient way to move long distances, as well as an excellent way to escape from predators. Indeed, the ability to glide may have been an important precursor to the evolution of flapping flight. Most researchers believe that before bats (and possibly birds and insects) evolved powered flight, they passed through a gliding stage. Soaring, in contrast, is a very specialized activity and soaring animals require specific mechanical characteristics, such as high aspect ratios or separated primaries. Soaring may be the most economical method of long-distance locomotion, and soaring animals come very close to getting something for nothing: they can range over dozens or hundreds of kilometers for no more than the energetic cost of holding their wings outspread.*

---

* Albatrosses eliminate part of even this small added cost by means of a partial locking mechanism in their shoulder joints.

Flapping and Hovering

*I*n contrast to gliding, powered flight gives an animal the freedom to fly up and down at will. Powered flight is thus immeasurably more versatile than gliding. Animals, of course, use flapping for power. Flying squirrels are limited to "downhill" gliding flights, but even the lowliest sparrow or mosquito can, by flapping, fly upward to its heart's content. The advantages that a flapper enjoys over a glider can hardly be overstated.* Flapping flyers can take off from the ground, ascend without updrafts, and fly long distances. These sorts of advantages open up an array of habitats and lifestyles that are inaccessible to gliders or terrestrial animals.

What does *powered flight* actually mean? Recall that drag is, in a sense, the

* Unfortunately, many biologists have gotten into the habit of calling powered—or flapping—flight in animals "true flight," as opposed to gliding. Aerodynamically, of course, gliding is just as much flight as powered flight. But biologists see such a quantum leap in the usefulness of powered flight over gliding that they have coined this misleading verbal shorthand to emphasize the distinction.

cost of producing lift. A flyer overcomes that cost by producing thrust. Indeed, power is force divided by time, and the "power" in question here is the thrust per unit of time needed to overcome drag. So flying animals flap their wings only to produce thrust, not to produce lift. Wings produce lift as long as an animal moves forward (that is how gliding works), but flapping allows the wings to generate a forward force while producing an upward force passively, almost as a byproduct. (Hovering is a special case: because the animal has no forward flight speed, its wing movements must produce both lift and thrust.)

The wings of animals that use flapping for power are fundamentally different from airplane wings. Airplane wings just produce lift; an airplane produces its thrust separately, with a propeller or a jet engine. Calling animal wings "wings" may even be a bit misleading, because as thrust producers, animal wings actually have as much in common with propellers as with airplane wings. In fact, helicopter rotors are probably more similar to animal wings than either airplane wings or propellers, in that helicopter rotors also produce both lift and thrust.

Although at first glance, a flapping bird or bee wing does not appear to have much in common with a helicopter rotor, they do have some general similarities.* Obviously, a helicopter rotor rotates continuously around a central shaft. A flapping wing rotates as well, it just does not rotate continuously. A bird's wing swings in an arc around its shoulder joint, but it reverses direction every half stroke. Also, helicopters and flying animals adjust their flight speed in a similar way. Helicopters hover with the rotor blades rotating in a horizontal plane. To move forward, the helicopter tilts the nose down, which tips the plane of rotation of the rotor forward. The steeper the tilt of the rotor, the faster the helicopter flies forward. Flying animals actually flap their wings in a path tilted from the vertical when they fly forward, down and forward on the downstroke, and up and backward on the upstroke. To fly fast, animals make the stroke more vertical by emphasizing the up-and-down component of the movements, similarly to the helicopter tipping its rotor forward. Conversely, to fly slowly, animals move their wings more horizontally (by emphasizing the fore-and-aft component), just as a helicopter keeps its rotor horizontal to hover. We shall see that one method flying animals use to hover is to flap their wings entirely horizontally.

* Steven Vogel has used the helicopter analogy for years in teaching, although as far as I know, he has never published it.

The Flapping Motion: How Do Wings Flap?

When a flying animal flaps its wings, the details of the wing movements are too fast to see clearly, and a casual observer may get the impression that the flapping movement is a simple up-and-down oscillation. In fact, animals flap their wings using several subtle but important departures from simple vertical movements. Much of our knowledge of the details of wing-beat patterns comes from observations of animals flying in wind tunnels, either freely or on some sort of tether. Some animals can be trained to fly freely in wind tunnels. For example, biologists have successfully photographed large and small bats, parakeets, pigeons, falcons, and hawks flying in wind tunnels [1, 2, 3, 4]. Many animals, however, are not amenable to wind-tunnel training. Moreover, techniques like high-speed cinematography or high-speed video provide only a limited field of view, and freely flying animals seem to fly everywhere except in the view of the camera. Tethering the animal avoids this problem (see, e.g., [5]), but introduces a new one, because assessing the errors or artifacts that a tether might cause for any given animal is very difficult. Researchers thus make strenuous efforts to prevent the tether from interfering with the animal's flight, with the result that the general patterns illustrated in tethered and untethered flight are often almost the same.

Details of the wing movements of a variety of flying animals can be seen in photographs and movies taken of animals flying both freely and tethered in wind tunnels. When played back at normal speed, high-speed films and videos show the movements in slow motion, and one of the most striking aspects of wing motion they reveal is that, as mentioned earlier, animals do not flap their wings vertically. In forward flight, the wing stroke is usually tilted about 30° from the vertical (or 60° from the horizontal). In other words, the wings move down and forward during the downstroke, and they move up and back during the upstroke. Birds, bats, and insects all flap their wings this way, although the exact angle may vary a good bit depending on the flight speed.

Furthermore, the wing tip does not follow the same path relative to the animal's body on the upstroke and downstroke. If we could fly in formation alongside a flapping animal and view its wing-beat in slow motion, we would see that its wing tip moved in a curved path. The wing tips of some animals follow a relatively simple path, such as the oval tip path of the albatross (Fig. 4.1). Some tip paths are more curved, as in the locust and fruit fly, and others form

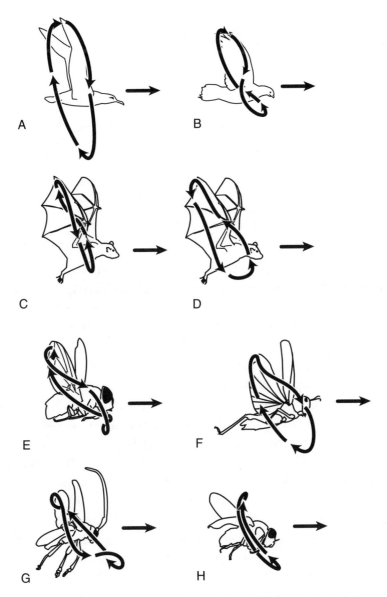

Figure 4.1. The arrows show the path of the wing tip relative to the body for a variety of flyers. A. Albatross, fast gait [18]. B. Pigeon, slow gait [67]. C. Horseshoe bat, fast flight [68]. D. Horseshoe bat, slow gait [68]. E. Blow fly [69]. F. Locust [21]. G. June beetle [17]. H. Fruit fly [70]. All figures redrawn from data in sources indicated. (S.T.)

Figure 4.2. Path of flapping wing tips relative to the air. Dashed curves show path of wing tip relative to body (same as Fig. 4.1), solid curves show wing-tip path relative to air. A. Albatross, fast gait [18]. B. Locust [21]. C. Blow fly [69]. All figures redrawn from data in sources indicated. (S.T.)

figure-eights, as in the pigeon. Some tip paths are quite intricate. June beetles and blow flies, for instance, have all sorts of complex loops (Fig. 4.1). Scientists do not yet understand why some animals use strokes with such complex tip paths. The loops and bends may play some as-yet-undiscovered role in enhancing lift or thrust production, or they may simply be a byproduct of the anatomy of the joint at the base of the wing.

Aerodynamically, the movements of the wing relative to a flying animal's body are not nearly as important as the movement of the wing relative to the air. If the wing tip left a visible path as the animal flew through the air, we would rarely find smooth, symmetrical waves, which researchers call simple harmonic motion. The larger the animal and the faster the flight, the closer the tip path approaches simple harmonic motion (Fig. 4.2). In other words, the tip paths of large, fast-flying animals look almost like a smooth, symmetrical wave,* but as the animals get smaller and their flight speeds get lower, the tip paths become increasingly distorted and asymmetrical. The upstrokes become vertical or even tilted backward relative to the direction in which the animal is flying.

The upstroke and downstroke also tend to differ in other ways. For example, the downstroke usually lasts longer than the upstroke. Again, this asymmetry is largest in slow flight, and may be small in fast flight. Table 4.1

---

* Such a wave is described mathematically as a *sine* wave, which represents perfect harmonic motion.

TABLE 4.1 *The Ratio between the Downstroke and Upstroke for a Variety of Flying Animals*

Animal	Downstroke:upstroke ratio
Albatross[a]	1.06
Vulture[a]	1.4
Bat (fast)[b]	1.31
Bat (slow)[b]	2.1
Dragonfly[c]	1.37
Honeybee[d]	1.3
June beetle[d]	1.5
House fly[d]	1.5
Locust[e,f]	1.9
Fruit fly[g]	1.7

Sources: a. J. M. V. Rayner 1991b [18]. *b.* J. M. V. Rayner et al. 1986 [13]. *c.* D. E. Alexander 1986 [5]. *d.* M. F. M. Osborne 1951 [72]. *e.* W. Nachtigall 1974 [73]. *f.* T. Weis-Fogh 1956 [21]. *g.* J. M. Zanker 1990 [70].

shows the ratio of the downstroke to the upstroke duration for a number of flying animals: a ratio of 1.0 means that the downstroke is exactly as long as the upstroke, while a ratio of 1.3 means that the downstroke lasts 30 percent longer than the upstroke. At the high end of the size and speed scale, the albatross has nearly equal upstrokes and downstrokes, but at the low end, the locust's cruising downstroke lasts almost twice as long as the upstroke, and the bat's downstroke in slow flight is more than twice as long as the upstroke.

Finally, a flapping animal holds its wings at different angles of attack on the downstroke and upstroke, with the downstroke at a higher angle and the upstroke at a lower (or even negative) angle to the air flowing over the wing. Birds even change the area of their wings on the upstroke and downstroke. All of these differences between the upstroke and downstroke are crucial for thrust production: the only way to get thrust by flapping is if the downstroke and upstroke are different. If the downstroke and upstroke were symmetrical, and used the same angle of attack and had the same duration, their horizontal components would cancel and the wings would produce no net thrust.

Gaits of Flying Animals

Just as terrestrial animals use different gaits—walking or running—for different speeds, flying animals also use different wing-beat patterns or *gaits* for

different flight speeds. Flying animals use these differences in wing-beat patterns in exactly the same way that horses or humans use different gaits. At a
slow flight speed, the *slow* gait uses less energy, just as it takes less energy to
walk than to run at low speeds. Above a critical speed, the *fast* gait is more
economical, just as running uses less energy than walking at high speeds.
"Slow" and "fast" refer only to the animal's flight speed, not the wing movements. In fact, an animal may move its wings faster in the slow gait than in
the fast gait. Most birds and many bats can use both slow and fast gaits. The
gaits are defined and named by what happens to their trailing vortices. In the
slow, or *ring vortex,* gait, the trailing vortices form rings after each downstroke, like a series of smoke rings (Fig. 4.3). In the fast, or *continuous wake,*
gait, the trailing vortices form undulating cylinders streaming back from
each wing tip [6], prompting one researcher to dub this the "roller-coaster"
gait [7] (Fig. 4.3). Some insects may use the slow gait, but most insects (and

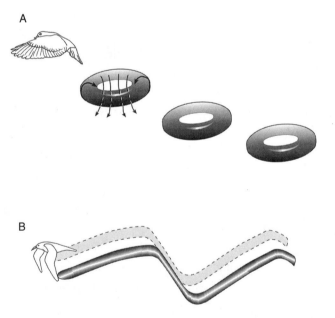

Figure 4.3. The two common flapping gaits in flying vertebrates. A. In the slow, or
ring vortex, gait, the animal's wings shed a doughnut-shaped vortex ring into the
wake at the end of each downstroke. B. In the fast, or *continuous wake,* gait, the tip
vortex streams off the wing continuously, forming an undulating cylinder trailing
behind each wing tip. (J.P.)

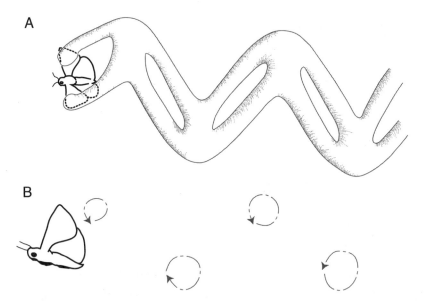

Figure 4.4. The ladderlike wake of many insects. A. Three-dimensional representation of wake structure. B. Cross section through wake to show the direction of rotation of the vortices. (Most wake measurements on insects have been made on tethered subjects, and some researchers are concerned that tethered wakes may differ significantly from wakes in free flight.) A and B redrawn from A. K. Brodsky 1994 [17], by permission of Oxford University Press. (S.T.)

perhaps some bats) appear to use a gait somewhat intermediate between the slow and fast gait. The wake of the gait used by insects is more complex than the slow or fast gaits, somewhat like a wavy ladder (Fig. 4.4). (To my knowledge, no one has given this insect gait a name.)

In nature, the wake of a flying animal is invisible, but the relationship between the flapping movements and flight speed gives a hint of which gait an animal is using. If an animal's wing tips move vertically (up and down) faster than the whole animal flies through the air, the animal is probably using the slow flapping gait. If the animal's flight speed is much greater than the flapping movements of the wing, the animal is most likely using the fast flapping gait. Thus, the slow gait actually requires faster wing movements, because the animal's flight speed is lower. The researchers who originally described separate flight gaits envisioned the different gaits as discrete and nonoverlapping, so animals would shift between them abruptly. Some scientists now think that these gaits may actually be extremes on a continuum. If so, then flyers could make gradual, rather than abrupt transitions between gaits.

The Downstroke: Most Lift and Thrust

Most of the useful aerodynamic forces are produced during the downstroke, when the wing moves down and forward. The wing's orientation at the midpoint of the downstroke is shown in Figure 4.5. Remember that the lift and drag of the wing are defined relative to the direction of the airflow over the wing, *not* relative to the directions in which the animal flies or gravity acts. Specifically, lift is perpendicular and drag is parallel to the direction of the airflow. The vector sum of the lift and drag is the resultant force (R). The resultant force is tilted forward during the downstroke, and this forward tilt provides thrust (Fig. 4.5). In other words, the resultant force includes both a vertical component (the upward force) and a forward component (thrust). The upward force supports the animal's weight, and the thrust pushes the animal forward. The thrust must be large enough to overcome the animal's drag, which includes the induced drag of the wings and the viscous and pressure drag of the wings and body. Even with all of these sources of drag for the thrust to overcome, the upward force is usually a much larger component of the resultant force than the thrust. For example, with a moderate lift-to-drag ratio of 8.0, an animal can produce eight times more lift than drag. The up-

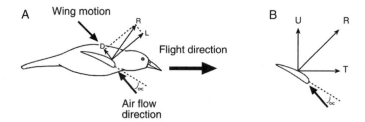

Figure 4.5. Movements of and forces on a flapping wing during the downstroke. A. The wing moves down and forward during the downstroke, so the lift (L) and resultant force (R, vector sum of lift and drag, D) are tilted forward. B. Because the resultant force is tilted forward, it has a forward component, thrust (T) as well as the weight-supporting upward component (U). (J.P.)

ward force will be slightly smaller than the lift, so with this lift-to-drag ratio, the animal would need thrust equal to just over one-eighth of its weight. Animals, therefore, produce a relatively small thrust to support a given amount of weight. In many cases, the upstroke contributes little or no useful aerodynamic force, and when this happens, the thrust and the upward force on the downstroke must be about twice the animal's drag and weight in order to compensate for the inactive upstroke.

Many descriptions of animal flight, both popular and technical, describe the forces on flapping wings this way: when an animal flaps its wings, the inner part of the wing (near the body) produces lift, and the outer part produces thrust. This statement is only partly true. As the wing sweeps down and forward in the downstroke, the wing near the root experiences a relative wind mostly from the animal's forward flight (Fig. 4.6). The tip, however, experiences a relative wind that is caused by a combination of the forward flight speed and the wing's flapping motion. The tip thus experiences a wind that is faster than at the root, and that approaches the wing from below. The speed and direction of the relative wind changes gradually along the wingspan from that at the root to that at the tip. (Animal wings usually have a built-in twist, so that the angle of attack anywhere along the wing is appropriate for the airflow direction at that location.) Because of these differences in the direction of the relative wind along the span, the resultant force near the wing root is nearly vertical (as it would be in a glide), but the resultant force is tilted forward near the tip. How much the resultant force is tilted forward near the tip is determined by the relationship between the forward flight speed and the wing-beat frequency and amplitude. The higher the frequency and/or amplitude, the greater the tilt; the higher the flight speed, the smaller the tilt. The amount by which the resultant force is tilted forward determines the amount of thrust.

Thus the inner part of the wing produces mostly upward force and no thrust, and the outer part of the wing provides most of the thrust. But the outer part of the wing also may produce plenty of upward force. Generally, the upward component of the force near the tip will be several times larger than the thrust component. Furthermore, the transition from no thrust (at the root) to lots of thrust (at the tip) is gradual along the length of the wing, so there is at least a little thrust production even on the inner half of the

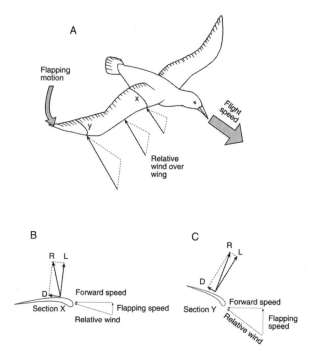

Figure 4.6. Differences in forces along the span of a flapping wing. A. The wing has little vertical motion near the base, so the air flows over it from nearly straight ahead. Farther from the base, the air approaches from an increasingly vertical direction, and the flapping motion also increases the speed of movement toward the tip. B. At cross section X (near wing base), the airflow over the wing or relative wind is nearly horizontal, so the lift is nearly vertical. C. At cross section Y (near the tip) the relative wind has a large vertical component due to the flapping motion, so the lift is tilted far forward. D: drag force; L: lift force; R: resultant force.

wing. A more accurate general statement would be that the upward force ("lift" in the colloquial sense) is produced along the whole span of a flapping wing, but thrust is mostly produced on the outer portion of the wing.

The Upstroke: Changes with Gait

The downstroke of the slow gait only differs from the downstroke of the fast gait in degree: as I mentioned earlier, animals move their wings faster in the

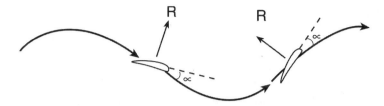

Figure 4.7. If a wing flaps with the same speed and angle of attack on the upstroke and downstroke, the forward and backward components cancel, and no net thrust is produced. (Actually, due to induced drag, a slight reverse or rearward thrust would be produced if speed and angle of attack were perfectly constant.) (J.P.)

slow gait. Indeed, frequency, amplitude, and angle of attack all tend to be higher in slow flight than in fast flight.* The upstrokes, however, differ substantially in slow and fast flight. Moreover, the upstrokes can differ even among animals using the slow gait. Insects and some bats produce useful aerodynamic forces during the upstroke—an *active* upstroke—while birds and other bat species produce very little force on the upstroke—a *passive* upstroke.

The upstroke and downstroke must be different for an animal to produce thrust. For example, if a flapping wing kept the same angle of attack relative to the air on the upstroke and downstroke, no thrust would be produced: the backward tilt on the upstroke would cancel the forward tilt on the downstroke (Fig. 4.7).† Clearly the upstroke needs to be modified in some way so that it does not cancel the thrust produced on the downstroke. The details of this modification depend on whether the animal is using the fast gait, the slow gait with an active upstroke, or the slow gait with a passive upstroke.

* This counterintuitive result comes about because the aerodynamic forces increase as the speed goes up, so smaller flapping movements suffice; lift increases with the square of the velocity. This decrease in movement does *not* mean the animal does less work as the speed increases. Drag, and hence, thrust, go up as speed increases, particularly in fast flight. Even though the movements are smaller, the flight muscles are pulling harder at higher speeds.

† A wing with a constant angle of attack over a symmetrical stroke ought to produce a net drag or reverse thrust, because the resultant force is tilted slightly back by the induced drag! Although Jeremy Rayner says that the horizontal components cancel, a net negative thrust is plainly visible in his own figure 2 [74].

Fast Flight

In high-speed forward flight, the wing's angle of attack is reduced on the upstroke, so that it produces much less force than on the downstroke (Fig. 4.8A). An albatross or a seagull flying rapidly uses this type of upstroke, which gives its wings a rather stiff appearance [8]. Some upward force and some reverse thrust are produced during this type of upstroke, but the down-stroke thrust is sufficient to compensate for the negative thrust of the up-stroke. In addition, birds and bats flex their wings slightly during this type of upstroke, which reduces the wing area (and hence the forces) a bit relative to the downstroke. At intermediate flight speeds, and especially for smaller birds, the flexion during the upstroke becomes more pronounced. At the be-ginning of the upstroke, the wing is sharply flexed, pulling it in close to the body and greatly reducing its effective surface area (Figs. 4.8B, 4.9). In some

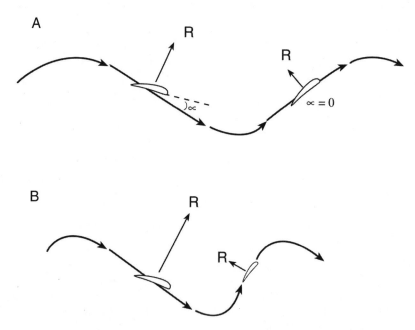

Figure 4.8. Thrust can only be produced if the downstroke and upstroke are not symmetrical. A. Reduction of the angle of attack on the upstroke reduces the overall aerodynamic forces, so the downstroke forces dominate the stroke. B. Reducing the area of the wing on the upstroke by flexing it (represented here by a smaller cross section) can reduce the upstroke forces as much or more than reducing the angle of attack. (J.P.)

cases, the area is reduced enough so that little force is produced on the up-stroke, and this may begin to resemble a slow flight upstroke.

In fast forward flight, the upstroke of a bat is essentially the same as a bird's. However, because of the difference in their wing structures, bats do not flex their wings in flight as much as birds, so their upstrokes become quite different from those of birds at slower speeds. The surface area of a bird's wing consists mostly of feathers, which can slide over each other as the wing is flexed and still maintain a smooth surface. Bat wings, in contrast, are mostly a thin membrane, supported by the arm bones and the enormously elongated finger bones. Given the stretchiness of the wing membrane, bats can flex their wings a bit, reducing the span by about 20 percent [2], but they cannot flex their wings too much or the wing membrane will go slack. Slack membranes are inefficient aerodynamically, because drag goes up, and struc-turally risky, because the trailing edge will flutter, which could tear the mem-brane or break a finger bone. So, at high speeds, a bat's upstroke looks about like that of a bird, but at more moderate speeds, bats do not flex their wings nearly as much as a bird of a similar size.

Slow Flight

The differences between birds, on the one hand, and bats and insects, on the other, are most pronounced in slow flight. (Remember that in slow flight, the speed of the wing tips due to flapping is greater than the animal's flight speed.) In the slow flight of birds, the wings are very strongly flexed during the upstroke (Fig. 4.9). Some earlier researchers believed that birds produce thrust on the upstroke in slow flight [8, 9]. More recently, Geoffrey Spedding has studied the wakes of birds in flight and found no evidence of thrust pro-duction during the upstroke of birds in slow flight [10, 11]. Insects and some bats, on the other hand, seem to generate thrust on the upstroke in slow flight [1, 12]. An animal must be able to do three things with its wings to pro-duce thrust on the upstroke: (1) move the wing backward faster than the for-ward flight speed of the animal, so that the wing actually moves backward through the air; (2) twist the wing so that the leading edge is pointing up, or even up and back; (3) flatten or reverse the camber (Fig. 4.10). If the animal can move the wing backward with the leading edge up, the wing assumes a negative angle of attack; the ventral (anatomically lower) surface of the wing functions as the top of the airfoil, so that the lift produced in this orientation

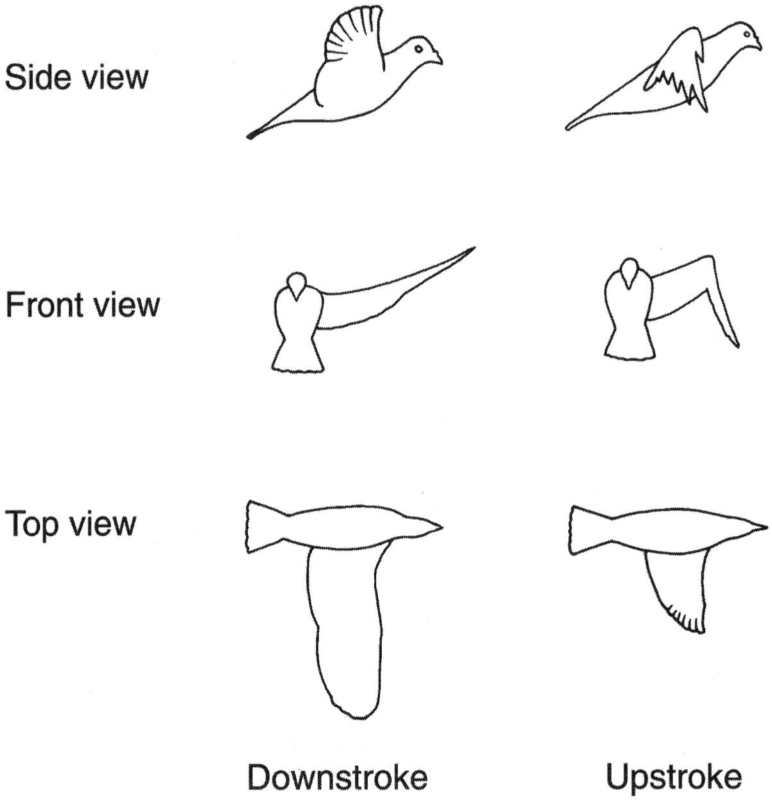

Side view

Front view

Top view

Downstroke   Upstroke

Figure 4.9. Flapping pattern of a pigeon from slow-motion film. Note how the wing is fully extended on the downstroke, but sharply flexed on the upstroke. Redrawn from R. J. H. Brown 1953 [8] by permission of the Company of Biologists Ltd. (J.P.)

is tilted strongly forward. The lift will be increased (and drag reduced) if the camber can be flattened or reversed. The amount of upward or downward tilt of the lift direction depends mainly on the speed with which the wing moves backward. An upward tilt is obviously useful, but a downward tilt may be tolerable, depending on the amount of thrust needed and the upward force produced on the downstroke.

Why are bats and insects able to produce thrust on the upstroke, whereas birds cannot? The answer may be related to the fact that neither bats nor insects can flex their wings along the span as much as birds. Bats must keep the wing membrane taut and can only flex their wings slightly. Insect wings are supported by rigid veins, which can only be adjusted at the wing base, so they cannot flex their wings at all (Chapter 1). Bats and insects are thus both constrained to use upstrokes with extended wings. In contrast, birds appear to have evolved an upstroke that minimizes aerodynamic forces, flexing their wings dramatically while keeping a relatively smooth surface because of their overlapping feathers. Bats and insects, forced to flap with extended wings, evolved the ability to rotate the leading edge of their wings up past the vertical. The structure of bat and insect wings also makes camber reversal much easier than on a feathered wing. Both of these features allow them to produce

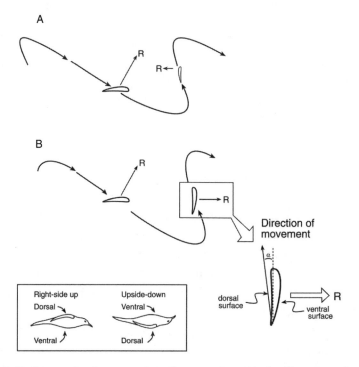

Figure 4.10. Some animals can produce thrust on the upstroke. The upper tip path is essentially the same as that of Fig. 4.8B: the animal flexes its wing to reduce wing area during the upstroke, and no thrust is produced on the upstroke. However, if the animal can move its wing upward and backward fast enough on the upstroke, the wing's ventral surface can function as the upper surface of an airfoil. (The ventral surface is the anatomical lower surface, see inset.) The lower tip path shows how the animal can adjust the angle of attack of its wing to produce lift on the ventral surface, which is directed mostly forward and acts as thrust. The diagram shows that the wing has reversed camber, which increases the amount of thrust but is not essential for thrust production on the upstroke. (J.P.)

useful aerodynamic force on the upstroke in slow flight. (Because of their small size and unique anatomy, specialized for hovering, hummingbirds' flight is probably closer to the insect pattern than to the typical bird pattern.)

Body Size: Gaits and Wing-Beat Frequencies

Body size has only an indirect relationship to flapping gaits. As a rule, large animals fly faster, and the largest flying animals may only be able to use the fast gait. Small animals may be limited to the slow gait because of their slow flight speed and low lift-to-drag ratios. Many birds and bats fall in the middle of the size range, and these animals can use either gait.

Body size can affect a bird or bat's ability to use different gaits in at least two ways. First, smaller animals tend to have shorter and broader wings, with lower aspect ratios. For reasons not fully understood, animals with low-aspect-ratio wings seem to be limited to the slow gait. Thus, most *passerine* birds (i.e., common perching birds and songbirds) tend to use the slow gait most of the time, and smaller passerines, such as chickadees and wrens, may be entirely restricted to the slow gait. Similarly, most insects are not fast enough and their wings have aspect ratios too low to use the fast gait. In contrast, large birds with high-aspect-ratio wings—gulls, petrels, and large falcons—use the fast gait most of the time. Albatrosses may be entirely restricted to the fast gait. Their wings have the highest aspect ratios, and they cannot flex or twist them enough to use the slow gait. Most birds, however, can use both gaits. Bats probably follow a similar pattern to birds. Small bats use the slow gait, and some long-winged bats also use the fast gait; at least a few bats can use both gaits [13, 14].

A combination of aerodynamic and physiological factors makes fast flight disproportionately easy for large birds and slow flight easier for small birds. The lift-to-drag ratio improves as the Reynolds number increases, so larger animals have less drag to overcome for a given amount of lift production. Conversely, small animals have to overcome greater drag for a given lift production, and there is simply too much drag for most insects to fly fast enough to use the fast gait. Big animals may find fast flight easier, but what restricts them from flying slowly? Geometry, in the form of the *surface-to-volume ratio,* holds the key.

If a body keeps the same general proportions but increases in size, its increase in surface area will be proportional to the square of its length increase, and its increase in volume—or mass—will be proportional to the cube of its length increase. In other words, if it gets twice as long, it will have four times more surface area and eight times more volume or mass. Thus, as an animal gets larger, its mass increases faster than its wing area. This increases its wing loading, so it will have to fly faster to offset the extra weight increase. To fly slowly, an animal must flap its wings faster to make up for its slower flight speed, and a larger animal will need to flap disproportionately faster. Moreover, the surface-to-volume ratio affects its muscles as well. Muscle strength depends mostly on the cross-sectional area of the muscle (Chapter 6), and cross-sectional area also follows the "square-cube" relationship of the surface-to-volume ratio. Just as with wing area, as an animal's body size increases, its cross-sectional area does not increase as fast as its body mass [15] (Fig. 4.11). In other words, the muscles of a big animal will produce less force *per kilogram of animal* than a small animal's muscles. Big animals' muscles also tend to be slower, further reducing power output (power being equal to force times speed) [16]. Slower, less powerful muscles will limit how fast a larger animal can flap its wings. Thus, by its effects on wing loading and muscle cross-sectional area, the surface-to-volume ratio makes slow flight more difficult for larger animals.

Specialization may be involved as well. Fast flight is more economical for large animals, so many have become specialized for fast flight. This very specialization may impair their slow flight abilities. Other large flyers have essentially given up on flapping and turned to soaring, and the specializations for soaring also interfere with slow flight abilities.

At the other end of the scale, very small flying animals have such low lift-to-drag ratios that they cannot fly fast enough to use the fast gait. They can, however, produce enough power for slow flight, and most can even hover. For example, large insects and small birds and bats can hover and use slow flight. Some small birds and bats may be able to use fast flight under some conditions, but small flyers probably use the slow gait most of the time. Some large insects may use a modified fast gait, but the exact details of this gait are not yet known [17]. Medium-sized birds and bats can probably use both slow and fast gaits, but they can only hover for a couple of wing-beats; larger birds and bats, and those with high-aspect-ratio wings, may be entirely limited to the fast gait [18]. Hovering requires the most power of any type of flight,

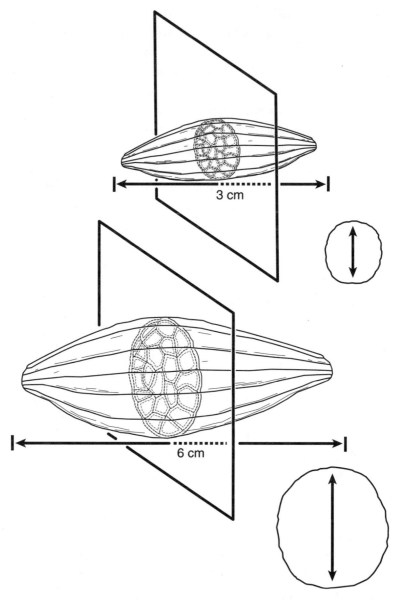

Figure 4.11. Effects of size increases on area-mass (*surface-to-volume*) relationships. The bottom muscle is twice as long (and twice as wide) as the top muscle. The lower muscle has four times as much cross-sectional area as the upper one, so the lower muscle can produce four times the force of the upper muscle. However, the lower muscle also weighs eight times as much as the upper muscle, demonstrating that weight increases faster than strength in muscles. (S.T.)

because there is no airflow over the wings from forward flight. Thus wing movements alone must produce all of the aerodynamic forces.

Just how fast do animals flap their wings? Flapping frequency is related to size, with larger animals using lower wing-beat frequencies, and smaller animals using higher wing-beat frequencies. Given that flying animals cover an enormous range in body size, from mosquitoes weighing a few one-thousandths of a gram to condors weighing over a dozen kilograms, a large range in flapping frequencies should be no surprise. Size is not the only factor, however; a number of other features, such as wing loading, aspect ratio, and muscle properties, also affect wing-beat frequency. The highest animal wing-beat frequency is only about a thousand times faster than the slowest, rather than the ten-million-fold difference we see in the mass of flying animals. Some tiny gnatlike midges (distant relatives of houseflies) hold the current high-frequency record, at over 1000 beats per second [19]. Mosquitoes have wing-beat frequencies of around 500 to 600 beats per second. Honeybees and houseflies are a bit lower, at 150 to 200 beats per second; June beetles flap at about 60 to 90 beats per second, which is probably typical for large beetles [20]. Large moths, dragonflies, and grasshoppers all flap at about 25 to 35 beats per second, while large desert locusts flap at 17 to 20 beats per second [5, 20, 21]. Hummingbirds operate in the same range as large insects, approximately 30 to 90 beats per second [22]. Most birds and bats flap much more slowly. Typical songbirds flap at approximately 10 to 20 beats per second [23]. R. H. J. Brown's pioneering high-speed films of flying birds showed that pigeons flap at about 14 or 15 beats per second in slow flight and 15 to 18 in fast flight; his data also show a gull in fast flight flapping at 10 or 12 beats per second [24]. We find the largest birds at the low end of the frequency scale. My impression from casually observing large hawks, eagles, and vultures is that they only flap 2 or 3 times per second. The Andean condor is the largest living animal that uses powered flight (although it much prefers soaring), and films of condors show that they flap from 1 to 3 beats per second [25].

Can Bumblebees Fly? Unsteady Aspects of Flapping

The often-asked question, "Didn't some engineer prove that bumblebees can't fly?" is intentionally absurd, because the questioner is quite aware that bumblebees are perfectly capable of flying. The question implies, however, that aerodynamicists (and probably scientists in general) do not know as much as they should about animal flight. In the version of the story tracked down by John McMasters [26], a prominent aerodynamicist was once dining with a biologist. The latter brought up the flight abilities of bees and flies and asked about their aerodynamic capabilities. The aerodynamicist made some quick, back-of-the-envelope calculations and came to the conclusion that bumblebees should not be able to produce enough lift to carry their weight. The aerodynamicist assumed that insect wings were easy-stalling flat plates with limited lift coefficients. But insect wings are cambered—increasing their lift coefficients—and quite stall-resistant because of their small size. Unbeknownst to either the aerodynamicist or the biologist, insects also make use of lift-enhancing "tricks," which were only discovered in the past twenty years or so. In any case, the poor fellow figured out a major part of the problem on his own after looking at an insect wing under a microscope. The original story was so much more entertaining than the correction that it spread widely through the engineering community and eventually appeared in the popular press.* Though this entertaining story tells of absurd results, it speaks to the fact that conventional aerodynamics *does not* always fully explain the forces measured by researchers on flying animals.

"Quasi-Steady" Approximation

In physics, a *steady* motion is one where there is no acceleration. An object in steady motion does not speed up, slow down, or change directions: in other words, its speed and direction do not change over time. A mathematical description of unsteady motion is enormously more complex than a similar description of steady motion. Some of the mathematics of unsteady aerodynamics could not be solved in a usable form until powerful computers became available. Thus, until very recently, engineers focused almost exclusively on steady aerodynamics. From a practical standpoint, this was not a

---

* For an entertaining investigation of the source of this story, along with a readable account of an engineer's view of insect aerodynamics, see John McMasters' article [26]. The "bumblebees can't fly" story was known to engineering students in Germany in the 1930s, but a similar story was being told in other parts of the world (including Asia and North America) at about the same time, which suggests it had multiple origins.

major problem, because most airplane wings spend almost all of their flight time in steady conditions. Even in the situations where the airflows are unsteady (say, initiating a turn or a climb, or changing the throttle setting), effects caused by accelerations are rarely substantial.*

With conventional steady aerodynamics in mind, a small number of biologists and engineers began to study the aerodynamics of animal flight. The pioneering collaboration of the biologist Torkel Weis-Fogh and the engineer Martin Jensen produced a classic series of research papers in the 1950s and 1960s that have become a benchmark against which all later studies are measured [21, 27, 28]. Weis-Fogh and Jensen studied flight in the migratory desert locust.† They made precise and detailed measurements of locust wing movements as the locusts flew in a wind tunnel, and then set about calculating the aerodynamic forces produced by the wings over the course of a wing-beat cycle. One problem they faced was that the speed and direction of the airflow are not constant over the length of a flapping wing (Fig. 4.6). To account for this variation, Weis-Fogh and Jensen borrowed a technique from helicopter theory called *blade element analysis.* They treated the wing as a series of short strips, starting at the base and working out toward the tip. For each strip, or *blade element,* they calculated the local wind speed and angle of attack by geometrically adding the flight speed and the speed of the wing's flapping movement at that strip. The speed and angle of attack could then be used to calculate the lift and drag on that element. (Figure 4.6 shows examples of these speeds and angles for two locations on a wing; Weis-Fogh and Jensen did the calculation on three strips along the forewing and two on the

---

* The only type of airplane that spends much time in unsteady flow situations is the high-performance fighter. Most of the modern work on unsteady aerodynamics has been stimulated by the requirements of fighter aircraft, but the Reynolds numbers (and Mach numbers) are so high that this work is largely irrelevant to animal flight. At speeds near or above the speed of sound (Mach number near or above 1.0), the compressibility of air becomes important, which is never the case for flying animals.

† The African desert locust, *Schistocerca gregaria,* makes an excellent subject for flight studies because of its powerful urge to fly. When reared in crowded conditions, the adults of these large grasshoppers develop strong flight muscles and a compulsion to spend most of the daylight hours in flight. In nature, this produces gigantic swarms of locusts, which use weather fronts to move across the landscape. The biblical "plague of locusts" was undoubtedly such a swarm. Their behavior is very handy for scientists, because the locusts will often fly for many minutes or even hours in a wind tunnel.

hindwing of a locust.) Adding the lift and drag from all the elements gives the overall lift and drag at any instant.

The blade element analysis accounts for the difference in relative wind along the wingspan at any given time, but what about over the course of a whole wing-beat cycle? To calculate the lift and drag over time, Weis-Fogh and Jensen made blade element calculations at a large number of instants during the course of a wing stroke: using steady aerodynamics, they treated each instant as a brief period of steady motion and calculated the lift and drag for that instant. By averaging values over parts of the stroke period, Weis-Fogh and Jensen then calculated lift and drag over the entire upstroke and downstroke. Weis-Fogh and Jensen's approach is called *quasi-steady* because it treats an unsteady motion as equivalent to a large number of brief periods of steady motion. The quasi-steady approach assumes that any effects of acceleration and deceleration are small relative to the steady effects. If the accelerations and decelarations are relatively small, the quasi-steady approach should produce realistic results. Just how unsteady can the motion get before the quasi-steady approximation breaks down, and when it does, how can the unsteady motions be analyzed? These questions have been the subject of much animal flight research for the past two or three decades.

The Quasi-Steady Approach Breaks Down: Unsteady Effects

Weis-Fogh and Jensen found that their quasi-steady analysis described the required thrust and weight support in locusts [28]. However, when biologists started to apply the quasi-steady blade element analysis to a variety of flying animals, they found that a quasi-steady analysis often required the wings to have unbelievably high lift coefficients to produce enough lift to support the animal's weight. In other words, scientists could only calculate enough lift when they used abnormally high lift coefficients in the quasi-steady equations. The researchers were left to wonder if flying animals could really produce such high lift coefficients, and, if so, how they accomplished the feat.

Just how high were these lift coefficients? Recall that lift coefficients tend to go down at small Reynolds numbers: a crow might have a maximum lift coefficient of 1.5 in steady conditions, but insects normally have maximum lift coefficients less than 1.0 [29, 30]. Yet biologists were calculating lift coefficients of over 5.0 for a small bird, 2.6 for larger birds, over 3.0 for a large dragonfly and a bat, and 1.3 for a bumblebee [1, 31, 32, 33, 34]. These outrageously high lift coefficients were mostly computed for hovering animals, where some extreme conditions occur (see below). But some biologists also realized they had another problem: even when the lift coefficient averaged over a stroke cycle was just barely within the normal steady state range, at some point in the stroke the *instantaneous* lift coefficient must have been well above values allowed by steady aerodynamics, because there are certainly points in the stroke when the instantaneous lift coefficient is well below the average lift coefficient. A group of French biologists developed a highly sensitive method to measure the lift on a flying insect, which was rapid enough to measure changes during a single wing-beat cycle. Using the same species of locust as Weis-Fogh, they found that the instantaneous peak lift at certain points in the stroke cycle was nearly twice that calculated by Weis-Fogh and Jensen, even though the average over the wing-beat was about the same as that calculated by Weis-Fogh and Jensen [35]. Clearly, more was going on in flapping flight than a quasi-steady analysis can comfortably explain.

Researchers have attacked the problem of unsteady flapping aerodynamics from both empirical and theoretical directions. Empirically minded biologists have focused on the details of the flapping movements in an effort to find unsteady mechanisms that might boost lift above steady-state values. The most unsteady phases of the wing-beat occur during the transitions between the upstroke and downstroke ("stroke reversals"), so biologists have focused on wing movements during these transitions with useful results. Theoretical biologists have developed a general mathematical description or *model* of flapping flight that incorporates unsteady motion, yet is simple enough to make useful, practical predictions, which scientists call "testing the model." Researchers have always expected the empirical and theoretical paths to merge to produce a comprehensive understanding of flapping aerodynamics, but in many ways the merger did not really begin until the late 1990s.

Empirical Approaches

Weis-Fogh must have come to doubt the accuracy of the quasi-steady approximation fairly soon after completing his work on locusts because, within a few years, he began investigating unsteady lifting mechanisms. Shortly before his untimely death in 1975, Weis-Fogh proposed several possible unsteady lift-enhancing mechanisms [19, 36]. He discovered the *clap-fling* mechanism, which he found by observing some of the smallest flying insects. Weis-Fogh clearly expected that at the very low Reynolds numbers experienced by these tiny insects, the insects would be limited to low lift coefficients (and low lift-to-drag ratios), so that they would benefit greatly from unsteady lift-enhancing mechanisms. He was partly right: he observed the clap-fling mechanism being used by many species of small insect, but we now know that clap-fling or its variations are also used by larger animals when they need to rapidly produce a lot of lift.

The clap-fling mechanism increases lift production by eliminating the *Wagner effect*. When any wing first starts moving through the air, it must travel several chord-lengths before the lift-producing bound vortex builds up to full strength. This effect is partly a matter of acceleration and partly due to the interfering effects of the starting vortex being left behind. If we consider the beginning of the downstroke as the start of the wing's movement, then the Wagner effect reduces the lift production in the early part of the downstroke. This effect is magnified at low Reynolds numbers, when viscosity slows development of the bound vortex. For a small insect, it is quite possible for the Wagner effect to reduce lift production over most of the downstroke. The clap-fling mechanism overcomes the Wagner effect in two ways. First, it eliminates the starting vortex. Second, it causes the bound vortex to form quickly at the very beginning of the downstroke. Figure 4.12 shows how the clap-fling mechanism works. At the top of the upstroke, the wings are "clapped" together over the animal's back. Then the wings are opened like a book, with the leading edges separating and the trailing edges staying together like the book's spine. This movement is the "fling." After the fling, the wings start down into

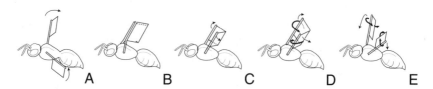

A   B   C   D   E

Figure 4.12. The clap-fling process discovered by Torkel Weis-Fogh. At the top of the upstroke (A, B), the wings are "clapped" together. At the beginning of the downstroke, the leading edges peel apart while the trailing edges stay together (C). This causes air to rush around the leading edges, forming a vortex at the front of the wing (D). As the wings move apart, these vortices become the bound vortices, producing lift almost immediately (E). Modified from T. Weis-Fogh 1975 [36] by permission of the Company of Biologists Ltd. (B.Hd.)

a normal downstroke. During the fling, air flows around the leading edge of each wing into the space between the wings. This airflow establishes a bound vortex on each wing almost instantaneously, and the bound vortex on each wing acts as the starting vortex for the opposite wing. Both vortices thus contribute useful force and minimize their interference. With little or no Wagner effect, lift production is high right from the beginning of the downstroke.

Weis-Fogh first observed the clap-fling mechanism in tiny, almost microscopic, parasitic wasps [19]. He correctly predicted that many small insects would use it in normal cruising flight. Small insects are not the only users of clap-fling, however. Because clap-fling can increase total lift production even in larger flyers, clap-fling is probably used in slow flight by some large insects, and it is used for maximum-effort climbing flight in vertebrates. For example, the slapping sound that a pigeon makes for the first few wing-beats after takeoff are caused by the wings clapping together: these birds use the clap-fling mechanism to improve their takeoff performance. Several variations on clap-fling are used by different animals [37], but they all work roughly the same way. Researchers have devoted considerable attention to the clap-fling phenomenon [38, 39, 40, 41], and one analysis predicted lift coefficients of 7.0 or 8.0 during the fling [42].

Delayed stall, or *dynamic stall,* is another unsteady mechanism that can increase lift production above quasi-steady values [7, 29, 37]. Delayed stall comes in two forms: *rotational* and *translational.* The rotational form occurs when a wing is rapidly rotated from a moderate angle of attack to one well above the normal stall angle of attack. During the rotation, the airflow stays attached, and during this brief period, the lift coefficient can dramatically exceed the maximum steady lift coefficient. Shortly after the rotation ends, separation and stall occur. Curiously, although engineers discovered it many decades ago [43], researchers have apparently ignored this phenomenon and its consequences for both animal and airplane flight until recently. Translational delayed stall occurs when a wing at an angle of attack well above the stall angle starts moving from rest. In this case, it is the Wagner effect itself that allows the bound vortex to continue to grow. For the first four or five chord-lengths of travel, lift can build up to values much higher than the maximum steady-state values. The wing will soon stall, unless the angle of attack is reduced to an angle below the steady stall angle of attack.

The conditions at the extremes of the stroke of a flapping wing seem tailor-made for using delayed stall. For example, if an animal supinated its wings (tilted the front of the wing up) just before the end of the downstroke rather than waiting until the beginning of the upstroke, the animal could get a transient increase in lift from rotational delayed stall. Similarly, if the animal started the downstroke with its wings at a very high angle of attack, it could get a lift boost at the beginning of the downstroke from translational delayed stall. A substantial portion of the lift may well be produced by delayed stall at the extremes of the stroke, especially by small animals. Indeed, Michael Dickinson and his colleagues recently made model tests based on the wing-beat pattern of tiny fruit flies, and they showed that both forms of delayed stall can contribute significantly to lift production [44]. The complex loops in the wing-tip path at the stroke reversals that we saw earlier (Fig. 4.1) suggest that insects may make use of delayed stall or similar unsteady effects at the extremes of their strokes.

Theoretical Approaches

While some researchers looked for unsteady "high-lift" mechanisms in flapping flight, others began to develop a general framework that could be used to analyze the aerodynamic forces and energetics of flapping flight. Jeremy Rayner and Charles Ellington have been at the forefront of this effort. Rayner used vertebrate forward flight as his starting point [45, 46], whereas Ellington began with hovering insect flight [47, 48]. Although they worked independently, both focused on the fluid mechanics of the wake behind the flapping wing, especially the vortex components. They based their analyses on Newton's Third Law, the law of equal and opposite reactions. They pointed out that any aerodynamic forces produced by the wings must be balanced by changes of momentum in the air. (Momentum is mass times speed, and the rate of change of momentum is a force.) Adding momentum to air produces vortices, so both of these theories predict the appearance of various types of vortices in the wakes of flying animals. In fact, these theories are usually referred to as *vortex* or *vortex wake* models. Vortices do, indeed, form in the wakes of flying animals [11, 17, 49]. If the shape and intensity of the vortices in the wake can be measured in enough detail, these models can predict the aerodynamic forces on the wings, which, in turn, can be used to calculate the aerodynamic power and energy requirements of a flying animal.

The form of a given wake depends largely on the gait. In the slow flight gait with an inactive upstroke (as in birds), the wake consists of a series of ring-shaped vortices, one for each downstroke [11]. The vortices probably start out somewhat rectangular, with the four sides consisting of a starting vortex formed at the beginning of the downstroke, a pair of tip vortices formed throughout the downstroke, and a stopping vortex that is shed by the wings at the end of the downstroke. These vortices quickly become oval or round, and would look exactly like smoke rings if they were visible (Fig. 4.3A). Geoffrey Spedding used some clever techniques to photograph this type of wake behind flapping jackdaws (small relatives of crows), as did Rayner and his colleagues with bats [13]. These researchers photographed birds or bats as they flew through chambers containing clouds of tiny, helium-filled soap bubbles. The photos were actually a high-speed sequence of stereo pairs, which allowed the biologists to map the three-dimensional movements of soap bubbles in great detail. The movements of the soap bubbles are a close approximation of the air movements, clearly showing the vortices in the wakes of the flying animals.

In the fast gait, the vortex wake is made up of a pair of undulating vortices trailing along in the path followed by each wing tip (Fig. 4.3B). The trailing vortices are larger and farther apart on the downstroke, and smaller and closer on the upstroke. Spedding calls this the "roller-coaster" wake [7], and he observed it behind a flying kestrel (small falcon) [49]. A third type of wake appears behind animals using slow flight with a thrust-producing upstroke (insects and some bats). In this gait, an undulating, ladderlike pattern of vortices forms behind the wings (Fig. 4.4). The longitudinal elements represent tip vortices and the transverse elements represent stopping vortices shed at the end of each upstroke and downstroke. Because these transverse vortices are shed alternately from the end of each upstroke and downstroke, these "rungs" across the wake alternate in their direction of rotation. This type of wake has been observed behind several types of insect in forward flight [17], and it has been predicted for some species of bats in slow flight [1].

A great advantage of the vortex wake models is that they are not concerned with the details of airflow and lift production on the wings themselves. Using some basic properties of the wing-beat—for example, wing length, wing-beat frequency, and amplitude—and measuring the size and strength of vortices in the wake, these models give a method for calculating reasonably accurate estimates of aerodynamic forces, which, in turn, can be used to calculate power output and energy use. The vortex wake models are based on the assumption that lift is produced by bound vortices on the wings, but the models give no insight into how bound vortices of the appropriate strength and timing are generated. There remains a gap between those researchers trying to learn how the wings use unsteady motions to enhance lift, then, and those working to put all flapping flight into a general aerodynamic framework.

Recently, however, researchers have made several attempts to close this gap [50, 51, 52]. Perhaps the most successful has been the combined use of a real insect (the tobacco hornworm moth, a large member of the sphinx moth family) and a mechanical wing-flapping machine. Ellington and his colleagues used this approach to demonstrate delayed stall in action [53, 54, 55, 56]. With this set of measurements, Ellington and his colleagues were able to link delayed stall to a remarkably complex flow pattern over the wing. They then showed how this flow pattern produced very high lift—equal to 150 percent of the animal's weight—and how it was related to the vortex wake. Michael Dickinson and his colleagues also built a mechanical wing flapper to study wing mechanics of tiny insects like fruit flies; they also measured a large lift contribution from delayed stall [44]. In addition to work by these biologists, a number of engineers have become interested in the problems of flapping flight. They have applied powerful computer-based techniques to calculate forces and airflow patterns around flapping wings. Some have used such simplified shapes and movements that a biologist might wonder whether the predictions have any applications to animal flight. Other researchers, however, have used very realistic wing shapes to model flapping aerodynamics. For example, Michael Vest and Joseph Katz even used actual airfoil shapes measured from real pigeons' wings. They found substantial agreement between their model and the actual forces that would be needed to keep a pigeon in flight [57]. Although these computer models can be very detailed, they consume so much computer time that it would be impractical to use them for more than one or two specific conditions. Vest and Katz, for example, modeled a single stroke cycle at one high speed and one low speed.

Hovering

Hovering appears to be the extreme end of the slow flight spectrum, and it differs qualitatively from forward flight, in that there is no airflow over the wings from the animal's forward progress through the air. In forward flight, the movement of the animal's body through the air provides airflow over the wings to help generate lift. Most of the animal's flapping power goes into overcoming drag, which is always much less than lift. Because there is little or no forward speed in hovering or very slow flight, all of the animal's weight must be supported by the power the animal puts into its wings. Thus, the animal's muscles must move the wings fast enough to produce not only thrust but lift as well. The result is that hovering (or very slowly flying) animals require a much greater power output than when in forward flight. Helicopters have the same limitation. They can carry a much heavier load in forward flight than in a hover, which means that they need to take off with a running start, like an airplane when overloaded. The huge power requirements of hovering place size limits on hovering animals. Earlier we saw that muscle cross-sectional area (relative to body size) decreases as animals get larger; in other words, an animal's power output per kilogram of body mass goes down as body size goes up. Thus insects and the smallest birds (hummingbirds, which Spedding calls "honorary insects") can hover for long periods, and small birds and bats can hover for shorter periods. Medium-sized birds and bats may manage to hover for a few wing-beats, but larger birds are unable to hover at all.

Avid birdwatchers will point out that several types of hawks can remain motionless over a spot on the ground while in flapping flight for many seconds at a time, which certainly looks like hovering. But these birds are cheating—although they may be motionless relative to the ground, they are moving relative to the atmosphere. They are actually flying into a headwind. They adjust their forward airspeed to exactly cancel the speed of the wind, which allows them to hang motionless relative to the ground. Perhaps "motionless" is not the most descriptive term, because when I have watched American kestrels and marsh hawks doing this, they give the impression of flapping furiously. They probably are using a slow flight gait (requiring high wing-beat frequency and amplitude) rather than a less frantic-looking fast gait.

Most biologists studying hovering have worked with insects, because they are more adept at hovering than almost all vertebrates. Hummingbirds are the only vertebrates with strong hovering abilities, and they overlap the size range of large insects and use very similar flight mechanics. Insects use two

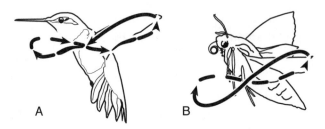

Figure 4.13. Horizontal wing-tip paths of a hovering hummingbird (A) and a hovering hawkmoth (B). The near-vertical body and horizontal wing movement are characteristic of normal hovering. (B.Hd.)

basic wing-beat patterns for hovering. In what is called *normal* hovering, the wings beat more or less horizontally. For an animal to beat its wings horizontally, it usually will have to tilt its body sharply head-up (Fig. 4.13). Bees, some flies, hawk moths and sphinx moths, and hummingbirds all use normal hovering, and it is probably the most common hovering method [47]. The second pattern is called *inclined-stroke-plane* hovering, and animals using this method beat their wings at an angle of 30° to 40° from the horizontal (as opposed to 60° to 80° from the horizontal in typical forward flight). Some small songbirds, small bats, dragonflies, and hover flies use this method. Animals hovering with an inclined stroke plane are easy to distinguish from those using normal hovering, because the former hover with the body nearly horizontal (Fig. 4.14). A few insects use a third type of hovering (*vertical-stroke-plane* hovering), but it seems to be a specialized form of hovering limited to animals with very low aspect-ratio wings, that is, butterflies and some moths [58].

Normal Hovering

The mechanism of normal hovering is actually quite simple to explain, but requires rather specialized anatomy to carry out. Beginning with the downstroke, the wing moves horizontally forward through the air at a positive angle of attack. Except for the movement being forward, rather than down and forward, it looks fairly similar to the downstroke in forward flight. The upstroke, however, is radically different from an upstroke in forward flight. At the end of the downstroke, the wing rotates along its length until the anatomically lower or ventral surface faces up. Then the wing sweeps back horizontally. At the end of the upstroke, the wing reverses the rotation along its length to bring the anatomically upper or dorsal surface back on top. The next downstroke repeats the cycle.

In addition to needing very powerful flight muscles, animals that use normal hovering require two further specializations. First, they need to be able to rotate the wing along its length by 180° or more. Biologists refer to this rotation as *supination* and *pronation. Supination* is rotating the leading edge up so that the ventral surface faces up, and *pronation* is rotating the leading edge down so that the ventral surface is down. A hovering animal, then, must be able to supinate the wings almost 180° at the end of the downstroke, and pronate them back at the end of the upstroke. This movement requires a combination of strength and flexibility in the joint at the base of the wing that is a major design challenge. This amount of flexibility is not typical of bird or bat shoulder joints, and the required combination of strength and freedom of movement is increasingly difficult to achieve as body size increases. This is yet another reason why hovering is limited to smaller animals. The other anatomical specialization is that the animal should be able

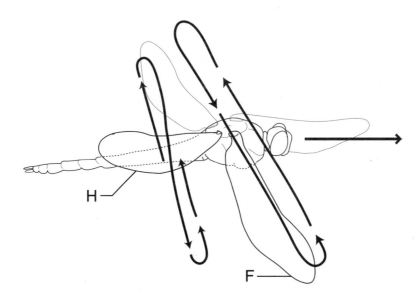

Figure 4.14. A hovering dragonfly, showing the horizontal body and tilted stroke plane of inclined-stroke-plane hovering. F: forewing; H: hindwing. Based on data from A. Azuma et al. 1985 [71] and R. Å.Norberg 1975 [33]. (S.T.)

to eliminate or preferably reverse the wing's camber on the upstroke. The wings are essentially turned upside down on the upstroke, and the ventral surface acts as the aerodynamic upper surface of the wing. If the camber were to remain unchanged from the downstroke, the amount of lift would be greatly reduced or eliminated, and the drag would be much higher, because the wing would have negative or inverted camber. High-speed movies of hovering bees, hummingbirds, and hawkmoths all show some camber reversal on the upstroke [58]. Inclined-stroke-plane hovering probably evolved as an alternative hovering mode that does not require such extensive specializations.

Inclined-Stroke-Plane Hovering

Bats, birds other than hummingbirds, and a small number of insect species use inclined-stroke-plane hovering (Fig. 4.14). The vertebrates undoubtedly use this type of hovering because they cannot completely supinate, or invert, the wing during the upstroke. Even if they could, it would be very difficult for birds to reverse the camber on their wings: much of their wings' camber comes from the cambered shape of the feathers themselves.* A key difference between normal and inclined-stroke-plane hovering is that almost all the aerodynamic force is produced on the downstroke in the latter. In other words, all the power used to support the weight is produced during the downstroke, and the upstroke does no useful work. If no upward force is produced on the upstroke, then the animal will have to produce twice as much on the downstroke. Lift production during the downstroke of inclined-stroke-plane hovering must make great use of unsteady lift-enhancing mechanisms because all quasi-steady analyses of animals using this type of hovering have resulted in unrealistically high lift coefficients (from 2.3 to over 6.0) [7]. The huge power demands of producing all the lift only during the downstroke are yet another reason why hovering in birds and bats is limited to brief periods, and then only by smaller animals. Aside from structural constraints, large birds and bats simply lack the muscle power reserves to hover, particularly when all the work must be done during half the stroke.

* Hummingbird wings are quite different from other birds', being formed almost completely from relatively uncambered primary feathers. Differential twisting of the primaries allows hummingbirds to change the camber of their wings.

Our knowledge of inclined-stroke-plane hovering is incomplete. Dragon-flies use only inclined-stroke-plane hovering, and they hover with almost the same stroke plane angle as in forward flight: 50° to 60° from the horizontal [33]. In contrast, many hover flies can use both types of hovering. Inclined-stroke-plane hovering in insects may be fundamentally different from that in birds because insects cannot flex their wings to reduce the effective planform area, which vertebrates commonly do in slow flight and hovering. The upstroke may generate some useful aerodynamic forces in insects, but biologists have not figured out the details of the wing and air movements during these upstrokes. Inclined-stroke-plane hovering in insects may be an adaptation to allow its users to hover with the body horizontal. Dragonflies hunt on the wing, and both dragonfly and hover fly males spend considerable time chasing rival males and passing females. These insects depend on their excellent vision to identify prey, rivals, and potential mates, as well as for tracking during a pursuit. Hovering with their bodies horizontal has two advantages. First, their eyes will be aimed forward, rather than up, giving them a more useful view. Second, keeping the body horizontal allows the insect to start a chase while looking directly at its target, rather than requiring a quick change in body orientation. Starting from a horizontal position may also allow quicker acceleration at the beginning of a chase. We do not know exactly how they manage it, but inclined-stroke-plane hovering has clear advantages for the observation and pursuit activities of these insects.

Flight Underwater

Underwater and *flight* may sound mutually exclusive, but air and water obey the same rules and follow the same general flow patterns, at least under biologically relevant conditions (see Chapter 2). One example of this similarity is that wings work perfectly well underwater.* Sir James Lighthill used the tail-flukes of whales as an example of underwater wings that flap to produce thrust [59]. He pointed out that most of a whale's weight is supported by its

---

* Wings used underwater are usually called *hydrofoils,* as opposed to airfoils. I prefer to simply call them wings, particularly when they are used in both air and water, because the term *hydrofoil* is also used to mean a boat or ship that uses underwater wings to lift its hull out of the water so that it can go faster.

buoyancy, so the flukes only need to produce thrust, not lift. The whale's flukes act like wings with symmetrical (uncambered) airfoils, using mirror-image upstrokes and downstrokes. The flukes thus produce lift up and forward on the downstroke, and down and forward on the upstroke. The upward and downward components cancel, leaving only a net forward force, or thrust (fig. 4.15).

In fact, most swimming animals as large or larger than minnows use lift-producing surfaces for locomotion. Many animals use a swimming stroke that is noticeably similar to flapping wings: for example, sea turtles, penguins, and sea lions all swim by flapping their "flippers" (forelimbs) (Fig. 4.16). Take the up-and-down swimming stroke of a whale tail and turn it on its side, so that the stroke moves right-and-left, and it becomes the tail-beat pattern of most fish. Most fish can also swim slowly by flapping their pectoral (right and left front) fins, and members of the boxfish family, such as cowfish and trunkfish, swim almost exclusively by flapping their pectoral fins. One of the most unusual flapping patterns is used by triggerfish, which are large fish common on coral reefs. Triggerfish swim mainly with the top (dorsal) and bottom (ventral) fins, which they flap right and left in unison. The first time I saw a triggerfish swim, it immediately reminded me of a bird flying on its side. It is a very graceful yet somehow disconcerting swimming motion.

A very small number of animal species can actually use the same wings to fly in air and swim underwater. Although the general flow patterns of air and water are the same, the increased density of water means that the forces on the wing will be 10 or 20 times greater underwater than for similar move-

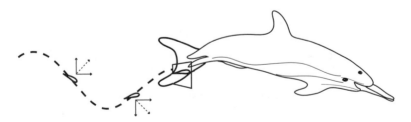

Figure 4.15. Dolphin tail flukes use a modified form of flapping to produce thrust. On the downstroke, the fluke has a positive angle of attack and lift is directed up and forward. On the upstroke, the fluke has a negative angle of attack, so lift is down and forward. The upward and downward components cancel (the animal is neutrally buoyant, so needs little or no lift), so the only force left is the forward or thrust component of both half strokes. (B.Hd.)

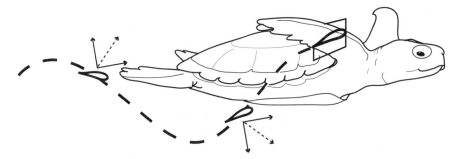

Figure 4.16. The front limbs of sea turtles are modified into paddles, and they operate with the same stroke as a dolphin. The upward and downward force components cancel, and the forward component drives the animal through the water.
(B.Hd.)

ments in air, even allowing for a much lower wing-beat frequency. This difference requires design trade-offs, because the lightness that is important in
air conflicts with the structural strength required underwater, a constraint
that helps explain the rarity of dual-purpose or *amphibian* wings for use in air
and water. The only well-known examples of such amphibian flyers are birds,
although some tiny wasp species may also be in this category [60]. One such
bird is the dipper, a small songbird of high elevations in western North America. The dipper flies in air, just like any other small bird, but it feeds on small
invertebrates in mountain streams. It walks into streams and continues walking along the bottom, fully submerged. It occasionally makes short underwater "flights" using its wings, but it probably spends more time walking
than flying underwater.

Perhaps the best amphibian flyers are the members of the auk family, such
as auks, murres, puffins, and auklets (Fig. 4.17). These seabirds nest on coasts
in high northern latitudes, and they both fly and swim with their wings.
Auks nest on shore, fly to food-rich patches of ocean, dive underwater, and
pursue prey by flapping their wings underwater. Auks have stocky, penguin-
like bodies and strong but heavy wings of low planform area, giving them
high wing loadings, over 200 kg/m², which is five or six times higher than
other birds of similar size. As a result, in air they must fly fast with a high
wing-beat frequency. Murres, the largest auks, have difficulty taking off
without a headwind, and their limited slow flight ability makes landing at
their rocky nest sites a challenge. Thanks to the scaling of muscle power,

Figure 4.17. A puffin, a member of the auk family, which can both fly and swim with its wings. (S.T.)

the smaller auks tend to fly more in air and have better slow flight capabilities, but they are still stocky-bodied and heavily wing-loaded compared to other birds of the same size.

Underwater, auks "fly" a bit differently from their aerial method. Underwater photographs show that they swim with their wings strongly flexed, so that the effective surface consists mainly of the primary feathers; the rest of the wing is held close to the bird's body. Why not use the whole wing? First, it takes much less area to produce the same amount of force underwater than in air, even though the wing moves more slowly. Second, the inner portion of the wing does not contribute much to thrust, as discussed earlier in this chapter—it mainly helps support the weight in flapping flight. Underwater, however, the animal's buoyancy makes its body weightless, so the bird needs no upward force. (They usually float, so to get the slight downward force they need to stay submerged, they swim with their bodies tilted downward.) Auks thus reduce the effective surface of the wing to little more than the thrust-producing outer parts. So far, no one has worked out the details of the auk's underwater flapping pattern, but it is clearly somewhat different from air-borne flapping [61]. In addition to being flexed, the wings go through a surprising amount of pronation on the downstroke, and the wing tip follows a nearly circular path. In any case, the underwater flapping pattern illustrates another trade-off: to fly in air, auks require far more wing surface than they need in water. That is, auk wings are overly strong and barely big enough for flight in air, yet far larger than necessary (and probably a bit fragile) for flight underwater. Auks are classic "jacks of all trades but masters of none"—they can get by in both air and water, but certainly not as well as full-time aerial or aquatic animals.

Interestingly, just as the smaller auk species are better flyers, the larger ones are better swimmers. Murres, the largest living auks, swim well enough to make a living feeding on fast-swimming fish. The Great Auks illustrate the limits that physics places on biology. Great Auks are now extinct, but they were by far the largest members of the auk family. Great Auks were flightless. The advantages of large body size that dominated their evolution must have outweighed flight ability so much that Great Auks lost the ability to fly in air (which may have actually improved their swimming ability by allowing stronger, shorter wings). Great Auks thus filled the same ecological role in the Arctic that penguins fill in the Antarctic. The name *penguin* was even once used to refer to Great Auks. Sadly, Great Auks were driven to extinction by hunting, first for their meat and eggs and later, on a commercial scale, for their feathers. They probably also suffered from competition for food with commercial fishermen. The gigantic colonies of Great Auks on islands off the coasts of Canada and Great Britain were mostly gone by 1800, and the last confirmed sighting near Iceland was in about 1840 [62].

Finally, many reference books and reviews state that the hatchetfish, members of the family Gasteropelecidae, are capable of flight in air. These tiny freshwater tropical fish do have long pectoral fins and a deep "chest" with strong muscles for moving the pectoral fins. Unfortunately for lovers of odd natural history, hatchetfish are *not* capable of powered flight in air. F. C. Wiest showed that hatchetfish do make spectacular leaps out of water. They do not, however, achieve a significant increase in distance by flapping [63]. Wiest found that they follow ballistic paths when they leap. In other words, just as you can calculate the distance a cannonball will travel by knowing only its speed and angle when fired, Wiest could accurately predict the leaping distance of the hatchetfish just from the angle and speed with which they left the water. If they had been getting useful thrust from flapping, they would have had longer, flatter paths than their actual ballistic arcs. The fish are thus not producing any significant lift or thrust with their pectoral fins. In spite of many reports that hatchetfish can fly in air, I know of at least one other, unpublished study that came to the same conclusion as Wiest's. Under carefully controlled conditions, hatchetfish show no sign of powered flight ability.

Birds Do It, Bees Do It, Why Can't We?

All flying animals use flapping to achieve powered flight. Yet no aircraft fly by flapping. Why don't we humans use flapping flight? By using wheels and axles to produce continuously rotating motion, we can build much simpler machines for producing thrust with rotating mechanisms than with flapping mechanisms. Our machines are largely based on rotating power systems, but animals have yet to evolve continuously rotating axles. On the other hand, animals generate power with muscles, which produce linear motion (shortening) rather than rotations. Understood this way, a seesaw flapping motion, with one set of muscles for the downstroke and another set for the upstroke, is a logical arrangement for producing aerodynamic power.

Can humans build machines that fly by flapping their wings? Such a device is called an *ornithopter*, and rubber-band-powered toy ornithopters have been around for over a century. Attempts to build engine-powered ornithopters have been rare, and have mostly had mixed results. One of the more remarkable examples is the mechanical scale pterosaur model built for a movie by a company called AeroVironments;* it has since been donated to the Smithsonian Institution. It was designed to fly by flapping. Unfortunately, the

---

* AeroVironments was started by Paul MacCready, the man who built the Gossamer Condor and Gossamer Albatross, which won the Kremer prizes for successful human-powered aircraft (Chapter 11).

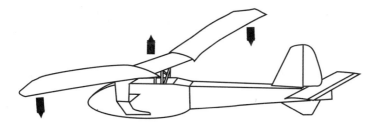

Figure 4.18. The successful ornithopter developed by James DeLaurier. (S.T.)

weight of batteries for the power and control systems, along with the constraints of a realistic appearance, prevented it from achieving true powered flight. It was flown successfully as a sailplane, and its flapping presumably reduced its descent rate, but it was never able to produce enough thrust to maintain level flight [64, 65].

The only well-documented engine-powered ornithopter to achieve true powered flight was the radio-controlled model built by James DeLaurier and his team of students at the University of Toronto. The team spent about seven years designing, building, and testing their craft to produce a successful ornithopter (Fig. 4.18). They developed a transmission system to convert the high-speed rotary motion of a model airplane engine into relatively slow up-and-down motion. They used this transmission system to raise and lower the center section of the wing. The main wing panel on each side pivoted on a support a few centimeters outboard of the fuselage. The model consists of a normal-looking airplane tail and rear fuselage, with a deep, short front fuselage housing the engine and transmission system. The wings of the successful version have a shape and aspect ratio quite similar to those of a typical radio-controlled model sailplane of similar size, although they are designed to twist and bend in controlled ways while flapping [66]. Total wingspan is about three meters. After experimenting with three different airfoils, and rebuilding almost every part of the craft at least once, the team made two successful, fully powered flights on September 4, 1991, which included climbs (and landings!) under power [65]. DeLaurier described this model as a technology test bed to prove the feasibility of building full-size, passenger- or payload-carrying ornithopters. He and his students have built a full-size, person-carrying ornithopter prototype, which they have been testing extensively on the ground. DeLaurier believes that they have overcome significant takeoff problems, and they are currently (2001) planning a flight attempt in the near future. At this point, there are no obvious advantages of an ornithopter over a conventional airplane, except for some possible stealth characteristics. At best, passenger-carrying ornithopters will be impractical, inefficient novelties for the foreseeable future.

FIVE

Staying on Course and Changing Direction

Steering

Any animal that moves needs a method to control where it goes, in other words, it must be able to steer. If it cannot steer, it will not be able to obtain the resources—food, mates, shelter—it needs. When an eagle sees a rabbit or a fly smells a dead skunk, the flying animal must be able to steer toward its goal. First, it must be able to change direction intentionally; if the eagle sees a rabbit off to its left, the eagle must be able to turn left. Second, the animal must avoid accidental or unintentional direction changes. As the eagle flies toward the rabbit, if gusts of wind blow it off course, it needs to be able to return to its desired direction. Changing directions intentionally is *maneuvering,* and avoiding or reversing unintentional changes is *stability.* Stability may not seem like an obvious component of steering, but consider driving an automobile. Clearly, to go around a corner—maneuver—I must make a large turn of the steering wheel. But even driving on a straight road, I need to make constant, small corrections, which represent a form of stabilization.

In some ways, stability and maneuvering are at opposite ends of a spectrum. For example, one way to achieve stability is simply to build in a strong tendency to move in a straight line. However, such a tendency will interfere with attempts to turn. This conflicting relationship affects all locomotion to some degree, but it is especially significant for flight. Because animals fly at relatively high speeds, a sudden unintended turn—or conversely, the inability to turn quickly—can lead to disastrously painful encounters in the blink of an eye. Surprisingly little research has been done on stability in animal flight, despite the great emphasis on stability and maneuvering ("control") in aeronautical engineering. As a consequence, much of our understanding of stability, and maneuverability to a lesser extent, is based on analogies with airplanes. Airplane designers consider stability and maneuverability to be mutually exclusive,* and so they have traditionally been traded off against each other, depending on a given aircraft's purpose. For example, passenger and cargo airplanes are stable at the expense of maneuverability, while fighters and competition aerobatic airplanes are maneuverable at the expense of stability. Most flying animals are extremely maneuverable (compared to airplanes) and have very little built-in stability. Yet they have evolved a variety of mechanisms that allow them to fly in a straight line.

Special Problems of Flight

Humans basically move two-dimensionally in a three-dimensional world. In spite of inhabiting skyscrapers and climbing mountains, our movements tend to be right or left and forward or backward. When I climb a hill or a stairway, I am actually just following the terrain. I am on a surface, and I still think mainly in terms of moving right or left, forward or backward. Consider maps: most people are comfortable with compass directions (north, south,

---

* In this chapter, I use the term *maneuverability* in its general sense, meaning a strong ability to change directions. Ulla Norberg and Jeremy Rayner have pointed out that this actually includes two rather different abilities: the ability to turn very quickly (in a short time), which they term *agility*, and the ability to turn very sharply (in a small distance), which they call *maneuverability* in a strict sense [29].

etc.), but interpreting hills or valleys with contour lines requires much more mental effort. Likewise, when a department store is on several floors, a map of the store shows each floor separately; we make little effort to provide a three-dimensional representation of the store, because we are not accustomed to understanding movements in three dimensions.

In contrast, flying animals must be able to move vertically as well as laterally (to the side) and longitudinally (to the front or back). While this requirement may sound obvious, it permeates an animal's flight-control systems with a thoroughness difficult for people (other than aviators) to truly appreciate. Indeed, vertical movement is what separates flight from terrestrial locomotion. Flight requires vertical control: without it, flight is literally a crash waiting to happen.

Flying animals typically move 10 or 20 times faster than a similarly sized animal moves on the ground. So a flying animal covers a lot more distance than a runner in the same amount of time. A running fox might have several seconds to notice and avoid an obstacle three or four meters away, but a flying crow might have less than half a second to turn to avoid an object the same distance ahead. Rapid response times are thus much more significant for flying animals than for terrestrial animals.

Curiously, by adding the ability to move in one new dimension—the vertical—flyers gain the ability to rotate about two new axes. As a walker, I am basically limited to left and right turns, or rotations about a vertical axis (yawing rotations). The terrain controls rotations about a side-to-side axis or a front-to-back axis, unless I fall down. Flyers, however, can rotate nose-up or nose-down—pitch rotations—or tilt so that either the left wing or the right wing is low—roll rotations. Controlling all these types of rotations means that stability and control requirements are considerably more complex for flyers than for walkers.

Takeoff and Landing

Clearly, taking off requires an animal to be able to control its vertical direction in order to fly up away from the ground or a perch. Moreover, it must move fast enough to generate enough lift, which means that larger—and heavier—flyers may need a takeoff run on the surface to build up speed. For example, geese and swans have particularly high wing loadings, so they run several steps to begin a flight from land. They appear to do the same thing on water as well. They are not exactly walking (or running) on water, however. They push down and back with their wide, webbed feet in a running motion. This helps lift their body out of the water, which greatly reduces drag and eases their acceleration to flight speed. Many large birds, especially those with relatively high wing loadings, such as flamingos and larger albatrosses, require a running takeoff unless they can leap from a high perch or into a strong headwind.

Landing requires exquisitely fine control of vertical direction and speed. Such control may be innate in some insects, but birds and bats must learn and practice some aspects of this skill. Some flying animals may spend many days at a time in flight (such as during migration) but sooner or later, all must land. For takeoff, the animal must simply accelerate to flying speed and aim up, but for landing the animal must descend toward its chosen landing spot, and adjust its speed so that it moves just fast enough to maintain lift, but slow enough to avoid injury at touchdown. One solution is to go into a hover just before touching down. This is how many insects land, but it is only an option for vertebrates if they are small, and it is energetically expensive. Some insects use an alternative tactic, which is to accept clumsy, crashlike arrivals; because of their small size, they have low inertia, and they rarely suffer significant damage in moderate collisions. Houseflies, for instance, often land on walls or ceilings by flying straight toward the surface with their front legs outstretched; when the front legs contact the surface, the legs absorb the shock and act as a pivot for the body to swing around and grab on with the middle and hind legs. Other insects, such as June beetles and bumblebees, simply fly toward their intended perch and grab on as they hit. In contrast, larger flying animals cannot afford routine crashes, so they generally try to stall their wings just before touching down. This approach greatly increases drag, slowing the animal to a safe touchdown speed, and simultaneously kills lift to avoid bouncing back into the air. The latter approach clearly requires a sensitive three-dimensional control ability to avoid painful, or even catastrophic, arrivals.

Stability

A flyer's stability is its tendency to return to an original direction after being pushed off course by some disturbance, such as a gust of wind. Flyers may have different amounts of stability. Simple balsa toy gliders are usually reasonably stable and give long, straight flights if carefully assembled. Leave off one of the tailpieces, however, and the glider loses all stability, tumbling out of control no matter how carefully it is launched. Between these extremes, a flyer can have more or less stability, and a variety of features can affect the stability. Engineers have developed complex and sophisticated methods of analyzing airplane stability [1, 2], but most are difficult or impractical to apply to animals because of the radical changes in wing orientation and shape that they routinely use, such as when flapping.

Advantages and Disadvantages

Why does the degree of stability matter? A stable flyer has at least two advantages. First, because a highly stable flyer tends to stay on a given heading, it flies on a straighter path, enabling it to fly a shorter distance from point A to point B than a less stable flyer. The stable flyer will use less muscular energy and may take less time. Second, the stable flyer does not require its control system to be as responsive or complex. A stable flying animal's nervous system simply has less to do: it need not detect and respond to every tiny gust or patch of turbulence. (Botanical flyers like the Javanese flying cucumber have evolved extreme stability because they have no control system at all; see Chapter 3.) Conversely, a less stable animal must put more effort into correcting deviations from its desired heading, and also requires more rapid control responses. Although the advantages of stability might seem desirable, very stable flyers have a crucial limitation: they resist *intentional* heading changes as well as unintentional ones. In other words, they have low maneuverability. Stability and maneuverability thus trade off against each other, as mentioned earlier, and the optimum balance depends on the task. In contrast to the enormous amount of work that airplane designers have put into the study of stability, very few studies have been done on stability of flying animals. Much of our understanding of animal flight stability comes from analogies with airplane design. Airplane designers consider high stability and high maneuverability to be the opposite extremes on a spectrum, and in some ways this applies to flying animals as well. However, as we shall see, the variable wing geometry required by flapping flight allows animals to sidestep some of the constraints on airplanes and achieve stability and maneuverability in the same elegant design.

Stabilizing Mechanisms

A number of design features contribute to stability. A well-designed tail is one of the most effective stabilizers (recall our toy glider). A tail adds "weathercock" stability, keeping the nose pointed into the wind just like a weathervane. Other stabilizing features are more subtle. The Javanese "flying cucumber" discussed in Chapter 3 has its wing tips raised a bit, so in front view, the wings form a very shallow V-shape. This dihedral gives a strong stabilizing effect; many gliding birds like vultures use noticeable dihedral. Dihedral stabilizes both yaw and roll. For example, if a flyer with dihedral rolls to the left, the left (lower) wing will be more horizontal and the right (upper) wing will be more vertical. Viewed from above, the low wing will look longer and the high wing will look shorter, so the upward component of lift on the low wing will be higher than the upward component of lift on the high wing. This tends to roll the flyer upright. The geometry of the yaw-stabilizing effect of dihedral is considerably more complex, but it is often stronger than the roll-stabilizing effect. Design features like tails and dihedral give a flyer *passive* stability: the shape of the flyer itself provides stability, and it is thus a built in and essentially permanent feature.

Flyers with little passive stability can still fly with apparently stable flight paths. If an animal can sense deviations from its desired direction, and if it can take corrective action fast enough, the animal can fly in a nearly straight line. This *active* stability requires the animal to sense tiny perturbations and respond to them very rapidly (usually within one wing-beat). This ability calls for extremely precise direction sensing, usually involving wind sensors combined with vision. Such a rapid response can be accomplished only with reflexes: automatic, "hardwired" responses built into the nervous system that do not require any input from the higher centers of the brain. This, then, is the opposite of the passively stable condition, because this stability depends on the animal's behavior rather than on its shape. Active stabilization requires a sophisticated, fast-acting, specialized nervous system to maintain the animal's heading in flight.

Flapping flight precludes the use of some stabilizing mechanisms, but opens the opportunity to use other, new ones. Some stabilizing methods (such as dihedral) only work with fixed wings, as when gliding. They cannot be used during flapping. However, flapping wings are already performing complex movements many times per second. Active stabilizing may only require relatively slight adjustments to the wing-beat pattern. For example, if a bird's flight path unintentionally tips down, the bird can correct this by flapping its wings a bit farther forward than normal. This, in effect, shifts the bird's center of gravity back, which causes the bird to tilt nose-up. These types of subtle shifts in the wing-beat pattern are typically controlled by a set of small, accessory muscles separate from the main power-producing (upstroke and downstroke) flight muscles, and these same muscles are used for maneuvering.

Stability and Evolution

In 1952, John Maynard Smith, a renowned theoretical population biologist (and former aeronautical engineer) published a short but insightful paper on stability in the evolution of animal flight.* Maynard Smith argued that primitive flying animals—those in the early stages of evolving flight—would not yet have developed the specialized nervous system required for active stabilizing, so they must have been passively stable. As animals evolved, their nervous systems acquired the sophistication and reflexes needed to do more and more of the stabilizing. Thus we should expect the more specialized (more highly evolved) animals to be less passively stable. Maynard Smith went on to argue the converse: we should be able to predict whether an animal represents a primitive or an advanced condition depending on its stability. According to Maynard Smith, stable animals should be primitive, unstable animals should be advanced [3].

Why would animals evolve from a body form that is passively stable to one requiring active stabilization? In other words, why evolve to be unstable? The important benefit is clearly maneuverability.† Most flying ani-

---

* In the scientific literature, the name "Maynard Smith" is treated as if it were hyphenated and alphabetized by "Maynard" rather than "Smith." The charming gentleman in question often introduces himself, however, as "John Smith."

† Interestingly, perfectly analogous active stabilizing systems using computers have been used in fighter jets for over two decades, and for the same reason. Fighter airplanes must be very maneuverable in combat, and one way to achieve this is to make them so unstable that a computer must monitor the airplane's movements and correct them many times per second. The pilot in effect "flies" the computer, which translates his or her commands into movements of the airplane (Chapter 11).

mals operate in a much more cluttered and obstacle-strewn environment than the average airport, so they must be maneuverable to avoid collisions and to squeeze into tight landing sites. Predation is also crucial. The ability to change direction quickly is a major survival skill both for prey species and for the predators that need to catch prey in order to eat. Increasing maneuverability means decreasing stability, and most modern flying animals have gone to the extreme; they are so unstable that they tumble out of control if anything interferes with their active stabilizing system. But even this can be an advantage: if you are about to tumble out of control the instant you stop actively stabilizing your flight, intentionally beginning such a tumble can be a very effective, rapid change of direction for emergency escapes.

Are there examples that bear out Maynard Smith's predictions? We have a problem: modern animals all have been evolving for a long time, and today's flying animals are, by and large, quite specialized for flight, and hence, passively unstable. Unfortunately, I know of no modern flying animal that exhibits the passive stability Maynard Smith described. Instead, we must look to the fossil record to test Maynard Smith's prediction about primitive flyers.

Fossils give us two cases that fit his prediction. Pterosaurs are extinct flying reptiles related to dinosaurs (see Chapter 7), and the early pterosaurs all had long tails with a flap of skin on the end that would have given these animals weathercock stability. Later pterosaurs had short, nearly vestigial, tails that could not have contributed any significant stabilizing effect. These later pterosaurs must have used active stabilizing mechanisms, rather than relying on passive stability from their tails. Birds show a similar but less extreme evolutionary pattern. *Archaeopteryx,* the oldest known animal with birdlike features, had a very long tail with a row of feathers running down each side. As in early pterosaurs, this long tail would have given *Archaeopteryx* weathercock stability, as well as simplifying the animal's flight-control requirements. Modern birds, in contrast, have short, stumpy tails. Although these tails are equipped with a fan of feathers, most birds can fly almost normally with plucked tails. Modern birds seem to use their tails primarily to adjust drag at slow speeds and during sharp turns, but not for stabilizing [4, 5]. (The enlarged, ornamental tail feathers of male birds like peacocks and pheasants are exceptions; their main function is for sexual display, but they may also contribute to stability.) Thus, the trend in both pterosaurs and birds is to start with some passive stability, but to evolve toward less stable, more maneuverable "designs." Our current knowledge of the early evolution of flight in bats and insects is too incomplete to say whether Maynard Smith's prediction holds for those groups as well.

Maneuvering

Whereas stability is avoiding unintentional changes in direction, maneuvering is intentionally causing a change in direction. Animal flight is not very useful if a flying animal cannot maneuver or choose its own direction. Vertical maneuvers—climbing and descending—are relatively simple to perform compared to horizontal maneuvers such as turning right or left. In fact, the most efficient way for a flyer to turn is quite different from anything in our everyday experience.

How Not to Turn: Flying Is Not Boating

Because airplanes and boats both have rudders, and because boats use their rudders to turn, people naturally assume that airplanes use their rudders to turn. Their assumption, however, is wrong. When a boat turns, the helmsman deflects the rudder in the direction of the turn. The rudder is actually an airfoil, which is at a positive angle of attack with the water after being deflected. A rudder deflected to the right produces lift to the left (Fig. 5.1), which pulls the back of the boat to the left and swings the front of the boat to the right. The front of the boat keeps swinging to the right until the front is pointing in the desired direction, at which point the helmsman centers the rudder. Figure 5.1 shows a subtle feature of this maneuver: the boat tends to point slightly to the inside of the turn as it curves, which means that it is actually sliding a bit sideways, or *sideslipping*. The amount of sideslip is usually small compared to the boat's forward speed, but ignoring sideslipping while, for instance, docking a 100,000-ton oil tanker can be disastrous.

An airplane can be turned using its rudder, but this has several drawbacks. First, in such a turn, the airplane is never pointing in the direction in which it is actually moving, which can be confusing for the pilot. More important, sideslipping generates tremendous drag, so that the craft either slows (and descends) or needs considerably more power to stay level. (Pilots do use sideslips in certain situations to lose height quickly.) Sideslipping can also produce asymmetric stalls, which could be catastrophic during a landing

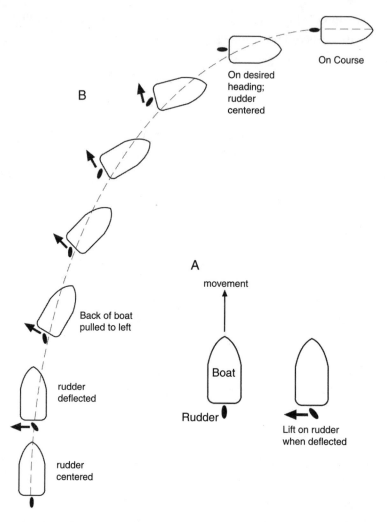

Figure 5.1. Turning a boat using rudder. A. The rudder is actually a small wing. Deflecting it increases its angle of attack, producing a lift force to one side. B. Lift on the rudder pulls the back of the boat to one side. Note that the boat is not aimed in the direction it is moving throughout most of the turn, and it slips sideways briefly even after the rudder is centered. (B.Ht.)

approach. Pilots do not use their rudders alone to turn under normal circumstances, because there is a much more efficient, safer way.

Using Wings to Turn

Since few flying animals have anything resembling a rudder, they must use their wings to produce turns. Wings can produce much more force, and with better leverage than a tail surface, and by varying the relative amount of lift produced by the right and left wings, flying animals can produce turns much more effectively than they could by using rudders. Indeed, the Wright brothers' greatest technological achievement was to discover and understand how birds turn, and to figure out how to adapt this method to a machine (Chapter 11).

The simplest way to see how to use wings for turning is to imagine a bird in a glide. As long as the left and right wings are held in the same orientation, the bird will glide straight and level (actually in a shallow descent; Chapter 3). What happens if the bird now tilts the leading edge of its left wing a few degrees up, so that the left wing is at a higher angle of attack than the right wing? Clearly, the left wing will now produce more lift than the right wing (Fig. 5.2), and the bird will roll to the right. Any change that increases the lift of one wing with respect to the other can be used, such as camber or area changes. The bird is now in a right bank, and the lift on each wing is now tilted to the right. This rightward tilt to the lift (Fig. 5.2) is what pulls the animal around in a turn.* Note that once the bank angle is established, the bird can, in principle, return the angle of attack (or camber, or area) of the left wing to that of the right wing, and the turn will continue.† In order to stop the turn, the bird must increase the lift on the *right* wing to roll back to level flight. In general, a turn consists of increasing the angle of attack of the wing

---

* All other things being equal, the outer (high) wing will produce more drag as well as more lift, which works against the turn. An airplane pilot overcomes this "adverse yaw" effect with a bit of rudder deflection. Although some birds can get a ruddering effect by twisting their tails at low speeds, most of the time, flying animals probably overcome adverse yaw by weight-shifting or fine wing or postural adjustments.

† In practice, a small amount of differential lift (from differences in angle of attack, camber, etc.) may be needed to maintain the bank. This depends on various aspects of the animal's anatomy, such as the location of the animal's center of gravity, the amount of dihedral or sweepback of the wings, etc.

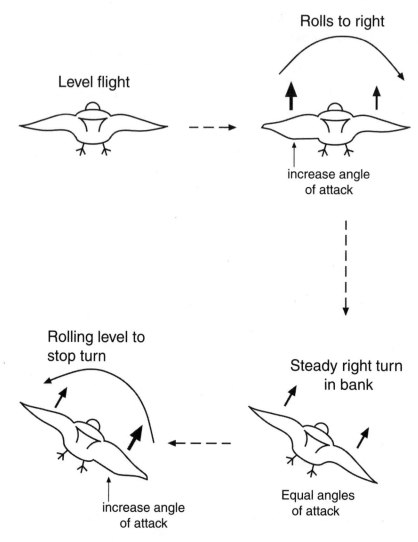

Figure 5.2. Banked turns. In this back view of a gliding bird, the bird tilts its left wing leading edge up, increasing the angle of attack and lift on that side. The unbalanced lift rolls the bird into a right bank, and the bird turns to the right. Once banked, the bird brings its left wing back to its original angle of attack. To complete the turn, the bird increases the angle of attack of the right wing to roll back out of the bank and stop turning. (B.Ht.)

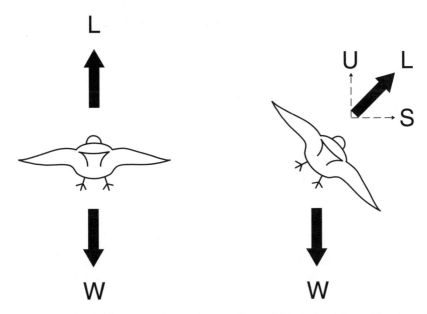

Figure 5.3. A banked turn requires an increase in total lift. A. Straight and level flight, with lift (L) equal to weight (W). B. When banked, lift is tilted to one side, and only the vertical component (U) is now opposing weight; the sideward component pulls the flyer around the turn. To prevent a descent, the total lift must be increased so that the upward component becomes equal to the weight. (S.T.)

to the *outside* of the turn, which causes the flyer to roll into a bank. Once in the bank, the angle of attack of the right and left wings is equalized. The steeper the angle of bank, the sharper the turn. To stop turning, the angle of attack of the wing to the *inside* of the turn must be increased to roll the animal back upright and return to a straight flight path. Of course, camber or wing area changes can be substituted for angle of attack changes, because they have similar effects.

To maintain a *level* turn, the flyer needs more total lift when banked than when its wings are level. When the flyer is banked, only part of the lift is supporting its weight; the rest is pulling it sideways, around the turn (Fig. 5.3). For the vertical (upward) component of the lift to balance the weight, the total lift produced by the wings must increase. A gliding animal could do this by increasing the angle of attack or camber of both wings (symmetrically) during the turn a bit more than during straight flight. If the animal made such a symmetrical change during straight flight, it would climb, which is why pilots sometimes describe a banked turn as "climbing toward the horizon."

Turns using wings in flight are thus fundamentally different from turning an automobile or boat. To turn an automobile, I turn the steering wheel in the direction I want to turn and hold it there until I have turned far enough; then I center the steering wheel. If automobiles turned like flyers, to start a right turn, I would turn the steering wheel briefly hard right, center the wheel and press harder on the accelerator pedal as I go around the turn, and then turn the wheel hard left briefly and ease off the accelerator to finish the turn. In a real sense, a flyer in a banked attitude turns all by itself, without any control input. Flying in a bank *without* turning is most likely impossible for flying animals.

Turns while Flapping

The simplicity of gliding flight is handy for visualizing how banked turns work, but most flying animals spend nearly all of their flight time flapping. How does an animal turn in flapping flight? The basic principle is the same: the animal rolls into a bank to tilt the lift to one side, stays banked to complete the turn, then levels out on its new heading. Flapping animals can roll into a bank with the same methods as gliders, but they can use some new methods as well. The angle of attack and camber of a flapping wing changes throughout the stroke even in level flight, but if the animal increases the average angle of attack or camber of one wing more than the other during the downstroke, the animal will bank just as if gliding. The animal only needs to have the average angle of attack or camber higher on one side during the downstroke, because most of the lift is produced on the downstroke (Chapter 4). Just as in gliding, the animal would use higher angle of attack or camber on the wing to the outside of the turn to roll into the bank and begin turning, and higher angle of attack or camber on the inside wing to roll level and finish turning.

In addition, flappers can do something that gliders cannot: they can change the stroke amplitude (the amount of up-and-down movement) of their wings. When a flying animal flaps its wings with a greater amplitude— farther up and down—the wing moves faster, particularly toward the tip. Imagine a crow flapping its wings. It might move its wing tips 20 centimeters up and down when flapping with a low amplitude, and perhaps 50 centime-

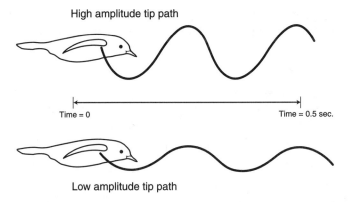

High amplitude tip path

Time = 0   Time = 0.5 sec.

Low amplitude tip path

Figure 5.4. Amplitude of the wing-beat affects the speed of wing movement through the air. When the wing flaps with high amplitude, the tip moves farther in a given time. Thus, the outer wing moves through the air faster when flapping at a higher amplitude. (S.T.)

ters up and down at a high amplitude. If it completes a stroke in three-tenths of a second, then the wing tip clearly must move much faster when the crow flaps at high amplitude than at low amplitude (Fig. 5.4). How do animals use this to turn? When a wing flaps with a greater amplitude, it moves faster, and when it moves faster, it produces more lift. Thus a flapping animal can use changes in amplitude in exactly the same way that it uses changes in angle of attack: increasing the amplitude of the wing on the outside of the turn initiates the turn, and increasing it on the inside wing completes the turn. Using differential amplitude to turn has an advantage over angle of attack or camber changes: the wing flapping with the higher amplitude produces more thrust as well as more lift, which tends to swing the animal's nose into the turn. Under some conditions, this differential thrust can speed up the turn.

Which of these methods do actual flying animals use? Few animals have been studied in detail while turning, but to some extent, the choice of method depends on the species, perhaps even on individual preference. Grasshoppers seem primarily to use differential angle of attack [6], while beetles seem to rely mainly on differential amplitudes [7, 8]. In my own work on dragonflies, I have seen dragonflies turn using differential angles of attack, differential amplitudes, and combinations of the two [9]. Moreover, dragonflies sometimes initiate turns by *reducing* the angle of attack or amplitude on the inside wing as well as increasing it on the outer wing. I have filmed more than one dragonfly that basically stopped flapping the inner wing while increasing the amplitude of the outer wing only slightly. This produces a very rapid roll into a steep bank, leading to a sharp turn.

Anyone watching large birds like ducks, geese, hawks, or vultures can easily observe banked turns, particularly when they glide. Bats and smaller birds may turn too fast for humans to easily see what they are doing, but these animals, too, use banked turns [10]. Birds and bats ought to be able to use differential angles of attack and amplitudes, and bats are particularly well suited to use differential camber. Surprisingly, biologists have hardly studied the wing movements of maneuvering birds or bats. In a study of pigeons, biologists found that these birds turn using a variation of one of the mechanisms used by insects. Pigeons roll into a bank by moving their outer wing downward faster than their inner wing. If the outer wing moved faster throughout the entire downstroke, the outer wing's amplitude would be higher. In most turns, however, the pigeon banks far enough to turn well before the first downstroke ends, so the bird speeds up the inner wing (which rather abruptly arrests the rolling motion) so that the inner and outer wings end up with roughly the same amplitude. Essentially the same process happens on the opposite wings to roll back upright [11]. Researchers were surprised to find that pigeons maneuver with large, constantly changing differential forces, rather than with subtle, finely adjusted changes in force. These birds seem to fly with an alternating series of rather violent lateral pushes, first to one side, then to the other, and the balance between the alternate pushes determines whether the bird turns or flies straight.

True flies—including houseflies, hover flies, bluebottle flies, horse and deer flies, and mosquitoes—are clearly among the most aerobatic of flyers, and they take the differential flapping to an extreme. When researchers looked at film and videotape slowed to make the wing movements visible, they found that flies sometimes turn by completely folding the inside wing back over the abdomen, so that only the wing to the outside of the turn continues flapping [12, 13]. When the fly does this, it turns so rapidly that to a human observer, the fly appears to make an instantaneous sharp turn. A fly can afford to turn this way because of its size: its inertia is so low that it can complete the turn in about one-fiftieth of a second (two or three wing-beats). Even though its lift is cut in half, the fly does not have time to fall any significant distance. If a pigeon tried to turn by folding back one wing, because of its larger size—and higher inertia—the bird would probably roll onto its side or back and fall a meter or two, while curving only slightly.

Non-wing Turn Mechanisms

Although changes in wing movements are normally the most important actions that flying animals take to turn, they can also move other body parts to enhance turning. For example, many insects bend the abdomen (back of the body) in the direction of a turn, and grasshoppers sometimes extend the large hindleg to the inside of a turn as well [6, 14]. Biologists call these movements *ruddering* [6], and typically describe them as mechanisms to increase drag on the inside of a turn to swing the front of the animal into the turn. I suspect that these movements are at least as important for *weight-shifting:* the animal uses these movements to shift its center of gravity to the inside of the turn, which increases its rolling tendency.*

Extending a hindleg can do more than shift the grasshopper's weight. When a grasshopper extends its hindleg, the leg often gets in the way of the hindwing on the inside of the turn, and reduces its stroke amplitude. Banging a wing into a leg seems a rather crude way to reduce the stroke amplitude, but in at least some situations, moving the leg requires fewer signals to and from the grasshopper's brain than adjusting the wing-beat directly.

Hovering

Turns while hovering are fundamentally different from turns in forward flight. First, the animal must turn about a vertical axis (yawing) rather than a horizontal one (rolling). Second, the turn must be produced by differential thrust rather than differential lift. The details of differential thrust production in hovering are still not entirely clear, but hovering dragonflies can easily turn to face in the opposite direction in two wing-beats (one-twenty-fifth to one-thirtieth of a second). For readers acquainted with the mechanics of rotating machinery, a dragonfly performing such an about-face rotates its body at almost 1000 RPM; the *propeller* of a light airplane at cruise speed rotates at between 2000 and 2500 RPM! Again, to a human observer this kind

---

* Weight-shifting is exactly how hang-glider pilots turn their craft.

of heading change would appear to be instantaneous, and the dragonfly can only bring it off because of its low inertia.

Vertical Maneuvering

Turning, however complicated, is only half the picture. Flyers must also be able to climb and descend at will. But, if we know how an animal turns, we already know how it changes vertical directions. To go up or down, the animal needs to change the total amount of lift the wings produce, but in this case both wings work together. To climb, the animal increases the lift on both wings, and to descend, the animal decreases the lift on both wings symmetrically. If the flyer increases the lift on both wings, it will accelerate upward until a combination of its weight and the vertical component of its drag becomes equal and opposite to the new, higher lift value, at which time the flyer settles into a constant climb (Fig. 5.5).

Animals use the same mechanisms to climb or descend as they use to turn. They can climb by increasing the angle of attack, camber, wing area, or amplitude on both sides simultaneously. Moreover, changes to the wing-stroke pattern that increase thrust will also cause them to climb, because increasing thrust increases the speed, which in turn increases the lift. The reverse of all these mechanisms decreases lift, which allows the animal to descend. In practice, all these mechanisms are interdependent. For example, increasing the angle of attack or the amplitude affects thrust as well as lift. Flying animals must be able to balance all these effects to achieve the desired climb or descent (or turn), and it is a tribute to their exquisitely specialized nervous systems that they seem to perform most of this balancing by reflex.

Maximum-Performance Descents

An animal's rate of climb is limited by how much power it has available for flapping, but falling has a physical, rather than biological, limit. An animal's size and shape determines its *terminal velocity,* which is its maximum falling speed. When an object falls, it accelerates until its drag equals its weight, at which point it falls at a constant, steady speed: its terminal velocity. Small animals have high drag coefficients, so they have low terminal velocities; by the same token, large animals have higher terminal velocities. In most cases, flying animals descend much more slowly than their terminal velocities, because they glide down, but some animals encounter situations where a rapid descent is a matter of survival.

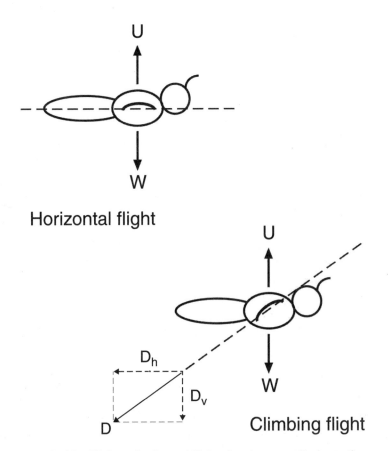

U

W

Horizontal flight

U

W

D_h

D_v

D

Climbing flight

Figure 5.5. Climbing flight. In horizontal flight, drag is perpendicular to the upward force. When the animal climbs, drag is tilted down, so it has both a vertical component (D_v) and a horizontal component (D_h). The vertical component adds to the weight so the animal must produce more lift to increase the upward force. (B.Ht.)

Can an animal descend faster than its terminal velocity? Yes, and the best-known example is the *stoop* of the peregrine falcon. Falcons are predators that make their living by attacking other birds in flight, and their preferred method of attack is to dive onto their victims at such a great speed that the impact breaks the victim's neck or back. In some cases, falcons just go into steep dives, fold their wings to decrease drag, and fall onto their hapless prey like feathered rocks. In other cases, the falcon does something amazing: it rolls onto its back with wings extended so its wings produce lift directed *downward*. This downward lift accelerates the falcon downward faster than gravity alone, and can, in principle, allow the falcon to fall faster than its terminal velocity. In practice, falcons probably use this trick mainly to reach their terminal velocity more quickly, because they normally hit their prey with their feet, and the falcon obviously needs to be upright to do that.

The falcon's trick is probably not widely used among most flying animals, though it could clearly benefit prey as well as predator species. Many insects keep the top of their heads oriented to light (by reflex), and since the sky is normally lighter than the ground, they stay upright. However, flies have a gyroscopic sensory mechanism (see below) that provides them with the equivalent of an artificial horizon; although no studies have shown this, these supremely maneuverable flyers may well fly inverted to accelerate downward on occasion. Bats, which are also highly maneuverable (and, like falcons, catch aerial prey on the wing), are also likely candidates for using the falcon's trick, but again, I know of no research in this area.

Sensing: Where Am I Now, and Where Am I Going?

For an animal to fly and maneuver effectively, it must have some way of determining its orientation in space ("Am I upright?" "Am I leaning to the left?") and its direction ("Am I moving forward?" "Am I sideslipping?" "Am I turning?"). Humans are primarily visual, so we tend to think that vision is the simplest and most direct way of sensing this information. Though vision is certainly important for many flying animals, many others have evolved other means to obtain this information.

Vision

Insects depend heavily on vision for some directional information. Insects in general do not seem to have any direct gravity sensors, so many flying insects rely on light to distinguish up from down. Even at night, except in extreme conditions, the sky is lighter than the ground, so flying insects can stay upright by keeping the lightest part of their visual field overhead [15, 16, 17]. Of course, if conditions are too dim to reliably tell the sky from the ground, this method does not work. Most insects appear to be incapable of flying with any stability in total darkness.

Insects also use vision for directional cues. In experiments where the "flow" of visual patterns past a subject animal could be adjusted by the researcher, bees, grasshoppers, and flies all showed some influence of vision on their ability to fly in a particular direction [18, 19, 20, 21, 22]. In these insects, the parts of their eyes looking downward are tied to stabilizing reflexes: if the patterns below the insect do not "flow" properly, the flight path is corrected. For example, if objects on the left do not appear to move back past the insect as fast as objects on the right, the insect interprets this movement as a turn to the left and responds by reflex with a turn to the right. Different insects use different parts of the visual field and sense different types of changes, but the principle remains the same.

Birds and bats have larger eyes and generally sharper vision than insects. Birds typically have excellent vision, and they undoubtedly use "visual flow" for directional information just as do insects. But birds have vastly more complex nervous systems than insects, so birds might have more flexibility in their responses to such information. The unyielding physical require-ments of flight, however, demand that birds make stabilizing adjustments very quickly, so such reflexes may be just as inflexible and stereotyped in birds as in insects. Bats do use visual information when it is available, but they are mostly nocturnal. On moonlit nights, they operate visually, like birds, but on cloudy or moonless nights, they probably cannot see enough detail to use "visual flow" for determining their direction. Their sonar (echo-location; see Chapter 7) is quite acute, however, and they use this sense to get some of the same directional cues that day-flyers get from vision.

Equilibrium Sensing

Unlike insects, vertebrates have a well-developed *equilibrium*-sensing sys-tem. The inner ear, in addition to sensing sound, contains gravity and rota-tion sensors. Humans make great use of these sensors. For example, with my eyes closed, I can easily tell if a surface I am lying on is level or tilted to one side. If I start to rotate, I can tell in what direction I am rotating.* These sen-sors might seem particularly valuable for flying animals, and they may in-deed be important in some situations (such as poor light, for example). Yet,

---

* The rotation sensors in our semicircular canals actually detect rotational *accelera-tions* and *decelerations,* so they are insensitive to steady, continuous rotation. In other words, if someone spins me around on a stool, after the first few revolutions, I lose a sense of rotating (at least until I stop).

these inner ear receptors can be fooled in many situations in flight. In a smooth, balanced, banked turn, for example, a vertebrate (like a person) has no sensation of leaning, even though the flyer (or airplane) is banked 20 or 30 degrees.* Indeed, pilots spend a lot of time learning to trust their instruments and ignore their equilibrium sensors. Thus, equilibrium sensors, at least of the vertebrate type, are less useful in flight than we might expect.

Sensing Airflow

Sensing the flow of air over the body can be useful to a flyer in two ways. On a large scale, if an animal can sense the speed and direction of the relative wind, it can keep itself aligned with its intended flight direction to avoid sideslipping and correct for the transient effects of wind gusts. On a much finer scale, if an animal could detect the fine details of airflows, particularly on its wings, it could, in principle, adjust its movements to optimize its wingstroke—to prevent a stall or smooth out eddies that might otherwise increase drag. Biologists have demonstrated a variety of large-scale wind sensors in flying animals, but they have not yet found any fine-scale mechanisms.

Insects: We know much more about insect wind sensors than about those of vertebrates. This surprising situation is largely because of the constraints of the insect exoskeleton and the relative simplicity of the insect nervous system. The exoskeleton forms a rigid body covering, so that every sensory receptor must have some structural specialization to allow it to detect anything outside the body. Often, these modifications take the form of a hair. Some biologists have become quite adept at determining what type of stimulus—wind, sound, odors—triggers a given hair, and how that hair's sensations are used by the insect's brain.

Most insects use either their antennae or wind-sensitive hairs on their heads (or both) as large-scale wind detectors. As with studies of flight mechanics (Chapter 4), desert locusts (large grasshoppers) have also been favorite subjects of researchers interested in wind sensing. In locusts, wind-sensitive hairs on the head trigger reflex steering responses [23, 24], which is

---

* I once had the opportunity to ride in an aerobatic airplane, in which I flew through several loops and rolls. Even when the airplane was completely inverted, I did not feel any sense of falling; the centrifugal force tricked my inner ears into telling me I was upright.

probably their function in other insects as well. Antennae are alternate wind sensors in many insects. Insects use their antennae for all sorts of sensing—smell, hearing, touch—and insects like aphids, flies, bees, grasshoppers, and moths use them for wind sensing as well [23, 25]. When insects use their antennae to detect wind, they typically hold their antennae in a characteristic orientation during flight. If wind strikes them from some orientation other than head-on, the relative wind deflects the antennae slightly. A nerve at the base of the antenna senses this deflection, and the insect uses this sensation to trigger a steering reflex.

The insects studied so far seem to use both antennae and head hairs for wind sensing. But why bother? Having more than one system may simply be redundant, so that if a predator happened to bite off the antennae, the head hairs could take over their functions, but this kind of redundancy is rare in biology. Rather, the two systems may detect slightly different aspects of the relative wind; for example, the antennae might be more sensitive to speed, while the hairs detect direction. Or the antennae might respond more to horizontal gusts and the head hairs to vertical gusts. They also probably operate on different scales; the hairs may respond to smaller, quicker disturbances, while the larger antennae respond to larger, more prolonged changes. In any case, they may have some overlap in function, so they can still act as each other's "backup" system.

Vertebrates: Birds and bats surely have similar sensing abilities to insects, but they are more difficult to study for several reasons. First, their tactile sensors are much denser and more numerous than those of insects, and they are not necessarily associated with obvious structural modifications—birds and bats have no constraining exoskeleton. Moreover, the complexity of vertebrate nervous systems means that researchers have much more difficulty demonstrating a physiological connection between a stimulus (like a change in wind direction) and a particular steering response. These animals may not even have specialized wind sensors. For instance, although humans are not flyers, a person can easily determine the direction of a brisk wind just from the sensation of the wind on his or her skin. Because birds and bats actually have an important use for such sensations, they ought to be at least as good as humans at detecting the speed and direction of wind on their faces. In addition, fur or feathers could make very sensitive small-scale flow detectors if coupled with a sensitive tactile nerve ending (as are many vertebrate hairs).* As yet, however, biologists have done little or no work trying to connect wind sensing with flight-steering behavior in birds and bats, and have not searched for small-scale flow sensors.

Specialized "Flight Instruments"

Two different lineages of insects have evolved special mechanisms that give them some ability to fly without reference to light. Dragonflies and true flies use very different methods to sense their orientation without using their eyes. Dragonflies literally use their heads. Dragonflies' heads are relatively large; they are almost covered on the top and sides with enormous eyes, and they have a set of heavy, powerful jaws on the bottom. This head is suspended on a short spindly neck that attaches to the upper back side of the head. The weight of the head tends to keep it upright, acting like a pendulum or plumb bob. The back of the head rubs against patches of sensory hairs on the thorax that sense any tilting of the thorax relative to the head. The head orientation thus gives the dragonfly an indication of the direction of gravity, in other words, which way is up. This sensing ability was demonstrated by a clever set of experiments. In the late 1940s, Horst Mittelstaedt developed a way to glue tiny magnets onto the front of the heads of dragonflies [26]. He then used a large magnet to manually tilt the dragonflies' heads without touching them. He found that for both flapping and gliding dragonflies, if he tilted their heads, they immediately made wing movements to correct what they perceived as a tilted body. For example, if he tilted a dragonfly's head to its left, it adjusted its wings to roll to the left, and vice versa. Why did the dragonflies try to correct in the same direction as their head was tilted? In free flight, if a gust tilted the body to the right, the head would tend to stay upright. When Mittelstaedt tilted the head to the left, he gave the dragonfly the same sensation it would feel if it had its head upright and its body tilted to the right, so the reflex response is to bank in the same direction the head tilts (Fig. 5.6). The dragonfly head system works as a gravity detector, which helps the dragonfly correct for transient gusts and turbulence. This system gives dragonflies a limited ability to fly stably

* Cats' whiskers are part of such a system: the long hair detects objects well before they reach the cat's face, and the length of the hair magnifies the effect of a light touch on the nerve ending at its base.

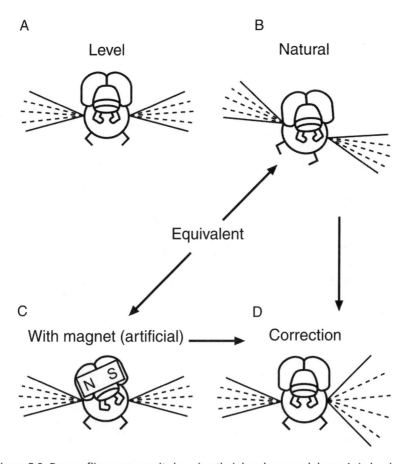

Figure 5.6. Dragonflies sense gravity by using their heads as pendulums. A. In level flight, the body and head are both upright. B. If a gust tilts the body, gravity keeps the head upright, and the dragonfly senses that its head is tilted. C. If the dragonfly's head is tilted artificially by using a strong magnetic field to tilt a magnet glued to the head, the dragonfly is fooled into feeling as if its body is tilted (N, S: North and South poles of the attached magnet). D. When the dragonfly's head tilts to one side, it beats the wings on the other side harder to produce more lift so that it can roll back upright. (B.Ht.)

without light, but it can be fooled in the same way as the vertebrate inner ear, particularly in smooth, steady turns.

True flies use what may be the most elegant and sophisticated sensory device (with the possible exception of some types of eyes) in the animal kingdom. Flies have evolved a gyroscopic stability sensor. A spinning gyroscope tends to stay in whatever orientation it is placed. Perhaps you have seen toy

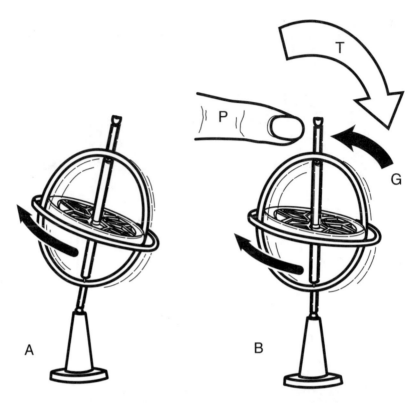

Figure 5.7. Properties of gyroscopes. A. Gyroscopes tend to stay in any orientation (although friction between the wheel and the frame may cause precession). B. Gyroscopes resist being turned. If some outside push (P) causes the gyroscope to turn in direction T, the gyroscope resists with force G. (S.T.)

gyroscopes that seem to balance at "impossible" angles as long as the wheel spins. A related property makes gyroscopes useful for several types of flight instruments: gyroscopes resist changing their orientation. In other words, if I push on a gyroscope to turn it, the gyroscope pushes back; furthermore, the faster I turn it, the harder the gyroscope pushes back (Fig. 5.7). Engineers have designed instruments that measure how hard a gyroscope resists when it is being turned, and such an instrument tells the pilot how fast an airplane is turning—it is a *rate-of-turn* indicator (Chapter 11).

The ancestors of flies had two pairs of wings, but in the fly lineage, the hindwings evolved into tiny sticks with knobbed ends, less than a tenth as long as the functional wings. These highly modified hindwings are called *halteres,* and they no longer have any aerodynamic function. (Males in two other small groups of parasitic insects, scale insects and strepsipterans, have forewings modified into halterelike structures; to my knowledge, however, no one has tested them experimentally to see if they have a stabilizing function.) When the animal flies, it flaps the halteres up and down at the same frequency as the wings (although with a simpler, arclike stroke pattern). Biologists have known for over a century that the halteres were somehow involved with stability, because removing a fly's halteres destroys the fly's ability to fly straight. At first, researchers thought that the halteres acted as a gyroscope and physically kept the fly on track by producing a force to resist turns, but J. W. S. Pringle showed otherwise [27, 28]. He calculated that the halteres were far too small to keep a fly on an even keel physically, and he demonstrated that the halteres are actually gyroscopic sensors.

Although a fly's halteres oscillate up and down, rather than rotating like a true gyroscope, an oscillating mass—like a tuning fork or the knob on the end of a haltere—has a similar tendency to maintain its orientation in space, and to resist turning. When the fly turns, the haltere resists changing its path, which causes the stalk to bend (Fig. 5.8). The fly has nerve endings in the stalk to detect this bending, which triggers reflex wing-beat adjustments to counteract the turn. The basis for the gyroscopic haltere action is fairly obvious for yawing rotations (rotations about a vertical axis; Fig. 5.8); what is less obvious is that halteres can also sense pitch and roll rotations. The reflex corrective action they trigger seems to be strongest in pitch and roll, at least in some flies. Responses to visual information may be more important for controlling yaw, especially when flies chase mates or prey, so the yaw control reflexes of the halteres seem to be weaker than the pitch and roll reflexes.

The gyroscopic action of the halteres allows true flies to fly stably in complete darkness. The haltere sensations cannot be "fooled" by steady turns like the vertebrate inner ear or the dragonfly head sensations. Flies have become so dependent on the haltere-based stabilizing reflexes that flies can no longer fly without them. Flies' active stabilizing ability is so strongly linked to the halteres that if their halteres are removed, these insects spiral out of control when they attempt to fly. Unlike a human aviator—who can fly using visual cues when conditions permit, and who only needs to use gyroscopic instruments when visual cues disappear—flies require the use of their gyroscopic "instruments" whenever they fly.

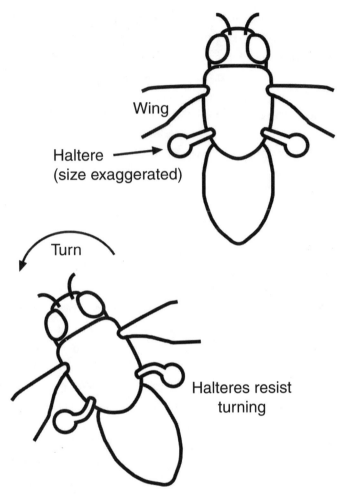

Figure 5.8. Flies' hindwings are modified into halteres that flap up and down with the wings. Like a gyroscope, the knob at the end of the haltere resists turning, and tends to keep flapping in its original orientation. When the fly turns, this tendency bends the stalk of the haltere, which the fly can sense with stretch receptors. (B.Ht.)

Maneuverability and Lifestyles

By engineering standards, flying animals are unstable, which gives them the potential to be highly maneuverable. Among animals, however, some are significantly better at maneuvering than others. For example, in a comprehensive review of all the major groups of bats, Ulla Norberg and Jeremy

Rayner [29, 30] found that differences in turning ability correlated nicely with the anatomy and lifestyles of different bats. Bats that fly in cluttered environments—forests—need to be able to turn sharply in a limited space. These bats tend to have low wing loadings, which allows them to fly slowly and turn sharply. They also tend to have wings with low aspect ratios; shorter, broader wings fit better in cluttered environments, and additionally, are quite effective for quickly rolling into a bank at low speeds. The drawback of low-aspect-ratio wings is that they require higher power, making their flight more expensive. Some bats that fly in more open areas have high-aspect-ratio wings, making their flight more economical. Many of these bats also have high wing loadings, giving them high flight speeds. High flight speeds and high-aspect-ratio wings are useful for rapid turns, because long wings can produce high torques for fast rolling: wing tips far from the body have a lot of leverage. Conversely, quick turns also benefit from concentrating wing mass as close to the body as possible to reduce the wings' moment of inertia. Thus, the most agile bats in this group have tapering, sharply pointed wings, giving tips far from the body but keeping most of the wings' area (and mass) close to the body. Other bats that fly in the open have high-aspect-ratio wings with low wing loadings, so they fly slowly. These bats may not be as maneuverable as the first two groups, but their flight is very economical. Some bats in this last group hunt over open water, where they can fly slowly and make few turns, and many of the bats that migrate long distances fall into this group as well.

Evolution has clearly favored maneuverability at the expense of stability among flying animals. Although they are not all equally maneuverable, even the clumsiest flying animal is vastly more maneuverable than any airplane. Perhaps we should expect this: first, almost all flying animals are either predators or potential prey or both, so high maneuverability is a vital asset. Moreover, flapping movements just do not seem to lend themselves to passive stability. With the wings constantly moving through a rather complex stroke, evolving the capability to make rapid adjustments to the stroke seems to have been more economical and advantageous than building structures—for example, long tails—to improve passive stability.

Fueling Flight

What exactly is energy? In the deceptively simple definitions found in physics textbooks, *energy* is the ability to do work, and *work* is force times distance, but what do these definitions mean in practical terms? When I push on an object—apply a force—and the object moves, I have done work. If I push on a chair hard enough—apply a force—so that the chair moves some distance, I have done work. Biologists call this *external work,* in contrast to the *internal work* of the body's biochemical and physiological processes. External work is usually much easier to measure than internal work. Furthermore, if I push on a wall and the wall does not move, I have done no external work regardless of how hard I push (although my muscles have used energy doing internal work). So if external or mechanical work means pushing on an object hard enough to make it move, then energy is anything that contributes to the push.

Work and Energy in Biology

The most important type of energy for animals is the energy stored in the chemical bonds that hold food molecules together. Obviously, animals collect this energy by eating. Animals use elaborate physiological and biochemical processes to capture some of the energy in the chemical bonds of food molecules, and use it in their own chemical processes. When an animal uses some of this chemical energy to contract a muscle, the animal may push on something hard enough to move it, thus doing external work. Animals cannot capture or use all of the chemical energy in their food, however. According to the Second Law of Thermodynamics, no energy-conversion process can be 100 percent efficient, because some of the energy will always be lost as heat. Most animals simply lose this heat to their environment, but "warmblooded" animals actually use this "waste" heat to warm their bodies.

A surprising proportion of an animal's physiological activities are used for extracting energy from food, converting the energy to a more useful form, and using the energy to do work. For example, digestion, respiration, and circulation are devoted almost entirely to these processes. Animals consume energy in many biochemical and cellular processes, as well as for doing work on their external environment. An animal's overall net consumption of energy is its *metabolic rate*. The energy consumption results from work done at all levels—biochemical to external—and from all heat given off. The more work an animal does, the higher its metabolic rate will be; conversely, the less food available to an animal, the less work it will be able to do. An animal's physical activities determine its metabolic rate, which in turn dictates the animal's food requirements. This relationship between food, energy, and activity levels has been a major focus of physiologists and ecologists for decades. The effects of this relationship range from the mundane to the global, from when and how long to eat, sleep, and court to where a species can survive, or whether it should migrate. Because flying can consume more energy than any other activity, flight clearly has a major influence on an animal's energy budget.

Of all its activities, an animal typically puts more work into locomotion than any other type of activity. Any type of locomotion can thus produce a dramatic increase in an animal's metabolic rate. The form of locomotion an animal uses, however, makes a difference. In energy terms, walking is the most expensive, swimming is the least expensive, and flying is in between. (We shall explore the reasons for these differences later in this chapter.) Because of the rapid movements required for flapping flight, flying usually requires the most power, which is equal to force times speed. Flight typically increases an animal's metabolic rate ten or twenty times above its resting level. For example, a 25-gram cardinal's body might require the energy in two sunflower seeds (about 2.5 joules or 0.6 calorie) each hour when it is at rest. To fly for an hour, the same bird needs the energy from forty-four sunflower seeds. Thus, an animal's ability to travel will depend on both its ability to find and eat food, and its ability to store that food energy for later use. Even though flight raises an animal's metabolic rate enormously, long, nonstop flights are common throughout the animal kingdom. Hummingbirds fly hundreds of kilometers across the Gulf of Mexico, and many ducks, geese, shorebirds, and songbirds fly thousands of kilometers nonstop on their spring and fall migrations. These long flights raise many questions. For instance, how much fuel (food energy) does a hummingbird need to fly across the Gulf of Mexico? How much fuel does a snow goose need to fly from Hudson's Bay to the Gulf Coast, or a monarch butterfly to fly from Minnesota to central Mexico? Can energetics tell us anything about why flight might be preferable to walking long distances? As far as fuel goes, how do animals store food energy, and why is fat the universal choice for long-distance journeys?

Muscles and Energy

Muscles are what power our movements, and one of the key characteristics that separates animals from other organisms is that animals have muscles for movement. Muscles use energy (from food) to produce force or motion or a combination. When muscles are active, they use enormous amounts of energy; active muscle tissue consumes energy dozens of times faster than most other types of tissue. When an animal's muscles are inactive, however, their energy requirements are much lower and similar to other tissues. Thus, when an animal is actively moving about, most of its energy is consumed by its muscles, but when it is at rest, other tissues dominate energy consumption.

Because of muscle tissue's obvious functions and its central role in so many activities, muscle cells may be the most-studied and best-characterized of all cell types. Muscle cells (and by extension, whole muscles) work by shortening. Unlike machines, which usually get power from a continuously rotating engine or motor, muscles produce linear movement over a limited distance. In other words, a muscle works by shortening, and it can only shorten so much. Some muscles can shorten by approximately 10 or 15 percent (that is, their contracted length is 85 to 90 percent of their resting length), although a highly specialized type of flight muscle in bees and flies only shortens 1 or 2 percent. Muscles use the leverage of the skeleton to convert short muscle movements into large movements of body parts.

Muscle cells are usually in the shape of elongated cylinders, and a whole muscle usually consists of a large bundle of these cells. Although muscle cells are typically less than half a millimeter in diameter, they often run the entire length of a muscle, which could be a few millimeters in a housefly to over a meter in an elephant. The basic contractile element of a muscle cell is a complex, submicroscopic, cylindrical array of protein threads called a sarcomere. The sarcomeres are the "motors" of the muscle cell, the parts that actually shorten [1]. The sarcomeres are joined end to end in cablelike *fibrils* that run the length of the cell, and each cell is filled with more than a thousand fibrils. Each sarcomere is approximately two micrometers long, and a fibril can be from a few hundred to hundreds of thousands of sarcomeres in length, depending on the length of the cell [2]. An elaborate control system ensures that when a muscle cell receives a command from the nervous system to contract, all sarcomeres in a cell contract at the same time. This arrangement of sarcomeres in cells and cells in muscles leads to several important properties of muscle.

First, if all sarcomeres pull equally hard, increasing a muscle's diameter will increase the whole muscle's pulling force. This increase is because a wider muscle will have more sarcomeres side by side, pulling in parallel. Thus, muscle force or strength is approximately proportional to the muscle's cross-sectional area. Sarcomeres from a wide range of vertebrates do, indeed, seem to pull with equal strengths, although there are rare exceptions [3]. Second, the sarcomeres themselves set limits on how much a muscle can shorten: few sarcomeres can shorten more than 10 or 15 percent, and many are limited to even less shortening. Because the sarcomeres are arranged end to end in the fibrils, if the sarcomeres all shorten 5 percent, then the whole muscle will shorten 5 percent.*

Although sarcomeres from many different kinds of muscles seem to produce equal force, they do not all seem to shorten at the same speed. If all sarcomeres shortened at the same rate, say 10 percent in half a second, then the whole muscle ought to shorten by the same proportion in the same time. A 2-centimeter muscle would shorten 0.2 centimeters in the same time that a 20-centimeter muscle shortens 2 centimeters. The longer one would thus shorten ten times faster. Scientists have repeatedly found that small animals have sarcomeres that shorten faster, and large animals have sarcomeres that shorten more slowly [3, 4]. This size effect means that small animals can move their muscles disproportionately faster than large animals, which results in some crucial size effects. For example, the muscles of small animals will produce more power per gram than those of large animals, but the small animal's sarcomeres will use energy at a correspondingly higher rate.

The sarcomeres are the main energy-consuming components of muscle cells, and they consume chemical energy as long as they are actively contracting. Not all of the chemical energy goes into force production, however. Muscle cells are about 25 percent efficient, meaning that 25 percent of the energy they consume goes into force production and 75 percent is given off as heat. (Muscles are thus the major source of the "waste" heat that warm-blooded, or endothermic, animals use to raise their body temperatures.) Muscle cells must be able to supply chemical energy to the sarcomeres whenever they are actively contracting, and this requirement is so important that most muscle cells have two or three alternate mechanisms to supply that energy and ensure uninterrupted contractions. In addition, muscles make up well over half of the body mass of vertebrates (and probably most invertebrates as well). This, combined with the great appetite of sarcomeres for chemical energy, makes active muscles by far the greatest energy-consuming component of the body. In other words, when an animal is actively using its muscles, the muscles are the greatest contributors to the animal's metabolic rate.

---

* Actually, the whole muscle's contraction lags a bit behind that of the sarcomeres because of the stretchiness of the tendons and other tissue at the ends of the muscle. Biologists call this material the *series elastic component*, and it may delay the buildup of pulling force of the whole muscle by several milliseconds.

The Cost of Locomotion

Suppose I want to find the cheapest way to ship a package a long distance. My choices might be to send it by truck, airplane, or cargo ship. Also, suppose that shipping costs are set mainly by fuel consumption of the vehicle (which may not be very realistic for a commercial shipping firm). What I need to know is how much fuel each vehicle consumes to carry my one-kilogram parcel a given distance. The truck requires the most fuel to carry the package and so is most expensive. The airplane uses about half as much fuel, and the cargo ship only needs one-tenth the fuel of the truck to carry the package the same distance [5]. This example demonstrates what engineers call the cost of transport—the amount of fuel needed to move a given mass a given distance, or fuel consumption per unit mass per unit distance. Biologists use a slight variation called the cost of locomotion. Rather than measuring the amount of fuel consumed, biologists measure the amount of energy an animal consumes to carry a given mass a given distance. The cost of locomotion is measured in joules per kilogram per kilometer in the SI system. An odd feature of the cost of locomotion is that it does not include speed or time. In a comparison of the costs of locomotion of a snail and a falcon, neither the speed nor the duration of travel has any effect on the cost. Because power, being a rate, *is* affected by time, this definition of the cost of locomotion produces an apparent paradox: flight is both a low-cost and a high-power activity. The low-cost, high-power nature of flight affects most aspects of flight energetics.

Energy in Biology: Metabolic Rates

Biologists use a variety of methods for measuring an animal's metabolic rate,* and one of the simplest is to measure the rate at which the animal uses oxygen. Oxygen is used in the processes that break down food molecules and capture their energy. Oxygen consumption thus has a simple, direct relationship with energy use, so an animal's metabolic rate can easily be cal-

---

* Biologists often use *metabolism* as a synonym for *metabolic rate,* but *metabolism* has several other related meanings and is thus easily misunderstood.

culated from its oxygen consumption rate if we know what kind of food it eats.*

An animal's metabolic rate can vary greatly, depending on its level of activity. A duck will have a much higher metabolic rate when grooming than when sleeping, and a flying duck will have yet a higher metabolic rate than when grooming. In order to make comparisons among animals, we need a standard, baseline condition so that we compare the metabolic characteristics of animals and not their activities. For an *endothermic,* or "warm-blooded," animal, the *basal metabolic rate* is defined as the metabolic rate of an animal at rest in unstressed conditions, not digesting a meal, and at a comfortable environmental temperature (an endothermic animal's metabolic rate increases in both cold and hot environments). The "comfortable temperature" is a temperature where the animal does not have to spend any metabolic energy on processes like shivering or perspiring to warm or cool its body. Thus the basal metabolic rate is the lowest metabolic rate of a waking animal, and it represents the minimum "maintenance" level of metabolic energy use. The metabolic rate of an *ectothermic,* or "cold-blooded," animal does not have a minimum value at some comfortable temperature. Instead, the metabolic rate simply decreases as the temperature drops, until the animal reaches its lower lethal temperature and dies. Ectotherms thus do not have a true basal metabolic rate, so biologists measure the resting metabolic rate at some typical temperature and call it the *standard metabolic rate at the specified temperature.* The standard metabolic rate represents the maintenance level of energy use for an ectothermic animal at a particular temperature.

Effects of Size

An elephant uses more energy in an hour than a mouse—in other words, the elephant has a higher metabolic rate. However, if we collected enough mice to equal the mass of the elephant (approximately 150,000), and measured the sum of all their metabolic rates, the result might be surprising. Those 150,000 mice would have a combined metabolic rate of over twenty times higher than an elephant of the same weight [6]. Put another way, small animals actually require higher energy consumption on a gram-for-gram ba-

---

* Until recently, in technical articles, biologists commonly reported the metabolic rate in units of amount of oxygen consumed in a given time.

sis. Energy consumption dictates food requirements: a kilogram of mice need the energy in about 44 grams of sugar (10 teaspoons) to survive a day, but the elephant only needs the equivalent of about 2 grams of sugar (less than ½ teaspoon) per kilogram of elephant to fuel it for a day. Of course, the elephant has several thousand kilograms to fuel. Metabolic rate per unit body mass is called *specific metabolic rate,* and when biologists measure the specific basal metabolic rate across a wide range of animal body sizes, they find that small animals have higher specific metabolic rates than big animals. As animals get larger and larger, their metabolic rate also increases, but only about three-fourths as fast as their body mass increases. Thus, gram for gram, humming-birds have higher specific metabolic rates than geese, ants have higher specific metabolic rates than dragonflies, and little brown bats have higher specific metabolic rates than flying foxes.

Body size also has an effect on muscle power output. As we saw earlier, larger animals tend to have slower muscles than smaller ones. Since power equals work per unit time,* and a slow muscle takes longer to do a given amount of work, a slow muscle will have lower power. If a gram of muscle in a wren can contract twice as fast as a gram of muscle in a stork, the wren's muscle will produce twice as much power per gram than the stork's. Of course, the wren has many fewer grams of muscle than the stork, so the wren's total power output will be lower, even though its power per gram is higher. Also, remember that muscle force is related to the cross-sectional area of a muscle, which only increases two-thirds as fast as body mass increases (Chapter 4). As an animal gets bigger, not only do its muscles move more slowly, but it has proportionately less muscle area, which further reduces its available muscle power. Just as specific metabolic rate decreases as animals get bigger, power per gram of body mass also decreases as animals get bigger.

Effects of Activity

Basal and standard metabolic rates measure minimum maintenance levels of energy use. What happens when an animal does something more ac-

---

* Power is work divided by time (W/t). Since work is force times distance (F × d), power equals force times distance divided by time (P = F × d/t). But distance divided by time is speed (d/t = v), so power also equals force times speed (F × v). Power is thus affected by both time and speed.

tive than the conditions specified for basal metabolic rate? For example, eating and digesting a meal can raise the metabolic rate 30 to 100 percent above resting levels [7]. The effect of eating is barely significant, however, compared to the effect of muscular activity, particularly locomotion. Vigorous exercise can raise the metabolic rate as much as ten to forty times above the basal level. Locomotion is thus a major, often primary, energy-consuming activity.

Effects of Temperature

Endotherms use heat from metabolic processes to warm their bodies. Birds and mammals are normally endothermic, and many insects are endothermic when flying. The biochemical and physiological processes that transfer energy and convert it into different forms give off heat as a byproduct. Ectotherms produce metabolic heat so slowly that they rapidly lose it to the environment, and their body temperature stays the same as that of their surroundings. Endotherms, in contrast, have such a high basal metabolic rate that they generate considerably more heat. A wren typically has a metabolic rate ten times higher than a lizard of the same size. In addition, endotherms have physiological mechanisms to regulate how much heat is retained and how much is lost to the environment, so they can maintain a constant internal body temperature. Insulation—fur or feathers—helps retain heat, while panting and sweating help rid the body of excess heat. Endotherms are usually in environments where their body temperature is higher than the environmental temperature, so their main concern is retaining heat. When the environmental temperature gets higher than their body temperature, endotherms must actively dump heat or risk elevated body temperatures. Hot conditions can lead to a vicious cycle, because processes for dumping heat (such as panting) also raise the animal's metabolic rate, which produces even more heat. Most endotherms can cope with mildly elevated body temperatures. But if the environmental temperature gets too high, a vicious cycle begins that leads to metabolic collapse (heat stroke) and eventually death.

The Advantages of Endothermy

Why bother being endothermic? Heating the body this way means that our wren must eat ten times as much food as the lizard to maintain its high metabolic rate. An obvious benefit of being warm-blooded is that endotherms can be active in cold environments where ectotherms become too sluggish to move. A sparrow can easily search for food in near-freezing temperatures, but a bullfrog's metabolic rate drops so low that its muscles can barely contract. The frog must hunker down in shelter and wait for warmer conditions.

The sparrow also gets a more subtle benefit. The frog's organs must operate effectively over a wide range of temperatures, but the sparrow's organs are always warm, as if it lived in a tropical paradise. The sparrow's cells have evolved high biochemical efficiency in their constantly balmy conditions, but the frog's cells must be "jacks of all temperatures but masters of none." In other words, frog cells can never be quite as efficient as sparrow cells when they work over a broad range of temperatures. Thus, in return for a higher metabolic rate, endotherms gain faster and more efficient physiological processes, such as digestion, internal transport, nerve signaling, and muscle contraction.

Endothermy is related to flight in at least a couple ways. First, two of the three extant groups of flying animals—birds and bats—are endothermic, and endothermic animals have much higher resting metabolic rates than ectotherms. This difference translates at least partly into proportionately higher metabolic rates during activity, with flight being among the most strenuous of activities. Thus, endothermic flyers should have much greater energy requirements than ectotherms of similar size. Paradoxically, however, flight itself may reduce the disparity between endotherms and some ectotherms. Because flight requires such a high rate of energy use, some large insects become effectively endothermic during flight. They reap some of the benefits (and incur some of the high energy costs) of endotherms during flight, while maintaining an appropriately low, ectothermic metabolic rate when not flying. Smaller insects, however, have such a high surface-to-volume ratio that they usually cannot retain enough heat to become endothermic.

Some biologists have assumed that endothermy is necessary, or at least highly beneficial for flight. As we shall see in Chapter 7, this assumption has had an important influence on the arguments about the evolution of flight in birds. Given that most flying insects are completely ectothermic, however, the evolution of flight clearly does not require that animals be endothermic.

How Much Does It Cost to Go from Here to There?

With metabolic energy as a "currency" to measure cost of locomotion, we can compare the costs of different types of locomotion. Terrestrial locomotion—walking or running—is the most expensive form of locomotion. Given that humans are naturally terrestrial, many people may be surprised to learn that walking is so costly. The cost of locomotion for human running is about five times higher than for the flight of a typical bird, and ten times more expensive than fish swimming. In other words, I use five times more energy to move a kilogram of me a kilometer than a goose uses to fly a kilogram of gooseflesh the same distance. A trout, however, only needs half the energy of the goose to move a kilogram of itself a kilometer. The cheapest way for animals to travel is clearly by swimming; flying is next best and walking or running is a distant third. This general relationship holds for all sizes of animals at all biologically realistic speeds.

Swimming

Just why is locomotion so cheap for a trout or a carp? The main reason is that the water supports most of the body weight of such a swimmer, so all the animal needs to do to swim is to produce enough force to overcome the drag of its body. Most aquatic animals are nearly the same density as the water in which they swim, so they do almost no work to support their weight against gravity. However, swimming is only cheap for those animals well adapted to swimming completely submerged. When an animal such as a duck or a muskrat swims on the surface, it uses two or three times more energy than when swimming submerged, and as much as twenty times more energy than a fish of a similar size. Why does it cost a duck several times more energy to swim on the surface than when submerged? Because of the bow wave: any object moving on the surface of water pushes up a bow wave at the front, which streams alongside and trails back as the wake. Boat designers have long known that the bigger the bow wave, the harder it is to push a boat through the water. Just as lift production leads to the induced drag, the bow wave produces extra drag on any body moving on the surface of water. Thus, for our purposes "swimming" means underwater locomotion by animals with streamlined bodies, not the exhausting, yet slow and inefficient, locomotion of humans in swimming pools.

Flying

Flying animals move through air that is less dense and less viscous than water, so why does flying cost more than swimming? First, a flying animal must move much faster than a swimmer in order to produce enough lift. This higher speed increases the drag that a flyer must overcome. Furthermore, a flyer has an extra source of drag that a swimmer does not: the induced drag that comes from lift production. In a way, the induced drag represents the cost of supporting the flyer's weight in air. Thus, a flying animal must do more work to overcome its drag, making flying more costly than swimming.

Walking (or running or galloping) is so costly because it involves at least three processes that require muscular work. The first is simply supporting the body's weight. The second is overcoming the friction in joints and muscles, and the third is constantly producing accelerations and decelerations. The exact proportion of muscular effort that goes into these three processes depends on the anatomy of a given animal, but the third process probably accounts for most of the energy used by the muscles. When a person takes a step, first one foot pushes off, which accelerates the body. Then the other foot swings forward and hits the ground, and as the weight shifts onto that foot, the body decelerates. Some of the leg muscles actively tense to act as shock absorbers during this deceleration. Momentum carries the body over the grounded foot, at which time that foot pushes off to accelerate the body, and the cycle repeats. In energetic terms, walking is inefficient because of the acceleration and deceleration required with every step. Both the decelerations and accelerations need muscular effort and thus energy use. In swimming and flying, animals accelerate and decelerate relatively little over the course of a tail-stroke or a wing-beat, so little or no energy is consumed by this process. As an analogy, consider riding a bicycle. When a person rides a bicycle, the bicycle does not accelerate or decelerate much with each turn of the pedal. Thus, a person can ride a bicycle much faster than he or she could run using the same amount of effort.

Hovering is an exception to the "flying is cheaper than walking" rule. Indeed, hovering is the most energetically expensive activity in the animal kingdom. When a hummingbird hovers, its wings get no lift from the speed of its flight through the air, because it is *not moving* through the air. The hummingbird must move its wings fast enough to produce all the lift the bird needs by direct muscular effort. A hummingbird puts lots of muscular effort into accelerating and decelerating its wings, just as we do when walking. Furthermore, the wings must move very rapidly to generate enough lift to support the bird's weight. All of this rapid wing movement, with lots of accelerations and decelerations, requires the bird's muscles to contract very quickly; in other words, hovering requires much more power than forward flight.

Something odd happens to the cost of locomotion during hovering. The cost of locomotion is the energy needed to move a given mass a given distance, but in a hover, the distance drops to zero. Dividing the energy per unit mass by zero might seem meaningless, but what it means is that the cost of locomotion becomes infinitely large in a hover.* In other words, hovering is infinitely inefficient as a method of movement because no matter how much energy a hovering hummingbird uses, it never makes any progress. (Hovering may, however, be quite effective or even efficient for other purposes, such as quickly sucking nectar from flowers.)

The U-Shaped Power Curve

Flight is different from other forms of locomotion because of the way power requirements change with speed. Being terrestrial, we are used to thinking of power as being directly related to speed: to run faster takes more power, to walk more slowly takes less power. Land and water-borne animals and vehicles all share this characteristic, which we tend to take for granted. For example, if I walk the block from my office to my parking lot, I hardly notice the effort; if I jog, I arrive panting, and if I run hard, I may have to sit down and rest. Flight, however, is different. Flight actually requires *increasing* power as the flyer *slows down,* a relationship between power and speed that is totally foreign to our everyday experience.

To gain an appreciation for this phenomenon, imagine walking briskly down an imaginary sidewalk with some peculiar properties. On this sidewalk, your feet get heavy and stick to the sidewalk when you slow down, so that you

---

* Despite the fact that calculators and computers call dividing by zero an error, division by zero has a simple mathematical meaning. The result of dividing any number by zero is infinity.

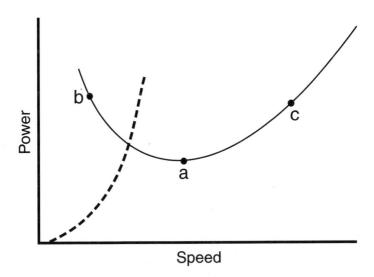

Figure 6.1. The U-shaped power curve (solid) is lowest at an intermediate speed (a). At both higher (c) and lower (b) speeds, a flyer requires more power. The dashed curve shows a typical power curve for walking. (S.T.)

need more and more effort to walk slower and slower. If you can summon the energy to walk faster, your feet get lighter and less sticky until you regain your original brisk pace. If you speed up beyond this pace, things work "normally": jogging requires a little more effort, and running requires a lot more. This strange hypothetical sidewalk is entirely analogous to the situation faced by all flying animals (and airplanes as well). At low speeds, they need more power to slow down because the wings must still be moved fast enough to support the animal's weight. Hovering, with its massive power requirements, is the logical extreme. Intermediate flight speeds require the least power. As a bird speeds up from this intermediate speed, it needs increasing power to fly faster, because pressure and viscous drag increase. If we plot the power required to fly over a wide speed range on a graph, the curve is high at low speeds, low in the middle, and rises again at high speeds. This is the U-shaped power curve, and it applies to anything that flies with wings (Fig. 6.1).

Drag and Power

What causes this counterintuitive relationship between speed and power? The power is used to overcome drag, and the total drag is dominated by different components at different speeds. The viscous and pressure drag behave in a straightforward way: at low speeds, these drag components are low, and as speed increases, they go up. Multiplying the drag by the speed gives power, so the power needed to overcome these drag components rises rather sharply at higher speeds (Fig. 6.2A).* The induced drag behaves very differently as speed changes: it is very high at low speeds, drops rapidly as speed increases, and flattens out to a shallow decline at high speeds. Induced power is so high at low speeds because it is the power that goes into lift production, and at low speeds, a flying animal has less airflow over its wings from its forward speed. Aerodynamically, the animal can increase the angle of attack of its wings to get more lift (and consequently increase their induced drag), while behaviorally, it can flap its wings faster to speed up the airflow over them. Both of these processes lead to increased induced power at low speeds (Fig. 6.2B).

Summing up the power to overcome viscous, pressure, and induced drag at many different speeds produces a graph that looks like Fig. 6.3. This graph shows the U-shaped power curve with several important features. The most obvious, of course, is that it shows that slow flight requires more power than flight at some intermediate speed. In addition, the lowest point on the curve indicates the flight speed that requires the least power. This point is called the *minimum-power speed* and given the symbol v_{mp} (Fig. 6.3). The graph also shows the speed of maximum range. The *maximum-range speed* (v_{mr}) is the speed that gives the greatest flight distance for a given amount of energy. In other words, if a robin has a fixed amount of fat to carry as fuel on a long-distance flight, the robin will get the most distance from that fuel if it flies at its maximum-range speed. At first glance, we might expect that the speed at which the robin uses the least power should give the greatest range. Why is the maximum-range speed faster than the minimum-power speed? Speed is the key: as the robin flies faster, it covers longer distances. If the robin flies a little faster than its minimum-power speed, its power consumption goes up,

* Anyone trying to make sense of the technical literature needs to know that engineers sort out the components of drag in an unusual way for power calculations. The viscous and pressure drag of the wings is called the *profile* drag and the viscous and pressure drag of the rest of the animal or airplane (mainly of the body or fuselage) is *parasite* drag. Their graphs thus show profile and parasite power curves, which I have lumped together as a combined curve for viscous and pressure drag.

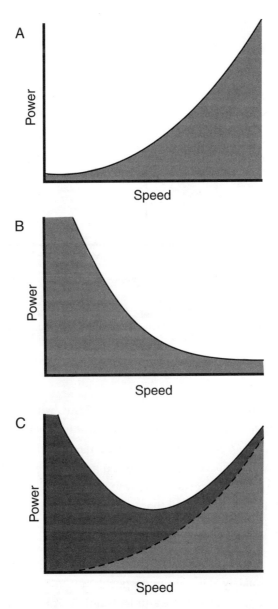

Figure 6.2. A. The profile plus parasite power increases as speed increases. B. The power needed to overcome induced drag actually decreases as speed increases. C. Combining the effects of profile + parasite power and induced power produces the U-shaped total power curve. (S.T.)

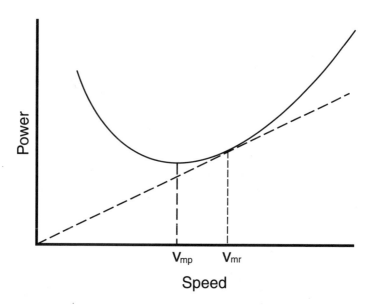

Figure 6.3. The lowest point on the U-shaped power curve represents the conditions for the lowest power requirement, which occurs at the minimum-power speed, v_{mp}. A line through the origin tangent to the curve, such as the dashed line, touches the curve at the power condition best for long-range fuel economy, which occurs at the maximum-range speed, v_{mr}. (S.T.)

but the distance it covers goes up a bit more. Thus, within a certain speed range, the total energy needed to fly a given distance actually drops as the bird flies faster. The upper end of this speed range is the maximum-range speed, and above this speed, increased speed no longer outweighs increased power. The maximum-range speed is easy to find with the power curve graph: a straight line through the origin that is tangent to the power curve touches the curve at the maximum-range speed (Fig. 6.3).

Which Speed Is Best?

Should an animal fly at its minimum-power or its maximum-range speed? The answer depends on what the animal is doing. For a warbler migrating from New England to Venezuela, the maximum-range speed is best. The warbler will be able to fly the longest distance on a fixed fuel load at this speed. In contrast, the same warbler is usually best off flying at its minimum-power speed when foraging for food or searching for mates. At this speed, the bird will maximize its duration; in other words, it will be able to spend the longest time aloft for a given amount of energy at its minimum-power speed.

The steep left-hand part of the curve demonstrates in yet another way that hovering is prodigiously expensive. When a walking cat (or person or beetle) slows down, it uses less power, and it uses the least when it comes to a stop. In contrast, a flying bat or chickadee uses more power to hover than to fly at high speeds. Flying animals operate in a situation much like the hypothetical sidewalk mentioned earlier, with the added complication that coming to a stop requires the most energy of all.

Size

The exact shape of the power curve varies from one animal species to another. Gulls and other animals with high-aspect-ratio wings will have less induced drag, which gives them flatter curves; stubby-winged birds such as ducks and sparrows have steeper curves. As flyers get smaller, their curve is shifted down and a bit to the left (Fig. 6.4). Bigger flyers have curves shifted up and to the right. In other words, as flying animals get bigger, their power requirements go up dramatically, but their flying speed increases more modestly. For example, a 20-gram swallow uses about one-tenth of a watt to fly at a minimum-power speed of 3.5 meters per second, while a 200-gram kestrel uses 1.6 watts to fly at its minimum-power speed of just over 5 meters per second (Fig. 6.4) [8, 9]. The kestrel weighs ten times more than the swallow, and it requires fifteen times more power to fly less than twice as fast.

Comparing power curves such as those of Figure 6.4 can give us insights into the performance of large and small flyers. Hovering is strenuous for all animals, but large animals are simply unable to produce enough power to hover: their muscles are not fast enough or strong enough. A gram of bee or hummingbird muscle produces more power than a gram of heron or eagle muscle. Small size (resulting in low Reynolds numbers) has two further effects on speed. First, the high proportion of viscous drag reduces the relative importance of high induced drag at low speeds. In other words, the viscous drag is so high all the time that the increase in induced drag at very low speeds is less noticeable (and has less effect on power needs). Second, the steep increase in viscous drag as speed increases places a cap on the maximum speed of small flyers. In a nutshell, it is easier for small flying animals to hover, but harder for them to fly fast. In the extreme case, the smallest

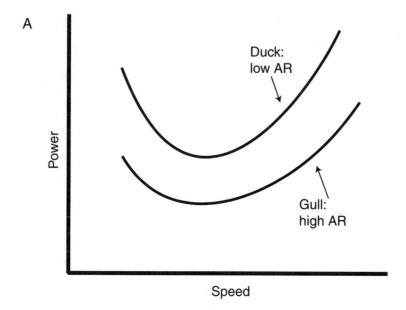

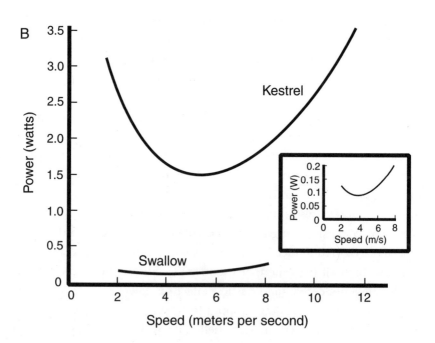

flying animals—gnats, midges, or thrips for example—may have a top speed of only a few centimeters per second, and this top speed may require nearly the same amount of power as hovering.

At the other end of the size scale, large flying animals are unable to hover, but their minimum-power and maximum-range speeds are relatively high [10]. The largest flyers also run into a power squeeze, but for very different reasons from the smallest ones. Because really big birds have proportionally less powerful muscles, they have difficulty producing enough power to flap continuously. Birds the size of larger storks or swans may be able to produce only enough power to fly at their minimum-power speeds, even when flying long distances [10, 11, 12]. When birds become so large that they can only fly at their minimum-power speed for a brief period, their only practical alternative for flying long periods or over long distances is to abandon flapping flight and soar. Thus, vultures, large birds of prey, cranes, storks, pelicans, gulls, and albatrosses are among the large birds that soar regularly, and the largest birds—eagles, condors, albatrosses—soar almost exclusively and avoid flapping flight as much as possible.

Many researchers have tried to measure the metabolic rate of animals flying at different speeds. Some have gotten results that show the expected U-shaped relationship between metabolic rate and flight speed [13, 14, 15], but others have not. Some curves seem to be flat at the high-speed end [16, 17], some, flat at the low-speed end [18], and some, entirely flat with no increase in metabolic rate at high or low speeds [19]. Fundamental aerodynamics dictate that the power *output* of a flying animal must be U-shaped, so why do some experiments appear to show that for some animals, their power *input* is not U-shaped?

First, an animal's total metabolic rate includes power for many activities other than flapping, and even the flight muscles only put 20 to 30 percent of

---

FACING PAGE:
Figure 6.4. Power curves for a variety of birds. A. Birds with higher aspect ratios (such as gulls) tend to have lower, flatter power curves than birds with lower aspect ratios like ducks. B. Large birds tend to have higher curves, slightly shifted to the right, relative to curves for smaller birds; kestrels weigh about 10 times more than swallows. Inset: the characteristic U-shape of the swallow's power curve is easier to see on a graph with an expanded vertical axis. Data for B from J. M. V. Rayner 1990 and 1995b [8, 9]. (S.T.)

their power into mechanical work (the rest goes into heat). The changes in metabolic rate due to flight speed may simply be swamped by other processes in some animals. Second, as Jeremy Rayner has pointed out, power curves for flying animals are quite shallow, and many animals may simply resist flying at speeds where the power requirements increase substantially. In other words, researchers may only be measuring the nearly flat, bottom region of the curve [20]. Rayner has also compared the curves published by several other researchers and shown that whether they are U-shaped, or flat on one end (called "J-shaped" or "L-shaped" by researchers), is at least partly a matter of perception. Finally, he suggested that some may indeed be a bit flatter on one end than the other, and such curves should be seen as "elements on a continuum," or slight variations on the basic U-shaped pattern [20].

Fuel for Flight

In an auto or airplane engine, fuel is burned to liberate its energy, and this energy is used to do work. The fuel is a mixture of hydrocarbon molecules in the form of gasoline or kerosene. Burning them is one way to break the chemical bonds that hold them together and release the chemical energy stored in those bonds. Plants and animals also run on energy that they extract from the chemical bonds of molecules. Plants synthesize their own fuel from sunlight and the carbon dioxide in the air, but animals must obtain their fuel by eating. Animals break down food molecules in a much more complex and controlled way than burning them in an engine, but the result is the same: energy is liberated and used to do work.

Most food molecules fit into one of three categories: proteins, carbohydrates, or fats. Animals usually can use all three for energy, but the three categories are not exactly equivalent. Proteins are generally more important for making structures and regulatory molecules (enzymes) than for fuel, but excess proteins may be used as fuel. In contrast, carbohydrates—starch or sugar—are used almost entirely for fuel. These molecules are usually the primary energy source in food from plants. They are the molecules that cells can break down most quickly for their energy, so carbohydrates are most valuable when an animal needs energy in a hurry. Cells need more time to extract the energy from fats than carbohydrates, and fats are also more difficult to transport from one place to another within the body—unlike carbohydrates, fats do not dissolve easily in water, the main component of blood. The body must package fat molecules in protein envelopes or partly disassemble them before transporting fat molecules in the blood.

Fat molecules have one major advantage over proteins and carbohydrates, however. Every gram of fat can be broken down to give over twice as much energy as a gram of protein or carbohydrate. Fat's advantage over carbohydrate is even greater, because every gram of carbohydrate must be bound to three or four grams of water when stored, so every gram of fat gives at least ten times more energy than a gram of stored carbohydrate bound to water. If an animal needs to store energy in a form that must be carried around, fat is the obvious choice. A given weight of fat will yield twice as much energy as protein and ten times more than carbohydrate. Fat is the universal choice for energy storage throughout the animal kingdom, and most animals can easily convert the proteins or carbohydrates in their diets into fats for storage.

If fat is best for being hauled about, what happens when an animal needs a quick burst of energy, such as a bobcat pouncing on a bird or a bird leaping away from a bobcat? Fat needs a bit of extra handling time before it can be broken down for energy, which could cause a serious problem if the bobcat or bird depended solely on fat. Most animals get around this problem by storing a limited amount of carbohydrates in some of the muscle cells themselves. This carbohydrate is available instantly, but may be used up in a few minutes. Those few minutes may be all the animal needs to switch its energy conversion processes over to fats. One of the reasons that carbohydrates can be used so quickly is that they can be used in the absence of oxygen; this is called *anaerobic metabolism*. Fats, in contrast, require oxygen to be broken down, which biologists call *aerobic metabolism*. Thus, an animal must supply oxygen to its tissues at several times the resting rate in order to use fat, which requires breathing harder and pumping blood faster. These physiological processes take some time for the animal to adjust, and fat cannot be used until the adjustments are completed.*

Some animals actually transfer energy molecules from one individual to another during flight, in a process startlingly similar to air-to-air refueling in

---

* The emphasis on the fitness aspects of "aerobic" exercise comes from the fact that fat is not used for energy until a person has shifted to aerobic metabolism, which may require exercising for several minutes or more.

airplanes. In air-to-air refueling, the "tanker" airplane flies in close formation to the other craft, transferring fuel through a hose; the other craft then disconnects and continues on its way without the necessity to find an airport, land, refuel, and take off again. An analogous process occurs in a few species of insect. When insects mate, the male transfers a packet of sperm called a *spermatophore* to the female's reproductive tract. In some species, the spermatophore contains a large amount of nutrients. Biologists originally assumed that the nutrients were to nourish the sperm cells, but in many cases, the nutrients end up in either the female's tissues or her eggs. The spermatophore nutrients thus may either fuel the female's muscles directly, or free up some of the female's own energy stores for locomotion by providing some of the energy needs of her developing eggs. A great many insects mate in flight—for example, mayflies, dragonflies, and some bees, wasps, and flies—and any that transfer nutrients in flight are thus performing the natural version of air-to-air refueling.

Is High Power Combined with Low Cost a Paradox?

Flight requires much more power than walking, yet we have seen that flight has a lower cost of locomotion than terrestrial travel. Superficially, the combination sounds mutually exclusive. How can both be true? Power is a rate, so it is affected by both time (work per unit time) and speed (force times speed). If two muscles lift a kilogram one centimeter at different speeds, they do the same total amount of work but the faster muscle does the work in less time, so it produces more power. The cost of locomotion, however, is independent of time. If the fast and slow muscles both consumed the same amount of energy to do the same amount of work, which should be approximately true, they have the same cost of locomotion, because they moved the same distance. Real animals, of course, are a good deal more complex than single muscles, but the same principle applies. For a whole animal's cost of locomotion, the total energy cost and the distance traveled are important, but the travel time or speed do not matter.

Consider a squirrel running and a blackbird flying. The blackbird requires two or three times the power of the running squirrel, so if they both flew or

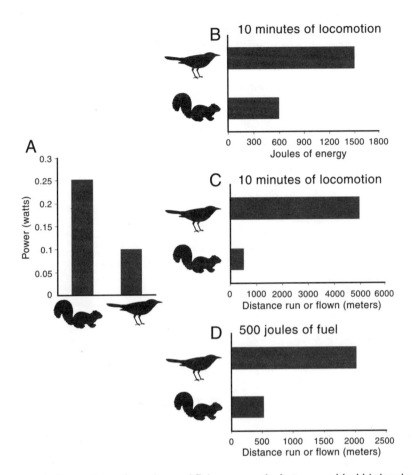

Figure 6.5. Comparison of running and flying energetics between a blackbird and a squirrel of the same weight. A. Power used at a speed that each animal would use for long-distance travel. B. Amount of energy they use to run or fly for 10 minutes. C. Distance they run or fly in 10 minutes. D. Distance they run or fly on a given amount of energy, 500 joules in this case. (S.T.)

ran for the same amount of time, the bird would consume two or three times more energy than the squirrel. Now consider their respective speeds: the blackbird flies perhaps ten times faster than the squirrel's top sustained running speed (Fig. 6.5). Finally, consider the distance they move using the same amount of energy. The blackbird can only fly for one-half to one-third of the *time* that the squirrel can run on the same energy, but the bird is so much faster that it covers three to five times more *distance* on the same amount of energy (Fig. 6.5). Speed is the two-edged sword of flight. A flying animal needs to be able to produce a lot of power to move its wings rapidly, but if it can, it can travel so much faster than a runner that the flyer needs less total energy to move any given distance.

SEVEN

Evolving Flyers

*T*he aerodynamic elegance and structural sophistication of an animal's wing must surely be one of the most marvelous results of evolution. A feathered bird wing, for example, maintains a smooth, contoured surface during large shape changes while flapping because of its overlapping feathers. The feathers themselves are extremely light for their strength, and the large quill feathers on the tip of the wing are precisely shaped so that each feather can act as an individual airfoil. An insect's wing, such as that of a dragonfly or bee or housefly, is completely different from that of a bird, consisting of a branching arrangement of rigid veins supporting a very thin but tough membrane. The insect wing cannot be flexed (to shorten it) like a bird wing, but its corrugated arrangement of veins, stabilized by the intervening membrane, gives the insect wing tremendous resistance to lengthwise bending, while simultaneously allowing the wing to twist (Fig. 7.1) [1]. This "twistiness" allows the wing to passively deform while flapping, so that some of the required pronation and supination occurs passively. And yet, the

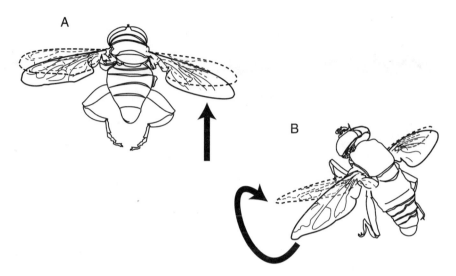

Figure 7.1. The arrangement of the veins of an insect wing have a strong effect on the wing's mechanical properties. A. The wings are very resistant to lengthwise bending. B. The wings have little resistance to twisting. (S.T.)

lengthwise bending strength of insect wings may be the highest, for their weight, of any known structures (including airplane wings).

Does Natural Selection Favor Flight?

Birds include more species than any other group of land vertebrates, and bats have the second-largest number of species among mammals. The vast majority of insects fly, and they outdo all others: there are more named species of insects than all other animal species combined, and there are certainly more individual insects than any other kind of animal. Flight has helped make this success possible in these animal groups, because powered flight confers several advantages. Flight is faster and more energetically efficient than walking or running. Flying animals may be able to use food or nest sites inaccessible to nonflyers. Perhaps the biggest advantage is that flyers tend to have lower mortality rates than walkers. Derek Pomeroy compared several groups of flying birds to nonflying birds like penguins, and compared bats to rodents, and the aerial groups had significantly lower mortality rates than the nonflyers in all cases [2]. Pomeroy was using data from other researchers, so he could not actually determine the causes of mortality, but he concluded that the ability of flyers to avoid predators was the only obvious advantage to which his findings could be attributed, given the timescale of his study. Put another way, flyers' ability to flee rapidly, plus the difficulty of aerial interception, means that predators have a harder time catching flying prey.

In spite of the great advantages of flight, true powered flight has arisen only four times in the 440 million–year history of terrestrial animals. Flight is a complex physical process, requiring precise anatomical specializations: wings must be aerodynamically effective and strong yet light, and their movements must be precisely controlled. Biologists' modern view of evolution by natural selection has been continuously plagued by one objection for over a century: the process seems ill-suited for producing complex structures. An eye or kidney with only half of its parts would be useless for carrying out normal eye or kidney functions. How can such a complex structure evolve when it must have come from a simpler, primitive structure that could not have been fully functional? Specifically, how could animals evolve exquisitely well adapted wings from vertebrate forelimbs or bumps on insect thoraxes? Darwin himself proposed the solution, still generally accepted to this day: structures may have different functions, and different advantages, in response to different selective pressures at various times in their evolution [3]. In other words, in early stages of their evolution, a proto-eye may have done something other than form visual images, and a proto-wing may have done something other than produce lift and thrust. At some point, by coincidence or random chance, the form of a given structure allows the animal to do something new with it. If the new function is advantageous, natural selection will act to refine and improve it. Biologists call this the principle of functional shift with structural continuity [4]. They often use the misleading term *preadaptation* to describe such functional shifts; but *preadaptation* does not refer to a directed or teleological process. It is, instead, driven by random chance, or changing environments, or both [3].

The Earliest Powered Flight

Insects were the first animals to fly, evolving flight sometime in the early-to-mid-Devonian period, more than 410 million years ago [5]. Insects are unique among flying animals because their wings are not modified walking legs. They have retained all their legs along with their wings, so many insects have excellent walking, running, burrowing, and climbing skills in addition to flight. In contrast, all the vertebrate flyers evolved wings from forelimbs, which has clearly constrained their running and climbing abilities. Pterosaurs were the first flying vertebrates, arising approximately 200 million years ago, followed by birds and then bats. *Archaeopteryx*, the earliest known animal with distinctly birdlike features, lived about 145 million years ago, and bats evolved flight at least 60 million years ago.

Putting dates on the first appearance of a group can be a risky business. The fossil record is very uneven, because some areas—such as swamps and seabeds—are good places for fossils to form, while others—most terrestrial habitats, especially uplands—are not so favorable. Under most conditions, fossilization is a rare event. Random chance is certainly a major factor: an animal carcass must not be destroyed by scavengers, it must encounter the proper conditions to become fossilized, it must not be destroyed by erosion, and, finally, it must be found by a paleontologist. For some times and places, the fossil record can be comprehensive, but for most, it has more gaps than information. Thus, when a paleontologist says that a species first appears in the fossil record at a particular time, that may or may not be the actual time when the group first arose. The group may have arisen earlier but earlier fossils have not yet been discovered, or perhaps no fossils formed from earlier animals. Scientists can sometimes use indirect evidence to help fill in the gaps, but there is always a bit of uncertainty about the exact time of origin of any animal group because of the gaps in the fossil record.

Insects, the First Flyers

The fossil record of insects before they evolved flight is currently almost nonexistent. Ironically, many of the older insect fossils are of isolated wings, and these ancient wings all appear to be fully functional, flappable wings. The very oldest fossils, a mere handful, are all of primitive, wingless species. Thus, we have no idea what the immediate ancestor of flying insects looked like. Furthermore, we have at best a vague idea of when these ancestors lived, because the oldest known winged insects were already competent flyers; their pre-flight ancestors lived at some unknown time earlier [5].

The Origins of Insect Wings

The lack of fossils showing the transition from flightless to flying species in insects has, if anything, inflamed speculation and debate among scientists about how flight must have evolved in insects. For well over a century, biologists have been discussing the origins of insect wings [3, 6]. The main arguments concern the origin of the structures—*winglets* or *proto-wings*—that eventually evolved into wings. Most scenarios fit into two general categories, "paranotal lobe" or "pleural appendage" theories. In the *paranotal lobe* scenario, small lateral lobes or plates evolved on the thoraxes of some insects. These proto-wings were originally flat, immovable extensions of the notum, a plate forming the roof of the thorax. Whatever their original function, these plates decreased falling speed when insects jumped from plants or rocks to escape from predators. Bigger plates make better parachutes, so the lobes got larger over time, until they became large enough to act as wings. These wings would not have been moveable, so the insects could only have used them for gliding. Eventually, some of these gliders evolved articulations to allow wing movements, perhaps for steering, which in turn allowed these insects to evolve powered flight. This scenario is reasonable in the absence of evidence pro or con, but many scientists remain uncomfortable with the step from gliding to powered flight. In addition to the articulation itself, the insect would simultaneously need to evolve a new set of muscles and much more sophisticated control mechanisms in the nervous system.

In the other scenario, called the *gill* or *pleural appendage* hypothesis, a series of small, moveable appendages ran along the side (pleuron) of the thorax and abdomen. Originally, proponents of this scenario suggested that these structures were gills, just as many aquatic insect nymphs (juveniles) have flat, moveable gills along their abdomens today. Most paleontologists, however, believe that early insects were terrestrial, so other uses have been proposed for these moveable pleural plates.

In recent years, most biologists and paleontologists have come to favor the pleural appendage scenario. The tracheal gills found on the abdomen of modern mayfly larvae are excellent candidates for proto-wings. They are flat, moveable plates that have tracheae (air tubes) arranged somewhat like veins in a wing (Fig. 7.2). Moreover, in some species they are more important as paddles for moving water over the body than as true gills or gas-exchange organs. Some

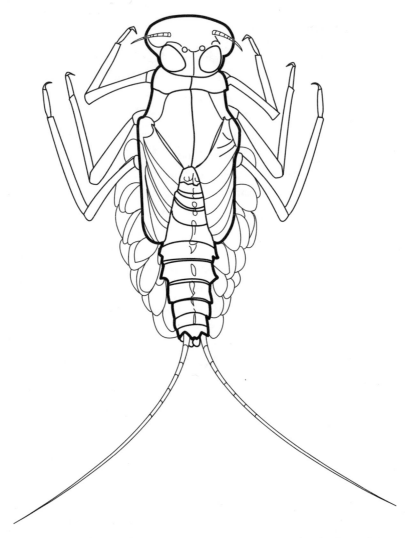

Figure 7.2. Mayfly nymph (immature or "larva") showing tracheal gills on abdomen. (S.T.)

may even be used as paddles for swimming. Thus these appendages already perform a function underwater that is closely analogous to flight in air. The complete scenario begins with aquatic insects that have these appendages running down each side from the thorax to the end of the abdomen. Perhaps the appendages produce respiratory currents; perhaps they are even used for swimming. At some point, the insects evolve an adult stage that spends at least part of its time out of the water. The appendages are retained by the adult, maybe for parachuting or to produce a little bit of thrust for running or jumping. This benefit leads to the enlargement of the appendages. Once the appendages get big enough, they can act as wings. The ones on the thorax—near the body's center of gravity—continue enlarging, while the others get smaller and are eventually lost. As the appendages become long enough to act as wings, the insects probably go through a gliding stage where they use the wing movements primarily for steering. If the ancestral forms used their appendages for swimming, however, the first winged insects may have flapped weakly from the beginning, at least enough to extend glides, if not to maintain level flight. Indeed, some modern species of stoneflies skim from rock to rock on the surface of streams, apparently never breaking completely free of the water. They are too weak to fly up away from the surface, and they use their small, weak wings entirely for thrust. Perhaps the ancestors of flying insects moved from swimming to flying by way of a similar "skimming" stage [7]. In any event, the muscles and articulation eventually became specialized for flapping, as the control mechanisms in the nervous system became better adapted for powered flight. These insects thus achieved powered flight. In either scenario, the first flying insects gave rise to the main insect lineage known as the Pterygota, or winged insects.

This evolution, including all the important transitions, took place during a prolonged gap in the fossil record of insects. In the early Devonian, approximately 400 million years ago, there were a few primitive wingless insects; thereafter, there is nothing until the early Carboniferous, 340 million years ago. By the time insects reappear in the fossil record, they had evolved fully competent, flapping flight, and the two main modern insect lineages had already diverged from each other.

The Function of Proto-Wings

The gills-to-wings, or *pleural appendage,* scenario is quite plausible, except for one drawback: the ancestors of flying insects may not have been aquatic. True, many Carboniferous fossils appear to be aquatic larvae, and many modern insects have aquatic larvae. Those ancient, primitive insects from the Devonian, however, were not aquatic. If the early pleural appendages were not gills or paddles for ventilation, what was their function?

Many researchers have suggested that lateral plates on the thorax originally evolved to increase drag and reduce falling speed—in other words, for parachuting. The aerodynamics of parachuting, however, should actually lead *away* from powered flight. If natural selection favored parachuting ability, perhaps to increase dispersal distance in winds, then small size would be favored. As size decreases, the Reynolds number also decreases, thus increasing the importance of viscous drag. Higher viscous drag decreases the animal's terminal velocity (maximum falling speed), so the insect would fall more slowly and be carried farther by winds. Wings, however, become less and less effective as the Reynolds number decreases. Moreover, at low Reynolds numbers, bristles or some sort of hairy fringe would be more effective than lateral plates for slowing a fall. In short, parachuting insects ought to end up looking much like dandelion seeds: tiny, with lots of bristles. Thus, a scenario based on selection for parachuting seems to lead in the wrong direction for gliding and eventually powered flight: the small wings of tiny animals have low lift-to-drag ratios and are particularly ineffective for gliding. Flight in insects thus must have arisen based on functions other than parachuting.

If moveable thoracic appendages did not start out as gills or paddles, what else might they have been for? One intriguing suggestion is that they might have been used for courtship displays [8]. This suggestion involves sexual selection, rather than natural selection. First described by Darwin, and still a major research focus for evolutionary biologists today, sexual selection occurs when members of one sex in a species choose mates based on traits not directly related to survival. These traits tend to be amplified in later generations, and in extreme cases, they may even interfere with the survival of the opposite sex. Birds provide many well-known examples, where males often have bright colors—wood ducks, cardinals—or large, elaborate tails—turkeys, pheasants, peacocks—that are used primarily to attract mates and intimidate rivals. Sexual selection could plausibly lead to moveable plates on male thoraxes, and could also explain their enlargement until they became aerodynamically effective. The drawback of this particular scenario is that females would not necessarily evolve the same structures, and even if females did have them, their plates would most likely have been small and poorly developed. Given that both sexes fly today, this hypothesis is not quite satisfactory.

A New Approach: "Bounded Ignorance"

Recently, several researchers have built and tested models of various possible configurations of insects with proto-wings. Their goal was to find out

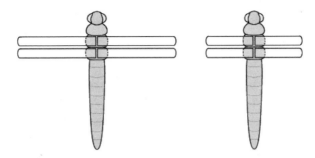

Figure 7.3. Model insects used by Kingsolver and Koehl to study aerodynamics and heat transfer. Modified from J. G. Kingsolver and M. A. R. Koehl 1985 [9]. (O.H.)

what factors are most influenced by small changes in size or shape; natural selection normally operates on precisely these sorts of small but influential changes. Joel Kingsolver and Mimi Koehl made models of insects with a range of body sizes and two or three pairs of stubby proto-wings of different lengths [9] (Fig. 7.3). Surprisingly, they found that short proto-wings (less than 20 percent of body length) have no effect on falling speed. Thus, proto-wings probably did not arise as parachutes to slow falls. They also found, predictably, that short proto-wings are not effective for gliding, and that proto-wings need to increase past a critical length before they make decent gliding surfaces. Furthermore, size matters: small, two-centimeter-long models needed relatively long wings—more than 60 percent of body length—to glide, while larger, six-centimeter models were able to glide when their wings were just over 20 percent of their body length. Robin Wootton and Charles Ellington tested some different models, comparing one with many short winglets along its length to a model with a single pair of wings on the thorax [10] (Fig. 7.4). Their many-winglet model glided as well as the single-pair model, and the many-winglet model was also reasonably stable. Results from all these model tests suggest that if early ancestors of winged insects were gliders, they were probably large, and they would have benefited from abdominal as well as thoracic proto-wings. Kingsolver and Koehl call this approach *bounded ignorance,* because, although we are ignorant of the actual form of the ancestral insects, experiments like these show the limits of what is physically possible.

Kingsolver and Koehl's aerodynamic experiments did not suggest a function for small, thoracic proto-wings, so they performed another set of experiments to find out if proto-wings could have been solar collectors [9]. Insects

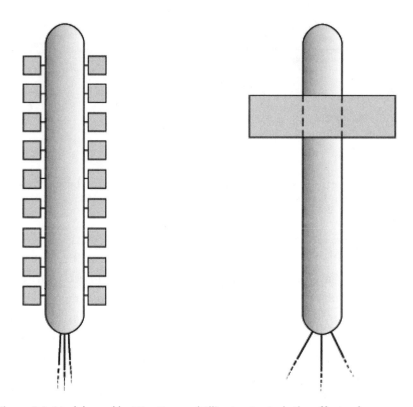

Figure 7.4. Models used by Wootton and Ellington to study the effects of many small winglets compared to one large winglet. Redrawn from R. J. Wootton and C. P. Ellington 1991 [10], reprinted with permission of Cambridge University Press. (O.H.)

are generally too small to warm themselves with metabolic heat, but if they could warm themselves with solar heat, their muscles would contract faster and they could be active at lower air temperatures. A particularly striking result from Kingsolver and Koehl's heating experiments was that as proto-wings increased in size on the six-centimeter model, they warmed the body faster until they were 20 percent as long as the body. But longer proto-wings did not increase heating. The aerodynamic experiments had shown that proto-wings needed to be 20 percent of the body length or longer for gliding, so at just the size where proto-wings exhaust their potential for heating, they start to become effective for gliding. Using wings as solar collectors is not merely theoretical: many insects bask in the sun to warm up, and butterflies and dragonflies often use their wings to help collect solar heat. Of course, their wings are bigger than necessary for solar collection, but then heating the body is not the wings' primary function. Thus, Kingsolver and Koehl have demonstrated a possible use for proto-wings too small to have any aerodynamic function.

In contrast, Wootton and Ellington's experiments show that small proto-wings could have had an aerodynamic function if there were many of them along the length of the body [10]. Even quite small proto-wings improved the glide angle somewhat, but they particularly improved stability and steering. The latter improvements may have been especially important for insects jumping or falling to escape predators. In addition, these proto-wings would have been most effective if they were adjustable, so they may have evolved the crucial movable articulation at an early stage. Wootton and Ellington's results suggest that leaping insects might have used many short winglets initially for stability, then evolved articulations for steering. Winglets would have enlarged to improve gliding, and eventually have been reduced to two pairs for flapping flight. Although Wootton and Ellington's results beg the question of the earliest function of the most primitive proto-wings, they do show that even small proto-wings could have had an aerodynamic function, which could then logically have led to the evolution of wings.

In his book *The Evolution of Insect Flight,* Andrei Brodsky argues that we probably cannot deduce the function of proto-wings before they were used for flight [11], contending that their original function must either have been minor or so specialized that descendents could do without it. Otherwise, once the wings had been taken over for flight, insects should have evolved replacement structures to take over the original proto-wing function. Although he discusses several other wing-origin scenarios, Brodsky completely ignores Kingsolver and Koehl's solar-collector suggestion, which sidesteps Brodsky's argument rather neatly: if proto-wings were first used as solar collectors, then later flying insects had no need to evolve replacement structures, because their wings still work fine as solar collectors. Brodsky, however, sees nothing in modern insects recognizable as a replacement for proto-wings, so he concludes that if proto-wings were not gills, then logic alone is unlikely to let us deduce the original function of proto-wings.

Atmospheric Changes, Flight Evolution, and Gigantism

The amount of oxygen in the earth's atmosphere has changed dramatically during certain periods in geologic history. Prior to the Devonian, for hundreds of millions of years, the level of atmospheric oxygen was about three-quarters its current level. Recent climate analyses agree that large changes in the oxygen level occurred before it settled at its current level. The most comprehensive study shows that the level of oxygen in the air started to rise in the late Devonian (about 350 million years ago), peaked in the Carboniferous (300 million years ago) at about double the current level, then gradually fell back to the low, pre-Devonian level in the Permian and Triassic (300 to 200 million years ago) [12]. If these analyses are accurate, such swings in oxygen amounts would have had significant biological and physiological consequences. The increase in oxygen began at about the same time insects first achieved powered flight, and the elevated oxygen almost certainly would have made it easier for the flight muscles of insects to produce power. Some biologists also believe that high oxygen levels led to evolution of very large body size in some insects because of the limitations of their respiratory systems [12, 13]. Unlike terrestrial vertebrates, which pump air in and out of lungs and distribute oxygen around the body with the blood, insects obtain oxygen by allowing it to diffuse passively down air tubes—tracheae—throughout the body. (Large, active insects may actively pump air through the largest tracheae, but diffusion is still the rule for the smaller branches.) Diffusion can be very efficient if distances are short—on the order of two or three millimeters, but it becomes disproportionately less effective as the distance increases. If the amount of oxygen in the atmosphere were to double, diffusion would be effective over longer distances, so insects could evolve larger bodies. Insects certainly did increase in size in the Carboniferous period. Fossil mayflies have been found with wingspans of 20 centimeters, some members of an extinct group called palaeodictyopterans had wingspans of over 40 to 50 centimeters (Fig. 7.5), and, most magnificent of all, some dragonflies had wingspans of over 70 centimeters! Although there is no proof that high oxygen levels caused gigantism in insects, the correlation is remarkably strong.

High oxygen levels may also have affected the effort of flying. An increase

FACING PAGE:
Figure 7.5. Two reconstructions of Palaeodictyoptera fossil species. A. *Goldenbergia.* B. *Stenodictya.* A. from A. K. Brodsky 1994 [11], by permission of Oxford University Press; B. from J. Kukalova 1970 [90].

A

B

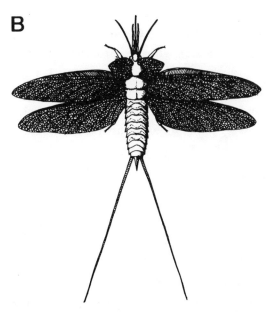

in the amount of oxygen in the air would have increased the density of the air, while having a negligible effect on the viscosity. This change essentially increases the Reynolds number for all air movements, and Robert Dudley has suggested that higher oxygen levels might have increased the aerodynamic effectiveness of small proto-wings and wings [13]. If insect proto-wings were still evolving into functional wings in the late Devonian, the increased air density may well have enhanced their aerodynamic benefits. Even if insects had already evolved functional wings before the oxygen increase occurred, higher oxygen levels might still have made flight physiologically and physically easier. The enhanced advantages of flight thus would have given flying animals an evolutionary edge, perhaps initiating and certainly accelerating the enormous diversification of insects during the Carboniferous.

Evolution in Pterygota

By the time winged insects appear in the fossil record, they had already diversified into the two main branches existing today, the Paleoptera and the Neoptera. The Paleoptera are considered more primitive, because they have a simple wing articulation that allows flapping but does not allow the wings to be folded flat over the abdomen. Among Paleoptera, several groups were diverse and successful during the Carboniferous. One of these was the palaeodictyopterans (Fig. 7.5), which had wings shaped like those of dragonflies, but with a much more complex vein network, reminiscent of modern mayflies or lacewings. Palaeodictyopterans ranged in wingspan from 1 to 56 centimeters and fed themselves by sucking juices from plants. Although they were the most successful insect group in the Carboniferous, palaeodictyopterans went extinct at the end of the Permian, about 245 million years ago. The only living Paleoptera are mayflies, dragonflies, and damselflies (Fig. 7.6).

The Neoptera have a more complex wing articulation, which allows the wings to fold over the abdomen at rest. This group is by far the most abundant and successful, containing all other flying insects alive today (as well as some extinct groups and some that have secondarily lost flight). Familiar neopterans include grasshoppers, cockroaches, cicadas, butterflies, bees, wasps, flies, and, most successful of all, beetles, which make up roughly one-third of all known animal species. However, these neopteran groups were not

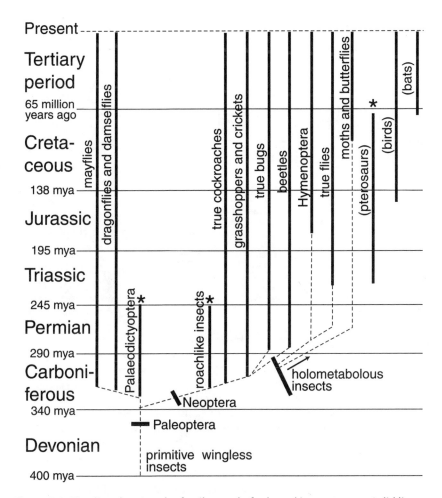

Figure 7.6. Timeline showing the fossil record of selected insect groups. Solid lines show approximate time of appearance and duration of groups in the fossil record; dashed lines show likely evolutionary relationships and inferred presence when fossil data are lacking, and stars indicate extinctions. (Groups of vertebrate flyers— pterosaurs, birds, and bats—also shown for comparison.) Times shown are for the beginning of each geologic period. (S.T.)

all present from the beginning. Most Carboniferous and Permian Neoptera were primitive roach- or cricketlike creatures. True cockroaches were not uncommon, and they shared this period with a few species of ancestral grasshoppers [14]. The first stoneflies and hemipterans (true bugs) also appeared in the Permian period (Fig. 7.6).

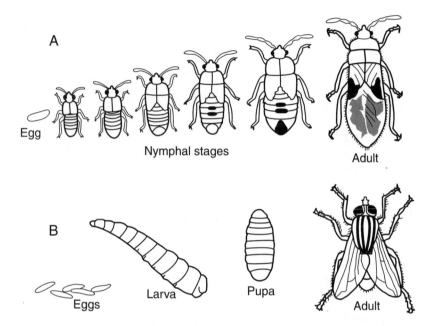

Figure 7.7. The young, or nymphs, of hemimetabolous insects like these chinch bugs (A) look like miniature, wingless adults, and their development is called *incomplete metamorphosis.* A nymphal stage lasts from one molt to the next, and the animal's appearance and size do not change during the stage. Holometabolous insects go through three radically different life stages, and the transformation from larva to adult is called *complete metamorphosis,* as demonstrated by flies (B). (Only one larva is shown, although holometabolous insects typically have several larval stages to allow for larval growth.) (S.T.)

Holometaboly, or *complete metamorphosis,* the last great evolutionary innovation of insects, also appeared in the Permian (Fig. 7.7). Holometabolous insects go through a quiescent pupal stage between the larval and adult stages, which allows vast reorganizations of anatomy. Holometabolous larvae, such as grubs, caterpillars, and maggots, look and live completely differently from their adult forms, which are beetles, butterflies, and flies, respectively. In other insects, such as grasshoppers and stink bugs (with incomplete metamorphosis), the young look much like miniature adults and eat similar food. All but the youngest juveniles have obvious but immovable "wing pads" where the adult wings will be (Fig. 7.7). Holometabolous larvae, in contrast, normally eat completely different food from adults and show no signs of wings or wing pads until the pupal stage. Thus, adults and young do not compete with each other for food or shelter. Holometabolous insects went on from the Permian to become the most diverse and successful animals of all time.

Only adult insects have wings. Juvenile insects must molt (shed their exoskeletons) periodically in order to grow, but adult insects do not continue molting. Some biologists have speculated that immature insects do not have functional wings because wings would make molting too difficult. Have insects made a virtue of necessity, adapting immature stages to a sedentary lifestyle because they could not molt with functional wings anyway? One bit of evidence suggests otherwise: alone among all known living insects, adult mayflies molt once after the initial molt to the adult stage, and they appear to have no particular problem molting with wings. More likely, ancestral pterygotes had juveniles with wings (or at least winglets) that were not terribly effective airfoils. Natural selection seems to have favored completely flightless juveniles, which put all of their efforts into eating and growing rather than expending energy in flight. The adults, with large, fully formed wings, could take care of all of the long-distance locomotion and dispersal.

The earliest major holometabolous group to appear was the Coleoptera, or beetles. Beetles evolved protective wing covers from the forewings, so the functional (hind) wings are well protected. With their wings out of harm's way, beetles are able to burrow in soil, tunnel in wood, swim underwater, and yet fly when the need arises. This ability to thrive in many habitats and still fly allowed beetles to successfully exploit more differing habitats and lifestyles than any other group, becoming the most diverse and successful of all insects.

As Robin Wootton has remarked, beetles are a "magnificent general purpose" group, in contrast to the Hymenoptera, which "are notable for doing a few things extremely well" [5]. The Hymenoptera include wasps, ants, and bees. The earliest hymenopterans were parasitic and predatory wasps, appearing in the Jurassic (about 190 million years ago). Ants, efficient general-purpose foragers and scavengers, appeared in the Cretaceous (130 million years ago). Although most ants in a colony are sterile, flightless workers, the reproductive individuals have functional wings and fly to find mates and new nest sites. Once a queen finds an acceptable nest site, she sheds her wings and does not fly again. The evolution of bees is tightly linked to the evolution of flowering plants, which appeared in the Cretaceous (Chapter

10). The earliest bees appeared in the late Cretaceous, about 70 million years ago.

Diptera or true flies (house, horse, deer, and black flies, bluebottles, mosquitoes, and midges) evolved and diversified throughout the Triassic, Jurassic, and Cretaceous. At least part of their success is due to remarkably adaptable larvae, which live in water, sediments, soil, manure, and decomposing organic material. Flies are certainly well named: they have extremely well developed flight abilities, including a unique set of gyroscopic balance organs, the halteres, formed from highly modified hind wings (see Chapter 5). The last major insect group to appear were the Lepidoptera, butterflies and moths. Their fossil record is rather poor. Some may have been present as early as the Cretaceous, but most of their diversification has been more recent.

Five major insect groups encompass over 90 percent of known insect species (and thus almost three-fourths of all animal species): Hemiptera, Coleoptera, Hymenoptera, Diptera, and Lepidoptera. They are all pterygote (winged) members of the Neoptera, so they have the more advanced folding wing articulation. Flight has clearly played a role in their striking success. In addition, Coleoptera, Hymenoptera, Diptera, and Lepidoptera are all holometabolous. The distinct separation of development into a relatively immobile larva specialized for growth and a highly mobile adult specialized for dispersal seems to have enhanced the success of the holometabolous groups.

Next To Fly: Pterosaurs

The flying reptiles called pterosaurs were the second group of animals to evolve flight. Most pterosaurs were about the size of modern seagulls. A few were as small as sparrows, but some of the later species were the largest flying animals that have ever lived. As with insects, paleontologists have not yet found any transitional fossils, or "missing links": pterosaurs fully capable of flight appear suddenly in the fossil record about 220 million years ago.

Pterosaurs are popularly associated with dinosaurs, probably because pterosaurs and dinosaurs are both extinct reptile groups with some spectacularly large members. In the past, many paleontologists thought that pterosaurs had evolved from dinosaurs. Most current evidence suggests, however, that pterosaurs were closely related to the thecodont* reptiles that were the dinosaurs' immediate ancestors. *Scleromochlus taylori,* a tiny thecodont that lived in Scotland about 220 million years ago, is currently the best candidate for the immediate ancestor of pterosaurs. *Scleromochlus* was only 20 centimeters long, but had a big head, a long tail, and large hind legs. This animal probably walked on all fours but ran just on its hind legs, and it must have eaten insects. Its relatively large skull was lightly built, with lots of openings for weight reduction, characteristics it shared with pterosaurs. *Scleromochlus* and pterosaurs also shared similarities in their shoulder blades, vertebral columns, and ankle joints [15, 16]. Interestingly, *Scleromochlus* also shares many similarities with the earliest dinosaurs, and it is probably closely related to the immediate ancestors of dinosaurs as well [16]. Thus, pterosaurs are not themselves dinosaurs, but pterosaurs and dinosaurs share a close common ancestor. Moreover, pterosaurs prospered and evolved alongside dinosaurs for over 150 million years.

Batlike Pterosaurs?

In 1817, Theodore von Soemmerring published the first description of a pterosaur fossil, and thinking that it was that of an aberrant bat, he drew his reconstruction with a very batlike posture and wing. His early reconstruction of a pterosaur has haunted the public and scientific perception of pterosaurs ever since. Soemmmerring's reconstruction is understandable, given that he was the first to try to describe a pterosaur, that few naturalists of the time accepted the idea of major groups of extinct animals, and that both pterosaurs' and bats' wings consist of a membrane supported by enormously elongated finger bones. Soemmerring showed his pterosaur with the laterally directed legs and reoriented feet of bats, and with the wing membrane stretching from the arm and finger along the trunk and legs all the way to the ankle (Fig. 7.8). This reconstruction also included a membrane stretching between the legs, again like the uropatagium (tail membrane) of bats. Even though other scientists developed less batlike descriptions of pterosaurs in the late 1800s,

---

* Thecodonts were once considered to be a single lineage of reptiles that gave rise to several other major groups. Paleontologists now consider them to be a diverse collection of not-too-closely related species, hence not a natural group. "Thecodont" is still a handy common name for the animals ancestral to pterosaurs, crocodiles, dinosaurs, and birds.

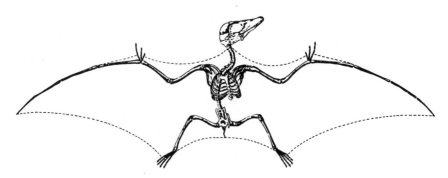

Figure 7.8. Von Soemmerring's reconstruction of a pterosaur, giving it a batlike leg posture and wing outline.

the popular literature, and even some scientific literature, continued to describe pterosaurs as batlike into the 1980s.

Bats perch by hanging upside down from tree limbs and the roofs of caves. Though many are surprisingly agile climbers, bats are generally awkward when crawling on level surfaces. Did pterosaurs also hang upside down and avoid landing on the ground? Until recently, some paleontologists thought they did, but most scientists now agree that pterosaurs got around on the ground reasonably well. What is still uncertain is whether pterosaurs walked on all fours or just on their hindlegs. Pterosaurs' ancestors were bipedal and used their tails to balance their forward-tilted trunks and heads. Early pterosaurs also had long tails and probably could have run on their hindlegs, certainly handy for an animal with wings for forelimbs. These early pterosaurs, however, could have used their forelimbs for walking because their arm and hand bones were only slightly enlarged—most of the wing was supported by the gigantic fourth finger (Fig. 7.9A, B). Later pterosaurs are more enigmatic: their arms and especially their hands seem too long to be used comfortably for walking, but their tails were too short to counterbalance their bodies if they just walked on their hindlegs (Fig. 7.9C). Birds also have short, stubby tail skeletons, but they manage to walk quite well on their hindlegs. Birds manage this by angling the thighs forward to get their feet under the body's center of gravity (Fig. 7.10). They hold their thighs at this unstable angle with extensive hip and thigh muscles (to the delight of "dark meat" lovers everywhere). Colin Pennycuick suggested that pterosaurs' hip bones were too small to anchor extensive thigh-positioning muscles [17],

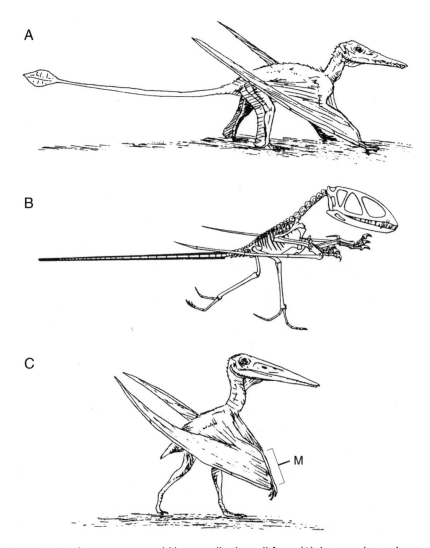

Figure 7.9. Early pterosaurs could have walked on all fours (A), but may have also run on their hindlegs (B). In later pterosaurs, the "hand" or metacarpal region (M) was so elongated that walking with the forelimb may have been awkward (C) although fossil trackways are all quadrupedal. A. and C. from S. C. Bennett 1995 [91], by permission of the artist, Brian Franczak; B from K. Padian 1991 [19], reprinted with the permission of Cambridge University Press.

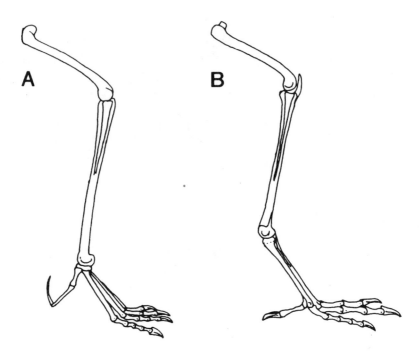

Figure 7.10. Reconstructed hindleg skeleton of a rhamphorhynchoid pterosaur (A) compared with the hindleg skeleton of a modern pigeon (B). From K. Padian 1985 [28], by permission of the Palaeontological Association.

but Kevin Padian replied (repeatedly) that pterosaurs' leg and foot bones are so strikingly birdlike that pterosaurs must surely have walked like birds [18, 19, 20]. Other researchers, including the pterosaur experts Christopher Bennett and David Unwin, have recently concluded that a number of fossil trackways—trails of preserved footprints—were made by pterydactyloid pterosaurs, and these animals clearly walked on all four limbs [21, 22]. Perhaps some pterosaurs walked on their hindlegs, but according to current evidence, most probably walked on all fours. In any case, large pterosaurs, with eight- or ten-meter wingspans and weighing as much as an adult human, do not seem likely candidates for a batlike existence confined to clambering about in trees and hanging upside down from branches.

Pterosaur Flight

Pterosaurs possessed some obvious adaptations for powered flight. They had large sternums (breastbones) for attaching powerful flight muscles, well-developed shoulder bones to carry the body's weight in flight, and air-filled bones to lighten the skeleton. Some even had a furcula (wishbone), perhaps to flex like a spring and help raise the wings during the upstroke. How competent were they at flying? The original batlike reconstructions, along with their classification as cold-blooded reptiles, suggested to many earlier biologists that pterosaurs were only gliders. Biologists now, however, generally agree that pterosaurs were capable of powered, flapping flight. Indeed, the shoulder joint is clearly specialized for the down-and-forward, up-and-back movement of normal flapping. Moreover, most pterosaurs were gull-sized or smaller, and the smaller species would not have been very effective at soaring. Even so, some scientists questioned the ability of pterosaurs to flap effectively until well into this century. Why?

Structural and Physiological Questions

Batlike reconstructions harking back to Soemmerring's again fueled the debate. If the wing membrane had been attached to the leg, its huge area would have given pterosaurs very low wing loadings. If they had been built this way, pterosaurs would have had low flight speeds and difficulty penetrating winds. Also, such a thin, extensive wing membrane would not have made a very good airfoil, because it would have deformed too much under aerodynamic loads, its only stiffeners being the wing finger in front and the legs in back. Perhaps most important, pterosaurs were classified as reptiles, and biologists could not imagine a sluggish, cold-blooded reptile producing the muscle power needed for flapping flight.

Gradually, and with some resistance, paleontologists have developed a new understanding of the structure of pterosaurs. This new interpretation is based largely on several remarkable pterosaur fossils from the Solnhofen limestone beds in Germany, which ironically were some of the first pterosaurs known to science. These fossils are so detailed that they preserved imprints of soft tissue, including wing membranes, and in some cases include actual fossilized wing membranes. Early paleontologists had noticed fine ridges and grooves in some preserved wing membranes, and in the 1970s, Peter Wellnhofer suggested that these ridges might represent some sort of stiffening fibers [23, 24]. Kevin Padian carefully redescribed and reanalyzed these fossils, arguing persuasively that the ridges were stiffening fibers. They run in almost the identical directions to the shafts of feathers in

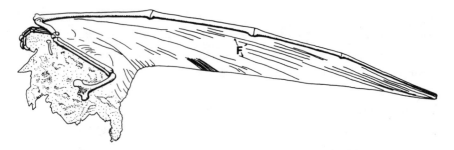

Figure 7.11. Drawing from an actual pterosaur wing fossil, showing fibers (F) that may have acted as stiffeners, like the battens of a sail. From K. Padian and J. M. V. Rayner 1993 [20], reprinted by permission of the *American Journal of Science*.

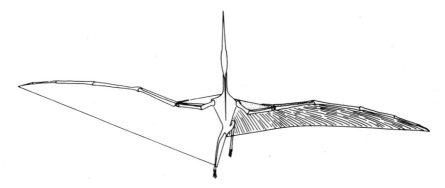

Figure 7.12. Possible pterosaur wing planforms. On the left, the batlike arrangement is supported mainly by tension in the membrane. On the right, in the birdlike reconstruction aerodynamic loads are supported by stiffening fibers oriented much like the feather shafts on a bird's wing. From K. Padian 1985 [28], by permission of the Palaeontological Association.

bird wings (Fig. 7.11), and would have been well placed to carry flight loads and maintain the wing's shape. These fibers thus eliminate the need to stretch the membrane between the wing finger and the legs to keep it taut. Although the wing bases of the fossils are too distorted to actually see the attachment, Padian thinks that the membrane was narrow and attached only along the pterosaurs' flanks, not all the way back to its ankles (Fig. 7.12). If Padian is right, pterosaur wings had high aspect ratios, and most pterosaurs had wing shapes and wing loadings quite similar to those of gulls and other seabirds (Fig. 7.12). Padian and Jeremy Rayner analyzed pterosaur wing structure based on Padian's reconstruction and concluded that pterosaur wings were well adapted to flapping flight, and may have been—like those of birds—capable of more flexing on the upstroke than those of bats [20].

Padian's description of pterosaurs with high-aspect-ratio wings is attractive aerodynamically, but many paleontologists believe he has misinterpreted the available wing impressions. Christopher Bennett reviewed the evidence from a variety of fossils. He concluded that pterosaur wing membranes were typically attached at least to the thigh, and sometimes all the way back to the feet [25]. The fibers Padian analyzed may well have been important wing stiffeners, but most other researchers think that pterosaur wings were broader—had lower aspect ratios—than Padian's reconstruction [21, 26].

The problem of pterosaurs being "mere reptiles" is partly one of semantics and perception, as well as physiology. The category of *reptile* was created to encompass living, scaly, cold-blooded, egg-laying land vertebrates. Living reptiles actually represent three separate branches of the vertebrate family tree: turtles, squamates (lizards and snakes), and crocodilians. Modern *phylogenies,* or "family trees," of vertebrates show that the turtle branch is actually more closely related to the branch leading to mammals than to the branch leading to crocodiles, pterosaurs, and dinosaurs. Thus, "reptile" might be a handy descriptive term for scaly egg-layers, but it is not really a natural grouping, in the same way that "fish" would not be a natural group if it included whales. Though they are both classified as reptiles, pterosaurs may not have been physiologically similar to modern lizards.

Another aspect of the perception problem is that modern reptiles have small brains compared to birds and mammals, and are generally considered to be much less intelligent. Some biologists have assumed that pterosaurs, being reptiles, would not have had enough brain capacity for active flight. Occasionally, fossils form casts of the inside of the braincase of an animal. To the surprise of many paleontologists, casts of pterosaur braincases were almost as big as those of birds of similar body size. Pterosaurs must have been active, reasonably intelligent animals.

Given that pterosaurs, dinosaurs, and birds are all on the same major branch of the vertebrate family tree, and given that birds are endothermic (warm-blooded), many paleontologists have concluded that pterosaurs may also have been at least partly endothermic. When A. G. Sharov, a Russian paleontologist, described what appeared to be fur from skin impressions of a pterosaur, the idea of endothermic pterosaurs gained even stronger support [18, 27]. Modern animals use fur for insulation, and endotherms are the main users of insulation, so at least some pterosaurs apparently were endothermic. Endothermy is supporting evidence for active flight, because it implies greater endurance and perhaps greater power output, but endothermy is not by any means a requirement. Insects are largely ectotherms, so they clearly demonstrate that endothermy is not required for successful powered flight. Moreover, the view of reptiles as sluggish and only capable of brief bursts of speed may not be accurate, as arguments about the flight ability of *Archaeopteryx* will show.

The Evolutionary History of Pterosaurs

As with insects, paleontologists have not found fossils that illustrate the origin or early history of pterosaurs' flying ability. In contrast to the huge volume of literature on the evolution of bird flight, little has been written about the evolution of flight in pterosaurs. Furthermore, the small amount that has been written breaks down into the identical arguments applied to birds: either small, agile, tree-dwelling predators evolved through a gliding stage to true flapping (the *trees-down* hypothesis) or small, fast-running bipedal predators evolved flight initially to aid jumping at prey, perhaps without even passing through a gliding stage (the *ground-up* hypothesis) [28]. Rather than repeat the arguments in detail, I shall postpone them until we look at birds, where there is actually some fossil evidence available.

The earliest pterosaurs differed significantly from later ones. Early pterosaurs had short necks, very long tails (often with a rudderlike flap on the end), and short metacarpal bones (Fig. 7.13A).* These pterosaurs make up the Rhamphorhynchoidea suborder.† Toward the end of the Jurassic, about 200 million years ago, rhamphorhynchoids gradually disappeared and were replaced by pterosaurs with long necks, short, stubby tails, and very long metacarpals. These later pterosaurs belong to the Pterodactyloidea suborder

---

* Metacarpal bones make up the skeletons of the palms of our hands; carpals are wrist bones.

† "Rhamphorhynchoidea" may be renamed in the future because some of its members—including the namesake genus *Rhamphorhynchus*—are probably more closely related to the pterodactyloid group.

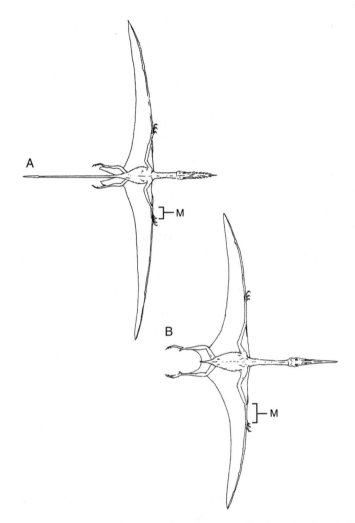

Figure 7.13. Differences in general body form between the rhamphorhynchoid pterosaurs (A) and the pterodactyloid pterosaurs (B). The metacarpal region (M) is much longer in pterydactyloid pterosaurs. From S. C. Bennett 1995 [91], by permission of the artist, Brian Franczak.

(Fig. 7.13B), which has given us the term *pterodactyl* as an alternative common name for pterosaurs. (Paleontologists avoid using *pterodactyl* because sometimes it is used for all pterosaurs, sometimes for Pterodactyloidea, and sometimes just for species in the genus *Pterodactylus* [20].)

The largest rhamphorhynchoids had wingspans of about a meter and a half, which made them about the size of a large seagull. Most were gull- or pigeon-sized, but at least one species had a wingspan of only 30 centimeters. They had long, toothy, beaklike snouts and appear to have been fish-eaters; some have even been found with fossil fish bones inside their rib cages. The suborder that appeared later, Pterodactyloidea, also included some small species, but most were the size of the largest rhamphorhynchoids or larger. Many pterodactyloids had long, toothy snouts and ate fish like the rhamphorhynchoids, but a few had hundreds of tiny teeth and probably strained plankton from the water like modern flamingos. The largest pterodactyloids had toothless, birdlike beaks. *Pteranodon ingens,* with its huge seven-meter wingspan, for many years held the title of the largest flying animal ever. *Pteranodon* had a beak longer than the length of the trunk of its body, and many had a backward-pointing crest almost as long as the beak (Fig. 7.14). *Pteranodon* was toppled from its position as the largest of all flying animals in the 1970s, when (appropriately enough, in Texas), Douglas Lawson discovered an enormous pterodactyloid humerus, or upper arm bone, as well as more complete skeletons of juveniles of the same species [29]. Based on comparisons with the juveniles and other large pterosaurs, paleontologists estimate that this new pterosaur had a wingspan of 11 or 12 meters when full-grown. That is larger than the wingspan of a small airplane! Lawson named his find *Quetzalcoatlus northropi,* after the Aztec winged-serpent god Quetzalcoatl and the Northrop company, builder of large, tailless airplanes. *Pteranodon* and *Quetzalcoatlus* were clearly soarers. Because of their great size, flapping would

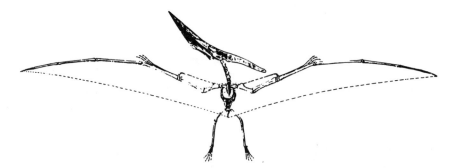

Figure 7.14. A skeleton of *Pteranodon,* a large pterodactyloid pterosaur. The drawing is based on a fossil specimen that had a wingspan of about 5 meters (over 16 feet). From K. Padian 1991 [19], reprinted with the permission of Cambridge University Press.

have been prohibitively expensive except for takeoff and emergencies. Unlike most other pterosaurs, *Quetzalcoatlus* was found hundreds of kilometers from the ocean or any significant bodies of water. Scientists have speculated that *Quetzalcoatlus* may have been a scavenger on dinosaur carcasses, but we have too little evidence to draw firm conclusions. Suffice it to say, *Quetzalcoatlus*'s gigantic size left scientists scrambling to revise estimates of size limits on flying animals and to figure out how such an apparently ungainly beast got around on the ground.

Assuming that pterosaurs flew much like birds, we can make a number of inferences about the ecology and life habits of pterosaurs using only skeletons and occasional fragments of skin. Jeremy Rayner plotted wing loadings against aspect ratios* for all major bird groups and found that birds with similar lifestyles formed cohesive clusters, even if they were not especially closely related [30]. Rayner pointed out that flight places very strong physical constraints on an animal, so to perform similar physical activities in flight, animals should evolve similar forms. Grant Hazlehurst extended Rayner's analysis to pterosaurs. Using all the pterosaur species for which he could get reasonable wing length and body mass estimates, Hazlehurst mapped pterosaurs onto Rayner's bird graphs [31]. Hazlehurst found that pterosaurs had higher aspect ratios and lower wing loadings than the average for birds [31]. Moreover, pterosaurs form a loose cluster at about the same place on the graph as long-winged seabirds like gulls and skimmers. Pterosaurs thus had similar wings to these birds and must have flown in a similar way. Along with the marine locations of pterosaur fossils, this is yet further evidence that most pterosaurs (at least, those discovered so far) were oceanic fish-eaters.

Hazlehurst noted one feature of pterosaurs that made them quite different from birds, as well as from bats and insects. Juvenile pterosaurs appear to have been fully capable of powered flight at a much earlier growth stage than birds. Young birds must reach 75 to 90 percent of their adult wingspans before they can fly. Many pterosaur fossils suggest that young pterosaurs could fly at less than half their adult wingspans. Indeed, pterosaurs seem to

* Rayner actually used a statistical technique called Principal Component Analysis to remove the effects of size. His graphs show these size-independent versions of wing loading and aspect ratio.

have had indeterminate growth, which means that they kept on growing, albeit more slowly, throughout adulthood. Such large size changes suggest that young and old pterosaurs may have flown differently. Young animals mostly flapped, perhaps using gait changes, but older animals flapped only with the continuous vortex gait and made increasing use of soaring. Young pterosaurs had lower wing loadings and so were probably slower and more maneuverable than their elders. If pterosaurs did actually start to fly at such a small size, then juvenile and adult pterosaurs would have hunted differently and eaten different types of prey; Hazlehurst speculates that pterosaurs may also have had much shorter periods of parental care than modern birds and bats.

The End of the Line

The popular perception of pterosaurs is that they represent a failed evolutionary experiment: "They're extinct, aren't they?" The older descriptions of pterosaurs with a simple, "primitive" wing, and a reptilian brain and physiology simply added to the impression that pterosaurs were a first, unsuccessful try at evolving birds. In addition to being inaccurate, these perceptions overlook one simple fact: pterosaurs thrived for over 150 million years. Assuming that the 145-million-year-old *Archaeopteryx* represents the earliest birds, birds have not quite equaled the pterosaurs' tenure on earth. Anyone wishing to know if birds are really more successful than pterosaurs will have to check back in a few million years.

Pterosaurs, like dinosaurs, appear to have decreased in diversity toward the end of the Cretaceous, although the later pterosaurs included the biggest ones, such as *Pteranodon* and *Quetzalcoatlus*. Also like the dinosaurs, the pterosaurs disappeared entirely at the end of the Cretaceous, about 65 million years ago. Many paleontologists believe that a large asteroid struck the earth about 65 million years ago and caused such widespread devastation and climatic change that most animal species went extinct. Why birds, mammals, and crocodiles survived but pterosaurs and dinosaurs didn't is still a mystery. Perhaps birds were already starting to outcompete pterosaurs in the late Cretaceous, and the asteroid impact just hastened the process enormously. Whatever the case, no pterosaur fossils have been found from more recently than the end of the Cretaceous.

Birds, Proprietors of Blue Skies

Archaeopteryx

Judging from the literature on the subject, scientists have put more effort into studying the origins of bird flight than into the origins of flight in insects, pterosaurs, and bats combined. Much of this work was stimulated by the discovery of *Archaeopteryx lithographica,* one of the most important transitional species ever discovered. The same Solnhofen limestone in Germany that produced pterosaur fossils with visible wing membranes also yielded up several stunning fossils of *Archaeopteryx,* complete with detailed feather impressions (Fig. 7.15).* *Archaeopteryx* caused a sensation when the early specimens were described in the 1860s and 1870s, both among the general public and among scientists. If not for the feathers, *Archaeopteryx* would have been classified as a dinosaur: it had toothed jaws, a long, bony tail, three unfused fingers with claws on each hand, and a small, flat sternum (breastbone) completely unlike the massive, keeled sternum of a modern bird. In fact, two specimens were originally identified as dinosaurs, and one as a pterosaur, because their feather impressions were so faint. Thus *Archaeopteryx* appears to be a bona fide "missing link" between dinosaurs and birds. Of course, things in biology are never quite this simple. Modern paleontologists consider *Archaeopteryx* to be a relative of the primitive ancestor of modern birds (and perhaps similar to it), but on a separate side branch off the main bird lineage. Nonetheless, *Archaeopteryx*'s age and combination of ancestral and birdlike features have aroused keen interest in scientists for over a century. The study of *Archaeopteryx* has revolved around three questions: (1) What did *Archaeopteryx* evolve from? (2) Could *Archaeopteryx* itself

---

* The first recognized fossil of *Archaeopteryx* is a magnificently preserved, solitary wing feather found in 1860. The feather is so well preserved that photos of the fossil can easily be mistaken for photos of a real feather. Two fabulous skeletons with obvious feather impressions were discovered shortly afterward, the "London" specimen in 1861 and the "Berlin" specimen in 1877. The latter is so complete and detailed that it is the specimen usually shown whenever *Archaeopteryx* is mentioned. Ironically, the first fossil of *Archaeopteryx* was actually discovered in 1855, but because it was incomplete and the feather impressions were so faint, it was misidentified as a pterosaur for more than 100 years [93].

Figure 7.15. The "Berlin" specimen of *Archaeopteryx*. Drawn by Nicola Hector, reprinted by permission of the artist.

fly? (3) What does *Archaeopteryx* tell us about the evolution of powered flight in birds?

The Family Tree of Birds

Great controversy and debate has swirled about the ancestry of *Archaeopteryx* for over a century. Soon after the first fossil of *Archaeopteryx* was discovered, the great anatomist (and anti-evolutionist) Richard Owen published a description in which he argued that *Archaeopteryx* was clearly a bird and quite unlike modern reptiles. Owen's point was that birds and reptiles are separate, and one could not have evolved from the other. In reply, Thomas Huxley, an equally famous champion of evolution, pointed out that *Archaeopteryx* is clearly a transitional form, not between birds and modern reptiles, but between birds and dinosaurs. Thus, Huxley argued, birds must be the direct descendents of dinosaurs.

Huxley's view was widely accepted at first, but gradually objections arose. One issue that bothered many biologists was that no bipedal dinosaurs had been found with clavicles, the collarbones that fuse to form the furcula (wishbone) in modern birds. Gerhard Heilmann, a Danish physician, wrote a whole book detailing the many similarities between theropod dinosaurs and *Archaeopteryx*. When Heilmann wrote his book in 1916,* no known theropods had clavicles, and he concluded in the end that the missing clavicles in theropods meant that birds must be related to the ancestors of dinosaurs rather than to the dinosaurs themselves [32]. Like many other scientists, he thought that some of the similarities between birds and dinosaurs were due to similar habits (*convergent evolution;* see Chapter 11) rather than descent from a common ancestor. Until recently, paleontologists had drawn the reptile family tree with one large branch called Archosauria that branched more or less equally into dinosaurs, pterosaurs, birds, and crocodilians [33].† The most ancient archosaurs, too primitive to be placed into one of these groups, were lumped together as "thecodonts." (Paleontologists now suspect that thecodonts may actually represent several lineages, with the lineage of *Scleromochlus* leading to dinosaurs and pterosaurs.) *Euparkeria capensis,* a thecodont discovered around 1910, seemed to most paleontologists of the time to be a logical bird ancestor: it was a small, predacious animal that could walk on all fours and run bipedally. It seemed to have the major properties expected of an ancestor to the main archosaur groups, birds included. Paleontologists accepted *Euparkeria* as the closest common ancestor of birds and dinosaurs for most of the twentieth century.

---

* Published in Danish in 1916; translated into English and published as *The Origin of Birds* in 1927.

† The crocodile group was at one time much larger and more diverse than today. It formerly included small, terrestrial, partly bipedal species, and some that may even have lived in trees.

In the 1970s, just as some scientists were starting to wonder whether birds had arisen from crocodilians [34], the Yale paleontologist John Ostrom made a startling claim. Ostrom had earlier discovered a dinosaur he named *Deinonychus,* a medium-sized, bipedal carnivore with an unusually large brain and an enlarged, sickle-shaped claw on one toe of each foot. Ostrom then turned to a detailed study of *Archaeopteryx* and came to the conclusion that *Deinonychus* and *Archaeopteryx* are closely related [35, 36]. Most paleontologists were skeptical at first, and some still reject Ostrom's claim. But as more evidence accumulated and as a new method of reconstructing evolutionary lineages—phylogenetic systematics—came into common use, the evidence has continued to increase in favor of Ostrom's conclusion. *Deinonychus* and *Archaeopteryx* both have three fingers and an unusual wrist bone that allows the hand to bend from side to side, as birds do to flex the wing on the upstroke or to fold the wing at rest. They also have similar hindlimbs, particularly the extreme reduction of the fibula (one of the paired shinbones in each leg), the hingelike ankle, and the arrangement of the toes with the first toe reduced, the middle three carrying the weight, and the fifth toe lost. Since Ostrom's original work, other paleontologists have produced ever more detailed comparisons, and most have come to the same conclusion: *Archaeopteryx* shows a very strong relationship to dinosaurs like *Deinonychus.*

- *Deinonychus* belongs to a large group of bipedal, carnivorous dinosaurs called theropods (not to be confused with thecodonts). Theropods include such famous dinosaurs as *Tyrannosaurus rex* and *Allosaurus,* as well as less celebrated animals such as *Giganotosaurus* (a carnivore even bigger than *Tyrannosaurus*) and the chicken-sized *Compsognathus. Deinonychus* itself belongs to a subgroup within the theropods recently christened the Maniraptora. Maniraptorans were active, intelligent predators, ranging from dog-sized to person-sized, and include *Velociraptor* of Hollywood fame. The most recent phylogenetic studies show that maniraptorans are the closest known relatives of *Archaeopteryx* (and hence, birds) because they had hollow, air-filled bones, a fully-developed furcula (which had either been lost or unrecognized on earlier specimens), and many other birdlike traits [37, 38].

Some paleontologists reject the bird-maniraptor relationship [39]. Among other things, they cite several anatomical discrepancies. For instance, according to conventional embryology, birds seem to retain the second, third, and fourth fingers of the hand, whereas paleontologists think that maniraptoran dinosaurs retain the first, second, and third fingers [40]. (Recent studies of developing bird feet and hands suggest that conventional embryology may be misleading in this case [41, 42], and that the same fingers might be retained in both birds and dinosaurs.) The main objection to the maniraptoran-bird relationship, however, is that maniraptoran dinosaurs are a Cretaceous group and lived long *after* the late Jurassic *Archaeopteryx*. Even though *Archaeopteryx* may represent a side branch off the main line of evolution leading to modern birds, it was closely related to the ancestor of all birds, so birds must have arisen before *Archaeopteryx,* and their ancestors must have lived in the Jurassic. Thus, a few scientists still think *Archaeopteryx* descended from a much more ancient dinosaur lineage or perhaps even from a thecodont [43, 44]. Lack of evidence, however, is not the same as negative evidence. Until the fossil record improves to the point that we can say either, "Yes, ancestral maniraptorans lived before *Archaeopteryx,*" or, "No, maniraptorans definitely evolved after *Archaeopteryx,*" this objection cannot be resolved. Unfortunately, fossilization is such a rare, hit-or-miss process that bird ancestors simply may not have left any fossils. The weight of evidence at this point, however, does seem to be on the side of an undiscovered, maniraptorlike, theropod ancestor of *Archaeopteryx*.

Scientific and popular articles have appeared as recently as 1999 describing many new theropod dinosaur species with possible links to birds [45, 46, 47, 48]. Some of these dinosaurs apparently had feathers, and others had similar shoulder joints to birds. As provocative and exciting as these new discoveries are, the ones that have been formally described are still from the Cretaceous period and far too young to have been ancestral to *Archaeopteryx*. Their main significance is that they prove that several important birdlike features were present in at least some dinosaurs, bolstering the argument that earlier dinosaurs with similar features were ancestral to birds.

It Had Feathers, But Did It Fly?

The scientists who first looked at the magnificent fossils of *Archaeopteryx* in the 1800s took as a given that it could fly. The feathers are arranged in such a birdlike fashion that it is difficult to imagine the living animal *not* flying. Nevertheless, in the decades following the discovery of *Archaeopteryx*, some biologists began to have doubts. The reptilian features of *Archaeopteryx* made them question its ability to fly under power. Structurally, *Archaeopteryx* lacks not only a large, keeled sternum for anchoring powerful flight muscles but a unique muscular specialization found in modern birds. The supracoracoideus muscle lowers the forelimb in other tetrapods, but has evolved into the main upstroke muscle in modern birds. Birds evolved a unique pulleylike arrangement (the triosseal canal and the acrocoracoid process) in the bones near the shoulder joint that redirects the pull of the supracoracoideus muscle almost 180° to change the direction of its pull on the upper arm from down to up.* *Archaeopteryx* lacks the triosseal canal and acrocoracoid process, so presumably it used its dorsal shoulder muscles to raise its wings. Lacking these avian features, and with such otherwise dinosaurian anatomy, most scientists until recently assumed that *Archaeopteryx* must have been physiologically similar to modern reptiles. Just as with pterosaurs, they assumed that a cold-blooded, "reptilian" *Archaeopteryx* would not have had enough muscle power or endurance for flapping flight. Thus, many scientists were willing to concede that *Archaeopteryx* could have been an accomplished glider, but they thought that its structural and physiological limitations prevented it from using flapping flight.

When John Ostrom made his *Deinonychus-Archaeopteryx* connection in the 1970s, he also made another startling suggestion: maybe *Archaeopteryx* didn't fly at all. He suggested that *Archaeopteryx* used its feathered forelimbs as insect nets, sweeping insects out of the air for food [49, 50]. Ostrom sought a mechanism that could serve as an origin for flapping movements somehow useful for a running animal, which would bolster his argument linking birds and theropods. He mistakenly believed that the sweeping, tennislike strokes for netting insects were similar enough to flapping that they could actually provide some aerodynamic benefit. Ostrom's suggestion drew such a tidal wave of rebuttal that he later withdrew it, but the controversy he stirred up led to a sharp clarification in our understanding of flight in *Archaeopteryx*, well beyond just wing movements.

The strongest evidence that *Archaeopteryx* could fly comes from its feathers. Several scientists have pointed out that the primary feathers of *Archaeopteryx* are asymmetrical, with the shaft nearer to the front or leading

* Using a chest muscle—the supracoracoideus—rather than a dorsal shoulder muscle to raise the wing apparently keeps the bird's center of gravity in a lower, more stable location.

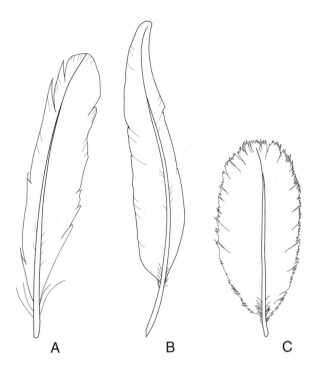

Figure 7.16. Comparison of *Archaeopteryx* and modern bird feathers. A. Primary feather from *Archaeopteryx* wing. Primary feathers are the large quill feathers that form the tip of the wing and often make up a significant portion of the wing surface area (Chapter 1). B. Primary feather from wing of a modern bird capable of flight. C. Wing feather from an emu, a flightless bird. (B.Hd.)

edge of the feather (Fig. 7.16). This arrangement makes each feather capable of acting as a little individual airfoil, and is also needed to maintain the whole wing's camber and to carry the aerodynamic flight loads [51, 52]. All modern flying birds have asymmetrical primary feathers, and birds that have become flightless, such as ostriches, never have asymmetric feathers. In fact, the feathers of flightless birds generally become soft and fluffy, because they no longer need to carry aerodynamic loads or maintain a streamlined body shape (Fig. 7.16). Thus, *Archaeopteryx* certainly used its forelimbs as wings.

Did *Archaeopteryx* flap its wings or merely glide from branch to branch? In recent years, most of the structural and physiological objections to powered flight in *Archaeopteryx* have been laid to rest. For instance, bats have a small sternum with a very small keel (compared with birds), and they are perfectly capable of flapping. The flight muscles of *Archaeopteryx* most likely were anchored to cartilage or soft connective tissue, as in many bats. Regarding the missing supracoracoideus pulley system, Jeremy Rayner thinks that a powerful upstroke muscle is mainly important for the slow (ring vortex) gait, and that aerodynamic forces alone are sufficient to raise the wing in the fast (continuous wake) gait. This suggestion is compelling, because the fast gait involves simpler flapping movements and would logically be a good choice as the primitive gait [53].

Was *Archaeopteryx* endothermic, like modern birds, or ectothermic, like modern crocodiles? The evidence seems thoroughly contradictory. For example, the feathers of *Archaeopteryx* have been used to argue both ways. One school of thought says that the feathers evolved initially to aid aerodynamics: perhaps as a fringe of enlarged scales on the forelimb to help steer during leaps from branch to branch. These scales eventually evolved into feathers to help form the wings and streamline the body. After this, their insulating properties became important for thermoregulation and allowed endothermy to develop [54]. Other scientists turn the whole argument around and say that since the wings of *Archaeopteryx* are so birdlike, it must have been an active flyer, and was therefore endothermic. Moreover, *Archaeopteryx* is so similar to some dinosaurs that its dinosaur ancestors must have been endothermic as well [50]. These scientists believe that feathers first evolved for insulation and were later partly taken over for aerodynamics; thus, to these scientists, feathers indicate that *Archaeopteryx* was endothermic. (Robert Bakker has long argued that all dinosaurs were warm-blooded [55, 56], but most paleontologists now think that only certain groups, such as some of the later theropods, are likely to have been endothermic.)

Both camps have recently received setbacks. The feathers-for-flight position has been shaken by a number of recent finds of what appear to be feathered dinosaurs in China [45, 47, 57]. The feathers may have been for insulation or for courtship displays, but they were symmetrical, and thus not for flight. The blow for the endothermic-dinosaur camp came when John Ruben and his colleagues demonstrated that the respiratory systems of dinosaurs and *Archaeopteryx* do not seem to have had the capacity needed to support the high oxygen consumption of endotherms. First, Ruben showed that maniraptoran dinosaurs do not have turbinates, which are structures in the nasal cavities of birds and mammals. The high body temperature of endotherms causes them to lose water vapor rapidly from the lungs, and the turbinates are important for minimizing this loss [58]. Second, Ruben and his colleagues showed that dinosaurs and *Archaeopteryx* did not have large and efficient enough air passages and lung arrangements to supply oxygen at the high levels needed by endotherms. They concluded that *Archaeopteryx* and its ancestors must have been cold-blooded [59, 60]. Although these studies seem quite conclusive, they are necessarily based on indirect relationships—we cannot measure the breathing rate of a fossil—and some researchers are still a bit skeptical of Ruben's conclusions. At this time, the question of how feathers evolved remains unresolved.

Regardless of its physiology, *Archaeopteryx* was most likely an active flyer. The shape and arrangement of *Archaeopteryx*'s wing is almost indistinguishable from the wings of modern birds such as woodpeckers and magpies. Why would *Archaeopteryx*'s wing be so similar to those of modern birds' unless it too was an active flyer? Furthermore, John Ruben used data from modern reptiles to show that *Archaeopteryx* could still have flown under power even if it was ectothermic [61, 62]. To correct the common "sluggish reptiles" misperception, Ruben noted that reptilian muscles can produce almost twice as much power as comparable bird or mammal muscles. The trade-off is that reptilian muscles can only work anaerobically, which limits how long they can work before they need to rest to recycle toxic physiological waste products. Many biologists assume that reptiles can only maintain powerful "burst" locomotion for a second or two, but Ruben pointed out that monitor lizards can maintain these bursts for many minutes at a time. He calculated that if *Archaeopteryx* had reptilian muscles, it still would have been strong enough to take off from the ground, and could have made powered flights of 10 to 20 meters with only a minute or two of rest between flights. Although a warm-blooded *Archaeopteryx* makes an attractive image, current evidence weighs against endothermy in *Archaeopteryx*, because it lacked several skeletal features associated with the uniquely efficient respiratory system of modern birds.

Whence Wings?

Since at least the time of Darwin, scientists have been arguing about the evolution of flight in birds. The two main suggestions, both first proposed in the late nineteenth century, have come to be known as the *arboreal* and *cur-*

sorial models, and both models have also been applied essentially wholesale to pterosaurs. The arboreal, or "trees-down," model was originally proposed for bats by Darwin himself, in *On the Origin of Species,* and starts with a tree-dwelling animal leaping from branch to branch. At some point, this tree-dweller extended its arms to the side when jumping, perhaps for stabilizing. Feathers, initially used for courtship displays or as solar collectors, might have given the arms a bit of extra area to extend leaps or soften landings. Then the animal evolved longer arms and larger feathers, increasing the arms' surface area and lift, and eventually this proto-bird became a true glider. Flapping then evolved from gliding; perhaps steering movements led to the fast, low-amplitude flapping of the fast gait, which strengthened to prolong flights until the animal was flying under power.

The cursorial, or "ground-up," model begins with a small, fast-running, bipedal animal that regularly jumped or leaped to catch prey (probably flying insects). When leaping, the animal stretched out its forelimbs (possibly already feathered) for balance or steering. Some versions of this model have the leaping animal begin gliding to extend its leaps, but other versions have the animal begin flapping to extend its leaps without going through a gliding stage (in an extreme version, the animal begins flapping while running, entirely for thrust and acceleration, and before it ever leaves the ground). In any case, the animal's flapping became stronger to lengthen its leaps, until it was able to sustain powered flight.

Prior to the 1970s, most scientists favored the arboreal model, but when John Ostrom proposed his dinosaur-bird linkage, he also revived the cursorial model [49, 50]. Biologists and paleontologists debated the two models, sometimes hotly, for about a decade [63, 64, 65, 66, 67]. The arboreal model is much more reasonable biomechanically and aerodynamically, because gliding is fast, and the simplest wing-beat pattern—the fast gait—requires high speed to provide any benefit. However, the fossil record seems to support the cursorial model. Assuming that *Archaeopteryx* evolved from theropod dinosaurs, paleontologists can find no evidence of arboreal or climbing adaptations in any theropods. This lack of evidence may be a bit misleading, inasmuch as many theropods were too big to have climbed trees, but the small ones like *Compsognathus* had no particular climbing specializations. Some of the cursorial proponents point out that the habitat in which *Archaeopteryx* lived had no true trees, just low shrubby bushes, but given that

Archaeopteryx was most likely already an accomplished flyer, this fact is probably irrelevant.

For many years, the arboreal model has been supported by scientists with a biomechanics or flight-oriented perspective, and the cursorial model tends to be favored by paleontologists. From the perspective of a scientist outside the argument, a significant part of the controversy stems from confusing the phylogeny—evolutionary history or family tree—of *Archaeopteryx* with the process of evolving flight. Ostrom proposed both a theropod ancestry for birds and a cursorial model for the evolution of flight, whereas several scientists who disagreed about the theropod ancestry also argued strongly for the arboreal model [54, 68, 69]. Although they are intimately related, the questions of ancestry and evolution of flight are not identical, and considering them separately might help resolve the debate.

In 1984, a conference was held in Eichstätt, Germany, to bring together all the scientists working on *Archaeopteryx* or the origin of bird flight [70]. At this conference, one of the major topics of discussion was the controversy over the arboreal versus cursorial models. Ulla Norberg and Jeremy Rayner both gave presentations strongly favoring the arboreal model. They showed how gliding from a high place to a lower one, then progressing to flapping, makes much more sense aerodynamically and energetically than any scenario starting with running and leaping from level ground. The discussion after Rayner's presentation was long and intense, and most participants seemed to agree afterward that the cursorial model, as then understood, was unrealistic [71]. The main problem with the cursorial model is that a runner has to run tremendously fast in order to leap high enough and fast enough to glide, and then it immediately decelerates once in the air. Few scientists could envision any advantage for a leap from a blistering run that causes the animal to immediately slow down. Furthermore, such a runner loses any ability to maintain its speed or accelerate once in the air, because its hindlegs—where a fast runner must have most of its muscle mass—are off the ground. Finally, running and jumping and gliding turns out to be energetically more expensive than just running, while gliding down from a height is much less expensive than running.

What emerged from the Eichstätt conference was a partial blending of the models. Rayner suggested that the arboreal model ought to be renamed the "gliding" model, because it only requires the animal to glide down from a height, which does not need to be a tree. Cliffs, large shrubs, or large rocks would work just as well, so actual trees are unnecessary. In a supplemental article based on the discussion after his presentation, Rayner looked explicitly at the performance requirements and energetic consequences for a "running-leaping-gliding" animal of the size of *Archaeopteryx*. His calculations showed that such a proto-flyer could actually lower its cost of locomotion if it could run fast enough—7 meters per second (over 15 miles per hour). This is two or three times higher than the sustainable running speed of any modern reptile, bird, or mammal, and faster than estimates of small dinosaur speeds based on fossil tracks and their anatomy. So, while not impossible, the strict cursorial model seems highly unlikely. On the other hand, *Archaeopteryx* was relatively small, no bigger than a pigeon, and small animals can be good climbers without any obvious specializations for climbing. Perhaps the proto-birds were runners that also commonly jumped or climbed to launching heights for short glides. Although many paleontologists continue to favor the cursorial model, because no one has found convincing evidence of climbing ability in theropods, nowadays they tend to concede that some combination of the gliding and cursorial models is probably the best estimate of how bird flight actually evolved [33, 38, 72].

A group of biologists recently proposed just such a model. The Oxford University biologists Joseph Garner, Graham Taylor, and Adrian Thomas pointed out that neither the cursorial nor the arboreal model fits convincingly with the fossil evidence. Using an evolutionary tree that included unfeathered and feathered theropod dinosaurs, *Archaeopteryx*, later fossil birds, and modern birds, these scientists showed that neither model predicts the correct order of evolution of various traits. For example, the cursorial and arboreal models both predict that bird ancestors should evolve asymmetrical flight feathers right along with aerodynamic wings. At least one theropod dinosaur, *Caudipteryx zoui* (see below), had elongated, symmetrical feathers on its hand, however. Based on this and several other traits, including weight-reducing adaptations and birdlike running, Garner and his colleagues proposed a new model, which they dubbed the "pouncing proavis" theory. This theory starts with a small, bipedal predator adapted to leaping onto prey from elevated sites. This leaping animal would have benefited from some ability to steer during its leaps. Enlargements or extensions of the hand and forearms could work effectively for this purpose, perhaps initially using asymmetrical drag increases and later relying on asymmetrical lift (Chapter 5). These extensions would have been located at about the same place on the arm as ailerons on an airplane wing (Chapter 11) and would have functioned in much the same way. If the symmetrical feathers on the hands of *Caudipteryx* pointed straight back, they would have served nicely as this type of steering device. Improvements in aerodynamic effectiveness for steering would start to provide enough lift to extend "leaps" into "swoops." Selection for longer and longer swoops would lead to actual flight, either through a gliding stage or directly to flapping [73]. The pouncing proavis model borrows from both of the earlier models, the elevated start from the arboreal model and the bipedal leaper from the cursorial model. Garner and his colleagues based their model on an evolutionary sequence from dinosaurs to modern birds that may or may not be supported by future fossil discoveries, but it reflects the most up-to-date evidence currently available.

Archaeopteryx was involved in one other brief controversy. In what must be one of the most bizarre episodes in modern science, a group of physicists and astronomers led by the famous astronomer Sir Fred Hoyle claimed that the *Archaeopteryx* fossils are forgeries. Hoyle sought to discredit transitional forms like *Archaeopteryx* because he believed that evolution occurred due to periodic invasions of viruses from outer space, rather than by Darwinian natural selection. He and his followers claimed that the *Archaeopteryx* fossils were actually fossils of small dinosaurs where someone had scraped away the stone around the skeleton, made it into cement, poured it back around the skeleton, and pressed feathers of modern birds into the cement to make the feather impressions.

Hoyle's prominence demanded that paleontologists defend *Archaeopteryx*, and they rose to the challenge with a vengeance [74, 75]. Paleontologists pointed out that several of the fossils of *Archaeopteryx* consist of the main slab and the counterslab that was split away to reveal the fossil originally, and the feather impressions on main and counterslabs are identical mirror images, down to the locations of microscopic cracks and mineral crystals. Moreover, the London specimen has been used so many times for making casts that its surface has been called a "chemical laboratory" and any artificial additions would surely have come to light in over a century of study. Finally, three specimens of *Archaeopteryx* had lain unrecognized for decades in different museums (one for a couple decades, one for over a century), and these three fossils all have faint feather impressions; how could someone have faked those feather impressions when no one knew they were *Archaeopteryx* fossils until a century after the original *Archaeopteryx* fossils were discovered? One paleontologist went so far as to try to make his own fake feather impressions and was never successful. He came to the conclusion that such a successful forgery would be of greater technical significance than the fossils themselves [75]! No reputable paleontologists or evolutionary biologists supported Hoyle's claim, and as Lowell Dingus and Timothy Rowe put it in their book *The Mistaken Extinction,* the authenticity of the *Archaeopteryx* fossils "is one of the very few points that everyone on both sides of the bird-dinosaur debate can agree upon" (p. 212) [33].

A final point about *Archaeopteryx* seems to have been largely overlooked by most scientists. Although the dinosaurian features of *Archaeopteryx* give lots of clues about its ancestry, the wings of *Archaeopteryx* are so aerodynamically similar to those of modern birds that *Archaeopteryx* may not really tell us much more about the evolution of flight than we could learn from a sparrow or a duck [61]. After all, *Archaeopteryx* itself is not directly ancestral to modern birds, and despite scientists' hopes that *Archaeopteryx* is similar to the actual ancestral proto-bird, its forelimb was already a well-designed airfoil. If scientists want to learn about how birds evolved flight, they need to look at animals with more primitive flight characteristics than *Archaeopteryx*. Some recent fossil finds in China may be just what they are looking for. Spectacularly well preserved fossils of three species seem to show possible links to birds. *Sinosauropteryx prima* was a small theropod that had a featherlike fringe running down its spine and along its tail, raising the possibility that feathers—or at least their precursors—were not uncommon among theropod dinosaurs [57]. A second species, *Protarchaeopteryx robusta,* was a small maniraptoran with well-developed, symmetrical feathers on its tail; other smaller, downy feathers are preserved near its legs and chest [47]. Finally, *Caudipteryx zoui* (mentioned earlier) had at least 14 symmetrical, elongated feathers attached to each hand and wrist. This animal also had a row of feathers running down each side of its tail (much as in *Archaeopteryx*) and some small, downy feathers are preserved near its hips [47]. These are exactly the sorts of specimens needed to fill out the family tree between *Deinonychus* and *Archaeopteryx*. Although these three feathered dinosaurs lived after *Archaeopteryx* [76], they illustrate the evolutionary changes that led to *Ar-*

chaeopteryx and are probably closely related to the actual ancestors of *Archaeopteryx*. Many more post-*Archaeopteryx* (Cretaceous) bird fossils have been found since 1990 than in all the preceding years [38], so perhaps our pre-*Archaeopteryx* fossil record will continue to improve as well.

Thanks to those recently discovered bird fossils, scientists know quite a bit about the evolution that occurred between *Archaeopteryx* and modern birds. Some of these early birds had more fusion of hand and hip bones and short tail bones, like modern birds, but retained teeth and claws on their fingers, like *Archaeopteryx*. Most of these fossil birds sported the massive, keeled sternums of modern birds. Their skulls became lighter and more airspaces appeared in the skeleton. Fossils, of course, tell us very little about physiology or behavior, so we don't know whether these birds were partially or fully warm-blooded, or if they had elaborate courtship behavior. Sometimes they do offer tantalizing clues. One rock slab from China contained two very similar, exquisitely preserved bird fossils, complete with feather impressions. Dubbed *Confuciusornis sanctus,* one has an unremarkable tail, but the other has two enormously elongated tail feathers, much longer than the rest of the body [45, 77, 78]. Were these tail feathers used as a courtship display by the male to attract the female? If so, did the courtship involve the elaborate behavior common in modern birds? Scientists can only speculate.

At the end of the Cretaceous, when the (nonbird) dinosaurs died out, some bird lineages died out as well. Others that had evolved earlier, such as those leading to shorebirds and ducks, survived, and many new lineages arose. Since they began their major evolutionary diversification in the Cretaceous, birds have become the most diverse and successful group of air-breathing vertebrates of the modern world.

Wings in the Dark: Bats

Bats were the last group of animals—and the only mammals—to evolve powered flight. The oldest fossil bats are from the Eocene epoch (40 to 50 million years ago). These fossil species have fully developed, functional wings and were clearly capable of flapping. Thus, bats must have evolved flight considerably earlier, possibly just after the end of the Cretaceous period (about 60 million years ago) [79, 80]. Scientists have not found a bat equivalent of *Archaeopteryx,* so, as with insects, scenarios for the evolution of flight in bats are based on indirect evidence. Unlike insects, however, flight is the defining characteristic of bats; asking how and when they evolved flight is thus synonymous with asking how and when bats themselves evolved.

Early Bat Evolution

Birds had evolved and diversified long before bats took to the air, but many bird lineages died out at the end of the Cretaceous. Perhaps this reduction in bird diversity opened up ecological niches that bats evolved to fill. In any case, the ancestral proto-bat was most likely a small, gliding, treedweller. Living colugos from the rain forests of Southeast Asia may give us some insight into bat evolution (although commonly called "flying lemurs," colugos are not true lemurs and are gliders, not flappers). Colugos may be the

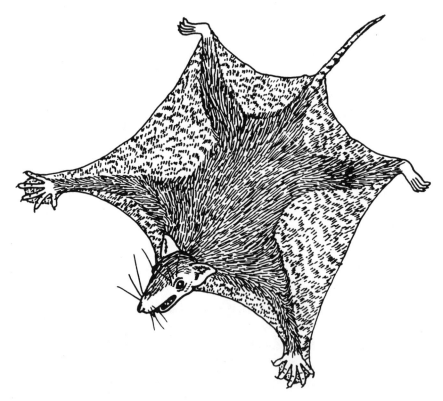

Figure 7.17. Gliding colugo. Note webs between fingers to increase wing area. From J. D. Smith 1977 [92], by permission of Kluwer Academic / Plenum Publishers.

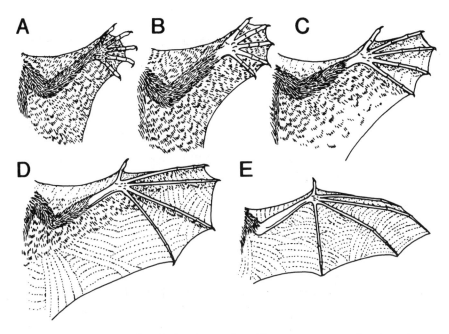

Figure 7.18. Sequence of evolution from colugo-like ancestor to modern bat wing. From J. D. Smith 1977 [92], by permission of Kluwer Academic / Plenum Publishers.

closest living relatives of bats. Unique among gliding mammals, colugos have webs of skin between their fingers (and toes) that act as parts of the wing membrane (Fig. 7.17); in other gliding mammals, the hands and fingers do not support the wing membrane. The wings of bats are supported largely by their greatly elongated fingers, so the proto-bat must have had a similar arrangement, and the mobility of the fingers undoubtedly improved the animal's steering ability. If natural selection favored longer glides, the arms and fingers of the proto-bat would elongate to increase wing area and improve its lift-to-drag ratio. Weak flapping to stretch the glide would not be difficult with such a wing. Eventually, true flapping flight evolved, at which time the animal would be recognizable as a bat (Fig. 7.18) [79, 81]. Note that this evolutionary sequence fits the arboreal model exactly.

Competition and predation pressure from birds probably pushed most bats toward nocturnal activity, or kept them there if their ancestors were already nocturnal. A few evolved larger body sizes, and some of these became partly diurnal, probably in isolated or specialized habitats [82]. The former became microbats, and the latter, megabats. Microbats (members of the Microchiroptera) are probably most familiar as small, nocturnal, cave-dwelling insect-eaters—although they can feed themselves in many other ways—and they are by far the most common and diverse group. Megabats (Megachiroptera) include flying foxes and fruit bats, and as their name suggests, most of them are relatively large (guinea-pig-sized or larger). They feed on fruit or other particularly nutritious plant parts, such as pollen or nectar. Their size, and perhaps their limited tropical distribution, gives them some protection from predatory birds.

Echolocation

Microbats initially flew from perch to perch to pick insects off tree branches; biologists call this *gleaning*. Early in their evolution, ancestral microbats evolved *echolocation*. Echolocation means producing sounds, and then using the echoes to locate surrounding objects.* Bats probably originally developed this sensing mechanism to avoid flying into obstacles in the dark. Once echolocation evolved, it was soon adapted for hunting insects in flight, which quickly became the predominant feeding method in microbats.

Some biologists think that body size played a role in the evolution of echolocation. The physics of light constrains eyes: in dim conditions, small eyes simply cannot gather as much light as large eyes, in the same way that large binoculars ("night glasses") work better in dim conditions than small binoculars of equivalent quality [80]. In other words, small eyes need more light to form sharp images than large eyes. Early microbats were small, and because they moved in cluttered environments and hunted tiny prey, they faced strong selection pressure to enhance other senses to compensate for limited vision. Their hearing certainly became highly acute, if it was not already. Early bats thus had the pieces in place to allow them to evolve echolocation. First, the bats would have learned to pay attention to the echoes of incidental sounds that they produced just by moving about. Eventually, they

---

* Microbats use a sophisticated echolocation system to track and intercept insects. Bat echolocation sounds are ultrasonic, meaning they are too high-pitched for humans to hear. High frequency sounds are easier to aim and allow bats to detect very small objects (like insects). "Sonar" comes from an acronym that refers to human-made devices for echolocation, and it is often used synonymously with echolocation.

would have begun producing more specialized sounds, and then developed more specialized detection mechanisms.

One reason that echolocation is universal in microbats but rare in other nocturnal mammals is that bats put almost no effort into producing these very intense, powerful calls. Bats emit their calls in synchrony with flapping: bats emit an echolocation call just as the chest gets compressed by the powerful downstroke muscles, so the effort of calling is essentially a byproduct of flapping. In the lab, flying bats have the same metabolic rate whether they are calling or silent, but the metabolic rate of perching bats goes up substantially when they call [80]. Echolocation arose in bats due to a unique combination of needs and abilities, and flight may have been a key requirement.

No "Archaeopteryx" for Bats

As mentioned earlier, paleontologists have not discovered any transitional fossils showing the evolution of flight in the lineage leading to bats. Thus, scientists must use comparisons among living species, along with general principles and informed speculation, to build scenarios for the evolution of flight in bats. Some scenarios have gradually been replaced as our knowledge of the relationships among mammal groups has improved. Other, more controversial suggestions stimulated research and analysis that clarified aspects of bat evolution.

Tree shrews are small, arboreal, somewhat squirrel-like, insect-eating mammals of the tropics that have a long fossil record and many primitive characteristics. They were originally grouped with the true shrews, but now biologists consider tree shrews to be different enough that they have been placed in a separate order from true shrews. Biologists place tree shrews near the root of the mammal family tree, so in a way they are fairly close relatives of most mammals, but obviously they must be much closer to some than to others. Modern tree shrews and microbats share many similarities, so earlier biologists naturally assumed that bats evolved from tree shrews. Both are small insect-eaters. Moreover, tree shrews are nocturnal and arboreal, and some even use a crude form of echolocation. These factors led biologists to suggest that the proto-bat evolved from a gliding, tree-shrewlike animal, and that it inherited nocturnal habits and perhaps even echolocation from its ancestors. As attractive as this scenario is, it runs counter to several modern lines of evidence. First, megabats do not use echolocation.* Second, the old-est fossil bats have few of the skeletal modifications associated with echolo-cation in modern bats [80]. Finally, although tree shrews are obviously re-lated to bats, recent analyses put colugos, and perhaps even primates, at least as close to bats [83, 84]. Biologists have yet to reach a strong consensus, but many think that the proto-bat was probably much more like a colugo than a tree shrew.

Given that the nearest relatives of bats—colugos, and possibly primates and tree shrews—are unquestionably arboreal, Darwin's original proposal of "trees-down" evolution of flight in bats seems self-evident. Although a few biologists have proposed cursorial, or "ground-up," hypotheses for the evo-lution of flight in bat ancestors [85, 86], these cursorial proposals were based on misunderstandings about the requirements of the arboreal model [87]. For bats, at least, the arboreal model is well supported and widely accepted [67, 79, 80, 87].

Over the years, some biologists have suggested that megabats (flying foxes and fruit bats) and microbats evolved separately and are more closely related to other mammals than to each other. This would, in turn, suggest that they evolved flight separately [88]. This suggestion caused controversy, because, if true, flight would have evolved in different ways and at different times in megabats and microbats. Moreover, it implies that the many structural sim-ilarities of megabats and microbats evolved convergently to accommodate flight, rather than by being inherited from a common ancestor. Most re-cently, this suggestion was inspired by a study showing differences between megabats and microbats in the arrangement of the visual pathways in their brains, as well as relatively minor differences in skeletal structures. However, when all the genetic and anatomical evidence is combined, it overwhelm-ingly favors a close relationship between megabats and microbats [80, 83, 89]. Thus, megabats and microbats both descended from a primitive ances-tral bat, and powered flight in mammals only evolved once.

One of the more common—and probably ancestral—feeding methods in microbats is hawking insects. *Hawking* refers to flying bats preying upon fly-

---

* Two species of megabats have evolved crude echolocation, but they produce the sound pulses with body movements, rather than with the voice as in microbats [80]. Megabat echolocation thus evolved independently from echolocation in microbats.

ing insects (and the term comes from the medieval sport of using falcons to hunt other birds in flight). Modern microbats, however, use a surprising variety of feeding methods. Many still glean insects from perches, some glean insects from foliage while flying, others fly low over streams and snatch fish from near the surface, and, yes, a few tropical species drink blood (usually from cattle or other large mammals). Along with megabats, some families of microbats eat fruit, nectar, or pollen. Many are rather generalized predators, eating anything they can snatch from the ground or foliage—insects, spiders, scorpions, frogs, small mammals, even small birds and other bats [84]. Most bats are flexible and opportunistic, and can use more than one foraging method, depending on what potential food items are most profitable. Contrary to popular belief, no bats are blind, and most microbats have good vision in moderate light. Many bats use vision and listening for prey noises, in addition to echolocation, to find prey. This foraging flexibility, along with their flying ability, clearly helped make bats one of the two most successful mammal groups.

Patterns in the Evolution of Flight

The great diversity in flying animals and the huge differences among major groups of flyers tend to obscure general patterns. The enormous size and anatomical differences between a mosquito and *Quetzalcoatlus* might suggest that they have nothing in common. However, all animals that have evolved flight share certain features. Most important, all animals that fly under power do so by flapping. Unlike machines built by humans, animals have not evolved rotating axle systems, so the only way that animals can produce thrust is by flapping (Chapter 4). Adaptations that permit and improve the flapping movements are thus among the key innovations required for true powered flight in animals.

Second, gliding is mechanically and energetically the easiest way to begin evolving flight. Scientists have developed logical and consistent models that use gliding as an intermediate step in the evolution of flapping flight for all flying animals. Other scenarios are possible, such as surface skimming in insects, or variations on the cursorial model in birds, but the fact remains that all nongliding alternatives require abilities that seem extreme compared to those required for gliding. In the absence of evidence to the contrary, most scientists expect gliding to have been crucial in the evolution of powered flight in most, if not all, flying animals.

Finally, while scientists have steadily improved our knowledge of the evolution of animal flight, large areas of ignorance remain. Though the gaps in the fossil record are frustrating, evolutionary biologists expect exactly these sorts of gaps. Evolution by natural selection usually changes transitional forms rapidly, so they appear only briefly in geological history. Evolutionary novelties—like winglets, for example—that are beneficial tend to spread rapidly through populations and evolve quickly into more effective forms. Moreover, changes that occur over tens or hundreds of thousands of years look instantaneous to paleontologists, to whom a million years may be a brief interval. Unless paleontologists are lucky, they rarely find transitional forms. *Archaeopteryx* is an exception. Thus many of the questions about the evolution of flight cannot be settled until scientists have similar fortune with other groups.

EIGHT

Migrating

*M*ost people have some familiarity with animal migrations. Wedges of geese flying south in the fall, songbirds appearing in the spring, salmon returning to their native streams to spawn: these are all familiar examples of migration. Animals usually migrate to exploit a habitat that is temporarily bountiful. When conditions deteriorate—temperatures drop or food runs out—migratory animals leave. They may return to a region with a milder climate, or they may seek out richer habitats by following weather fronts. Geese, for example, take advantage of mild summers in the northern United States and Canada, but they head south in the fall to avoid the harsh northern winter. Tracking animals from one habitat to another is not always practical, so how can we tell if an animal is migrating, as opposed to merely looking for the next rich patch of food? Biologists have discovered that migrating animals tend to show distinctive behaviors. For example, migrants tend to build up substantial fat reserves before departing, and they also move

along straighter paths and stop less often to feed [1]. Thus, we can tell if animals are migrating even if we cannot track them on their entire journey.

The Advantages of Aerial Migration

Why do animals migrate? The most familiar migrations, such as those of
songbirds or waterfowl, involve moving to a temperate zone or high latitude
for the summer, and then moving to lower latitudes or the tropics for the
winter. Surprisingly, some animals may only migrate a few dozen kilometers
or even just a few hundred meters. Most migrating animals go to all this
trouble in order to find hospitable conditions or avoid harsh conditions. The
summering habitat typically has an abundance of food and nesting sites that
would be unoccupied during the summer if migratory animals were not present. A small number of year-round residents means that there would have
been relatively little competition for these resources when animals were
evolving the migratory habit. However, the migratory species cannot survive
the harsh winter conditions of the summering region, so they must move
back to milder climes to pass the time until the next summer. This pattern,
with some differences in times and distances, probably applies to most species that migrate in the air or on land.

Migration gives an animal the ability to exploit temporary habitats or resources, yet to get out before the conditions get too nasty. For example, songbirds in the northern United States use the shortening day length in early
autumn as their cue to head south, leaving well before winter conditions actually occur. In general, the timescale for migrations can be weeks or months
(if driven by weather fronts, for example), seasons, or years. The distances
can be as short as a few hundred meters (bean aphids in England or milkweed
bugs on Baltic islands), to hundreds of kilometers (dragonflies, migrating locusts), to thousands of kilometers (songbirds and shorebirds of northern
North America) [1, 2, 3].

Flight is a great advantage for migration, regardless of the absolute distance an animal travels. For an insect, a journey of a few kilometers might
take days or weeks on the ground, but even a tiny aphid can fly that distance
in an hour. In California, the so-called convergent ladybird beetle (*Hippo-
damia convergens*) migrates up into the mountains to overwintering sites in
the fall and comes back down to the coastal valleys and lowlands by the

Figure 8.1. Migration routes of two North American seasonal migrants, the monarch butterfly and the snow goose. (S.T.)

spring, a journey only possible because the beetles fly. On a larger scale, snow geese fly from Hudson's Bay in Canada to the coast of the Gulf of Mexico in one nonstop, 2700-kilometer trip of just under 60 hours (Fig. 8.1) [4]. A migratory journey that is amazing in both absolute and relative terms is made by monarch butterflies. In the fall, monarchs from the northern United States fly southwest to the mountains near Mexico City, a distance of about 3000 kilometers. Monarchs do not make the trip nonstop, and they depend at least partly on favorable tailwinds. Even so, they make the trip in just a few weeks (Fig. 8.1).

Flight Speed

Some species of whales migrate from the Arctic to the equator and back, and seals and sea lions range thousands of kilometers from their breeding areas. True, it costs sea lions or salmon less energy to move a kilogram of their body 1000 kilometers than it costs geese, but geese have an enormous speed advantage. Drag underwater is much higher than in air, so only really large, powerful swimmers such as porpoises or tuna can have cruising speeds as high as 20 to 30 kilometers per hour. Most fish swim in the range of 1 to 10 kilometers per hour, particularly over long distances [5]. In contrast, medium-sized birds and bats fly in the range of 20 to 65 kilometers per hour [6], geese clock in at over 80 kilometers per hour [7], and even dragonflies can fly more than 12 kilometers per hour [8]. So, while swimmers may have an energy advantage, flyers have a substantial speed advantage. At 5 kilometers per hour, a fish would take about 22 days to swim 2700 kilometers, which takes snow geese just over 2 days. Walking, of course, is both the slowest and the most costly form of locomotion. For example, caribou in Canada make the longest known terrestrial migration, a round-trip of about 500 kilometers each way. If a caribou herd moves 20 kilometers per day, it would take at least 25 days to make that 500-kilometer trip. Geese could make the same trip in 11 hours, or less than half a day.

Barrier Crossing

The high speeds possible in flight permit animals to migrate across deserts, oceans, or ice caps. Examples of such barrier crossings abound. Knots, which are small shorebirds related to sandpipers, fly from the British Isles to Iceland, and from Iceland to an island in the Canadian Arctic each spring. They fly the 2000-kilometer Iceland-to-Canada leg nonstop, mostly over the Greenland ice sheet (at air temperatures of about –10°C) [4]. Many shorebirds and songbirds fly nonstop from eastern Canada and the northeastern United States straight south over the North Atlantic to Caribbean islands or northern South America (Fig. 8.2). This is a 3000- to 4000-kilometer trip, depending on where it terminates. Flocks have been tracked on radar, so we know that they do not fly down the coast, but rather nonstop over the open ocean [4]. Many other birds cross similar barriers during migratory flights. For example, ruby-throated hummingbirds cross the Gulf of Mexico nonstop; a New Zealand cuckoo migrates 3500 kilometers between New Zealand and Samoa nonstop, and falcons and many songbirds migrate 3000 kilometers over the Indian ocean between east Africa and west India. Brent geese mi-

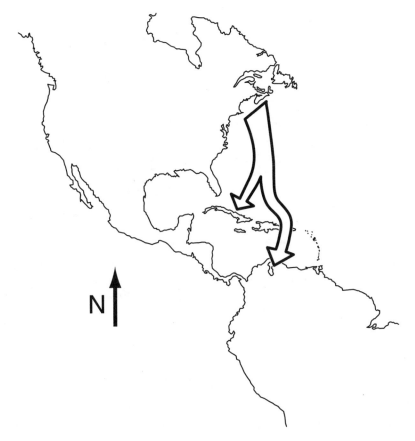

Figure 8.2. Route over the western Atlantic used by birds leaving northeastern North America in the fall. Some birds stop on Caribbean islands, others continue on to South America. (S.T.)

grate from Alaska to southern California, a distance of about 4000 kilometers over the northeastern Pacific [4]. Most of these overwater crossings are by birds that cannot swim, but even those that can swim, such as geese, apparently make the trip nonstop.

Many birds migrate between Europe and sub-Saharan Africa. This migration presents two barriers, the Mediterranean Sea and the Sahara desert. Many large birds use thermal soaring to conserve energy during migrations, but thermals do not form over the Mediterranean. Osprey and falcons give up soaring and fly across the Mediterranean by flapping, but eagles, storks,

Figure 8.3. Many birds fly directly across the Mediterranean by flapping, including songbirds, falcons, and ospreys. Others, such as eagles, storks, and buzzards, detour around the east or west end of the Mediterranean so that they can use thermal soaring along the whole route. (S.T.)

and common buzzards* avoid overwater flights and travel around the east or west end of the Mediterranean via Gibraltar or Turkey (Fig. 8.3). These birds would presumably be able to soar in thermals over the desert, but they probably skirt the edge of the desert in order to find food and roosts along the way. In contrast, small songbirds such as garden warblers and European robins fly directly across the desert. Because the southern Mediterranean coast is very dry in the autumn, it would provide little or no food, so biologists at first assumed that these birds flew across the sea and the desert in one flight. In fact, these songbirds do something much more complex [1]. Though they have no choice but to cross the Mediterranean nonstop, the Sahara is a different matter. Birds that run short of food or water tend to stop at oases for several days and build up their reserves. Other birds simply stop and rest in the shade during the day, and fly at night when the air is cooler. Most songbirds cross the Sahara in an intermittent series of several 10- to 12-hour flights.

Mountains are obviously much less of a barrier to a flying animal than deserts or oceans. Mountains that would completely block a terrestrial mi-

---

* These are Old World buzzards, or buteo hawks, not vultures.

gration may be a mere inconvenience to a bird. The reduced air density at high altitudes even allows the animal to fly faster, albeit with higher power consumption, so altitude probably has little net effect on maximum range [4]. As long as a bird or insect can produce enough muscle power and get enough oxygen in the thinner air over mountains, flying over a mountain range is not greatly different from flying over lowlands. Many birds migrate over the Alps, and birds migrating between Siberia and India cross the Himalayas. These latter migrants cross both deserts and mountains, but they typically stop to rest and forage in the deserts in the spring and in the mountains in the fall [9]. Birds cross some of the highest, coldest mountain ranges nonstop, but, by and large, birds treat most mountains just like any other land they migrate over.

Finally, predators constitute a very real barrier to long-distance travel by small animals. A flying animal has several advantages, however. First, it avoids running the gauntlet of all the terrestrial predators. A migrating swallow can safely ignore cats, foxes, weasels, snakes, speeding automobiles, and small boys with slingshots. Second, because flight is so much faster, a flying animal spends as little as one-tenth of the time traveling as it would if walking. The animal is thus exposed to predators during its journey for a tiny fraction of the time compared to walking, much improving its chances of arriving intact. Finally, catching prey in the air is a chancy proposition, and aerial predators seem to be less common than terrestrial ones. True, a great many flying animals are predators, but my guess is that fewer than half of them actually attack other flying animals while on the wing. Flying animals can see predators farther away, and can take evasive action in all three dimensions. Aerial predators thus need to be more specialized than terrestrial ones, which is why they may be less common.

The Energetics of Migration

In Chapter 6, we saw that an animal uses less energy to move a kilogram of itself a kilometer by flying than walking, but animals require more energy to fly than to swim. One interesting consequence is that energy requirements put different limits on body size depending on whether a migrant walks, flies, or swims. Some flying migrants are very small. Many insects migrate as far as the caribou mentioned earlier, and the autumn trip of monarch butterflies is as

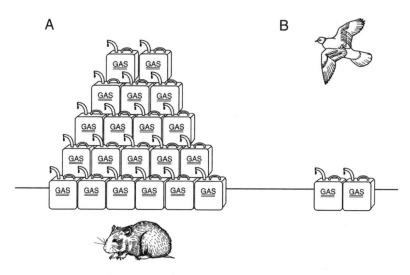

Figure 8.4. The stacks of fuel cans represent the amount of fuel (food energy) a hamster (A) and a songbird (B) of equal weight would need for a long-distance migration of the same distance. (J.P.)

much as *six times* longer than that of the caribou. Hummingbirds, of course, are the smallest birds. The ruby-throated hummingbird migrates across the Gulf of Mexico, and other hummingbirds migrate long distances as well.

Many small migratory birds double their weight before starting a long migration. In order to go the same distance by walking, a similarly sized hamster would need to be able to store fat equal to over ten times its lean mass (Fig. 8.4). Moreover, the extra cost of transporting the added fat itself would further increase the required fuel load, perhaps even doubling it. The hamster would need an enormous food intake to build up that much fat and lots of extra strength to haul it. Even if the poor rodent's feet could still reach the ground after increasing its body weight twenty times, these food and energy requirements and the long duration of the trip prevent small animals from making long-distance migrations on foot.

Economies of scale come into play so that larger animals can make long migrations on foot. Quite a few terrestrial animals of deer-size or larger migrate: pronghorn antelope migrate between high and low elevations over distances of about 100 kilometers; some of the large grazing animals of the African savannas move a bit farther (about 150 kilometers) on each leg of a three-stage journey. The caribou mentioned earlier walk about 500 kilometers between their winter and summer ranges. Large size is an advantage in several ways. For an animal to travel long distances on foot, it must be large enough to walk reasonably fast: consider the difference between the comfortable long-distance pace of a deer or human and that of a mouse or chipmunk. Moreover, animals' mass-specific metabolic rate (rate of energy use per kilogram of body mass) decreases with increasing body size; in other words, on a per-kilogram basis, the energy costs of body maintenance and locomotion go down as animals get bigger. It simply costs much less for a bison to haul a kilogram of itself a kilometer than it costs a kilogram of hamsters to walk the same distance. Large size gives the migrant a low specific metabolic rate (to keep energetic costs down) but enough size and strength to carry a substantial load of fat at a decent speed. Thus, ground-bound migrants are all big, but even the biggest walk over relatively modest migratory distances.

Migration Speed

The "U-shaped" power curve—a property of anything with wings (Chapter 6)—means that at the slow end of an animal's flight speed range, power requirements actually decrease as the animal flies faster. At some speed, the power needed to fly reaches a minimum, and then power starts to go up as the speed increases further. However, the minimum-power speed (v_{mp}) is not the most economical speed for long-range flight. Animals flying great distances should fly a bit faster, at their maximum-range speed (v_{mr}, Chapter 6). The maximum-range speed can easily be measured from a graph of the power curve (Fig. 8.5), although calculating the power curve in the first place for a real bird is not trivial.

To calculate minimum-power and maximum-range speeds for a variety of birds ranging from swifts to swans, Thomas Alerstam used an approach developed by Colin Pennycuick, a prominent ornithologist and flight biologist (Chapter 4) [10]. Alerstam then used radar data on migrating birds to compare their actual migratory speeds with his predicted minimum-power and maximum-range speeds. All the birds except the largest species flew at speeds closer to the maximum-range than the minimum-power speeds [4]. Actually, the birds of pigeon size or smaller flew even faster than Alerstam's predicted maximum-range speeds by 8 to 16 percent, suggesting that Pennycuick's method underestimates maximum-range speeds for small birds. The only birds that flew substantially slower than their predicted maximum-range

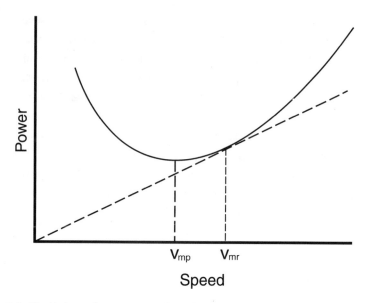

Figure 8.5. The U-shaped power curve for a flying animal, showing the points on the curve that give the minimum-power and maximum-range speeds (same as Figure 6.3, see Chapter 6 for details). (S.T.)

speeds were the whooper swans. Alerstam pointed out that these swans are so large that they may be physically incapable of flying fast enough to maintain their maximum-range speed. They thus compromise and fly at a speed between their minimum-power and maximum-range speeds. Perhaps the most telling aspect of Alerstam's study is that when swifts and sparrows forage or fly locally, they fly at speeds very close to their minimum-power speed of 5 or 6 meters per second, but when migrating, they fly almost twice as fast (9 to 11 meters per second). The migratory speeds are at or just above their predicted maximum-range speeds. For local flights, these birds fly at a speed that minimizes immediate power consumption (and maximizes flight duration), but during migration, they fly at quite a different speed so that they can travel the maximum distance on a given fuel load.

Fuel for Migration

The most portable form of stored energy that an animal can carry is fat. An animal can extract about twice as much energy from a gram of fat as from a gram of starch, sugar, or protein (Chapter 6). Thus most animals preparing for long migratory journeys build up a supply of body fat. The amount of stored fat can reach astonishing levels. Up to 50 percent of the body mass of a small bird setting out on a migratory journey is fat. This amount of fat is often described as a body fat proportion or body fat index of 0.5. Migrating monarch butterflies can have a dry body mass* of well over 50 percent fat, and migrating aphids can be up to 30 percent fat. Large birds generally carry a lower proportion of fat than small birds, partly because their energetic cost of locomotion is lower, but also because too much added weight can increase power requirements to prohibitive levels. In contrast, some very large birds may use tactics that help them avoid the need to build up large fat stores.

Range

Knowing how much fat a migratory animal carries, and knowing how fast the animal "burns" this fuel in flight, we can easily calculate how long the animal can fly before it needs to stop and eat. Small birds such as warblers and thrushes consume fat at about 0.7 percent of their body mass per hour, and larger birds such as ducks and geese consume fat at about 1 percent of body mass per hour. A warbler that starts out with a fat load of 50 percent of its total weight (body fat index of 0.5) should thus be able to fly for about 100 hours.† We know that the blackpoll warbler takes from 70 to 100 hours to fly over the Atlantic from New England to the northern coast of South America (Fig. 8.2). The rate of fat consumption is just an estimate for any individual bird, but if the rate is at all accurate, these warblers have very little reserve fuel for unexpected problems. Similarly, the knots that fly from Iceland to the Canadian Arctic start out with body fat indices of approximately 0.4, and they arrive at their destination with almost no body fat. In contrast, the ruby-throated hummingbirds that cross the Gulf of Mexico nonstop apparently carry enough fat for a flight almost twice as long as their normal crossing [4].

---

* Insects can have wildly varying levels of body water, so biologists normally subtract the mass of the body water before specifying the mass of other body components.

† The calculation of duration for a given amount of fat is more complex than simply multiplying the rate of energy use at the start of the trip by the amount of fat; as the bird flies, its weight continuously decreases, so its power and lift requirements change throughout the flight.

Refueling Stops

If a migrant's route of flight takes it over hospitable territory, and if it can afford the time, it might be better off making a series of short flights interspersed with feeding stops. This tactic reduces the need for increased eating and other physiological adjustments required to store large amounts of fat. The total energy cost of the trip may also be substantially reduced because the animal never has to "pay" for carrying a heavy load of fat with extra muscle power. Instead, the animal pays for this tactic with greatly increased migratory duration. In addition to two or three days to secure a feeding territory, birds that fly moderate distances between stops may require two or three days of feeding before they even begin to store significant fat—even birds that only fly 6 to 10 hours per day (or per night) still need to put on 10 to 15 percent fat before each series of flights. Altogether, a small bird could easily lose the better part of a week at each stopover. This heavy time penalty may explain why many birds, such as snow geese and European shorebirds, make long, nonstop flights over apparently hospitable territory. The shorebirds migrate along easily accessible coastlines, but they still begin their journey with a body fat index of 0.4 and fly most of the way nonstop [4].

Soaring

Many large birds migrate by thermal soaring. A soaring bird requires very little power compared to a flapping bird; soaring may allow large birds to migrate that would otherwise be incapable of long-distance flapping flight. These soaring migrants only fly on days when thermals are common, and their flights consist of a series of climbs in thermals followed by glides to the next thermal. When animals travel by soaring, their average ground speed depends on both the strength and the distance between thermals. If thermals are strong (i.e., have high vertical airspeeds), the animals climb rapidly and spend less time in the thermals. If the thermals are close together, the animal should fly between them at high speed to cover ground quickly; even in a steep glide, the animal will not lose much height in the short distance between thermals (Fig. 8.6). Alternatively, when thermals are widely separated, animals need to glide between them as efficiently as possible. In this case, the best speed is the speed that gives them their maximum lift-to-drag ratio,

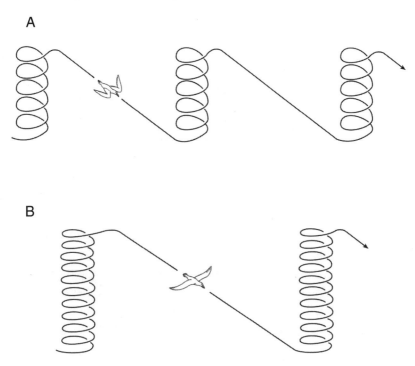

Figure 8.6. Cross-country travel in thermals. A. Strong, closely spaced thermals give high average ground speeds. The bird climbs faster in such thermals and can glide quickly (and steeply) between thermals because they are close together. B. Weak, more scattered thermals give low average ground speeds. The bird spends a long time climbing in each thermal and must glide slowly (at its maximum lift-to-drag ratio) to avoid losing too much altitude between them. (J.P.)

which will be relatively slow compared with flapping flight speeds and gliding speeds between closely spaced thermals.

Some birds, such as storks, eagles, and buteo hawks, are almost entirely dependent on soaring for long-distance migration. Other birds, such as the common crane, appear to use thermals opportunistically to reduce their overall flight costs. For example, some cranes migrate from the Mediterranean region to northern Scandinavia in the spring. Colin Pennycuick followed flocks of cranes in a small airplane as they flew over Europe. (This must have been quite a piloting challenge for Pennycuick, as his stall speed was much higher than the top speed of the cranes between thermals. He mentions flying on the edge of stall and making S-turns to keep from overrunning the birds.) His description of the behavior of one flock that he followed for several hours gives a graphic picture of the cranes' migratory tactics. In regions of strong thermals, the cranes flapped very little and stayed between 500 and 1300 m above the ground. When they entered an overcast area with weak thermals, the cranes continued to soar but gradually lost height to around 300 m. They then emerged from the overcast, found stronger thermals, and were able to work their way back up to 1300 m. Eventually, they flew into a region of weak thermal activity and began flapping. The cranes flew out of Pennycuick's sight at an altitude of about 150 m, flapping and flying at a speed slightly greater than their average cross-country soaring speed [11]. The migration routes of cranes often take them over long stretches of water (for example, the Baltic Sea) where they have no choice but to flap. Cranes appear to be large enough to take advantage of thermals but small enough to fly under power for long distances when necessary.

Pennycuick also studied white storks from a motorglider* as they migrated from Africa to Europe. Storks are somewhat larger than cranes and are obligate soaring migrants. In other words, they must soar in order to fly long distances. Pennycuick found that the average cross-country speed of migrating storks is about 43 kilometers per hour [12]. This is quite a respectable speed, especially considering that the storks probably spend only a little more energy soaring than they do when roosting. On the other hand, storks migrate about 7000 kilometers between European and African sites, which would take approximately 160 hours of flight time at 43 kilometers per hour. One hundred sixty hours would be almost one week nonstop, but soaring is not possible round the clock, because thermals only form during daylight hours when sunlight reaches the ground. Moreover, large birds require strong thermals, which may only occur during the hottest part of the day. Storks would thus be grounded by rain, fog, or overcast conditions, and maybe even during the early morning and late afternoon on sunny days. In reality, the storks' migratory journey takes several weeks. The honey-buzzard, a hawk that preys on the nests of bees and wasps, follows a very similar flight pattern and migratory route to the storks. Prior to the autumn migration, honey-buzzards increase in mass from about 600 to about 900 grams (giving them a body fat

* A motorglider is a modified sailplane with a small engine to boost it to thermalling altitude and to allow a powered return to an airport if no thermals are found. Pennycuick spent many hours thermal soaring along with the storks.

index of 0.33) [4]. Because of their specialized diets, honey-buzzards probably eat very little during their migration, but their fat stores would only give them enough energy to make about one-fourth of their trip by flapping. Soaring permits honey-buzzards to migrate and saves them a huge amount of energy, but, again, at a significant increase in travel time.

Some soaring migrants do not even bother putting on fat for their trips. Their migratory tactic is surely the epitome of living off the land while traversing hospitable territory. Old World common buzzards migrate over 10,000 kilometers between southern Africa and summering grounds in eastern Europe and western Siberia. These birds lay down very little extra fat before migrating, and so they must get almost all of their energy needs from hunting along the way [4]. By soaring, they minimize their energy (food) requirements while traveling. These hawks spend a leisurely two months or more on their migratory journey.

Modes of Migration

Classical Migration

Animals use at least three different patterns or modes of aerial migration: to-and-fro, or "classical," migration; multigeneration round-trip migration; and nonreturn migration. Classical migration, as used by many songbirds and waterfowl, is familiar to most people. In this mode, animals migrate between summer breeding grounds and more tropical overwintering sites. An individual normally makes one or more complete round-trips. We know from bird-banding studies that many species of songbirds migrate from North America to South America and back, or from northern Europe to Africa and back. Some birds return north to the same patch of forest, or even the same tree, to breed, year after year. Because most migratory birds spend the winter in tropical regions, where there are fewer biologists, we do not know as much about their winter behavior. They probably move around more, simply because overwintering birds are not tied to a nest in order to raise their young.

Some species of juncos and warblers migrate seasonally from high to low elevations in North America, rather than from north to south. Other birds, such as Arctic terns, migrate on a yearly, rather than a seasonal, schedule.

Some animals migrate entirely within the tropics. A species of kingfisher, for example, has populations that migrate into dryer regions to breed during the rainy season, and back to wetter areas during the dry season. Some African fruit bats make similar migrations, covering over 1000 kilometers to follow rainfall patterns [13]. Thus, classical migration is limited neither to north-south routes nor to birds.

Multigenerational Round-Trips

Seasonal migrations that require more than one generation to complete are another mode of migration. Among aerial migrants, this mode is used mainly by insects, and the best-known example is the monarch butterfly. When monarchs move north from Mexico to the north central and north-eastern United States in the spring and summer, they produce at least two new generations in the process. In late summer, monarchs start flying back toward overwintering sites in Mexico, and this southward leg of the journey is made by a single generation of butterflies (Fig. 8.1). This generation spends the winter at a small number of sites in the mountains of central Mexico. In the spring, they mate and head north, and the cycle repeats.

Many other insects make similar migrations. For example, like the monarch butterfly, the large milkweed bug, a relative of stink bugs, requires milkweed plants for food. The milkweed bugs thus migrate over almost the same route as the monarchs, timing their movements to arrive as milkweed plants become available. Unlike monarchs, milkweed bugs overwinter in the southern United States, but biologists do not yet know how many genera-tions occur during their migration. Many agricultural pests follow a similar pattern, again timing their movements to match crop availability. For ex-ample, fall armyworm moths move from the southern United States to the northeastern and north central United States as spring and summer crops be-come available, apparently with the help of south winds from spring and summer storm systems. The armyworm moths reproduce rapidly and build up enormous populations in agricultural areas. In late summer and early fall, some armyworm moths take advantage of seasonal winds from the north to head back south. The moths do not have as urgent a drive to fly south as birds and monarch butterflies, so every year a substantial number of moths wait too long and are killed by cold weather. The summer populations are so huge, however, that only a small percentage need to arrive successfully at over-wintering areas for the migratory habit to persist. Many aphid species follow a pattern nearly identical to the armyworm moth's.

A wide variety of insect species make multigenerational round-trip migra-tions. In most cases, biologists have not yet determined whether these in-sects follow the monarch or the armyworm pattern. Many species of moth, butterfly, and dragonfly fly north in the spring and south in the autumn, whether in North America or Europe. Many others, whose migratory pat-terns have yet to be discovered, undoubtedly do the reverse in the Southern Hemisphere. California ladybird beetles that overwinter in the mountains certainly produce more than one generation in lowland areas during the summer. If a species produces more than one generation during the breeding season (as is common among insects), then seasonal round-trip migrations must, by definition, occur across several generations.

Nonreturn Migration

The third mode of migration requires no well-defined round-trip. Animals usually make this type of migration in order to take advantage of temporary or ephemeral resources. Banded stilts are Australian wading birds common in salt marshes. These birds had been known to European biologists since the early 1800s, and they are common in southern Australia. Nevertheless, be-cause of the special breeding conditions that banded stilts need, they had not been observed breeding until 1930. These birds only breed when sporadic, heavy rains cause temporary lakes to form on salt pans in remote, arid re-gions. A population explosion of brine shrimp occurs in these salty lakes, and the brine shrimp provide food for the adult stilts and their chicks. The stilts migrate from their usual coastal salt marshes to these temporary lakes over several weeks. Huge numbers of stilts congregate at the lakes: about 80,000 birds nested at Lake Grace in 1930 when stilts were first seen breeding, and over 300,000 nested at Lake Barlee in 1980 (reviewed by Dingle [1]). Banded stilts have a very rapid reproductive cycle for birds as large as small egrets or ducks: they have been known to lay and hatch two sets of eggs in less than 12 weeks. Speed is important, because the young birds starve when the lake dries up unless they are fully fledged. These remarkable birds apparently go several years without reproducing. What triggers their migration and how they find the temporary lakes is unknown. Perhaps the birds can somehow tell which weather patterns are related to unusually large, rain-bearing frontal systems that will create the temporary lakes.

Migratory locusts, particularly the North African desert locust, seem to epitomize one-way, weather-driven migration [14]. (These are the same insects used by Torkel Weis-Fogh and Martin Jensen in their pioneering studies of flapping flight.) Migratory locusts have two forms: the *gregaria* or swarming form and the *solitaria* or nonswarming form (Fig. 8.7). The two forms, which occur because of a developmental switch based on crowding, are different enough in shape and color that they are often mistaken for two separate species. Juveniles that grow up in uncrowded conditions become solitaria adults, and juveniles from crowded conditions become gregaria adults. Gregaria locusts, commonly shown denuding the landscape in nature films, form the gigantic swarms of biblical fame that periodically plague northern

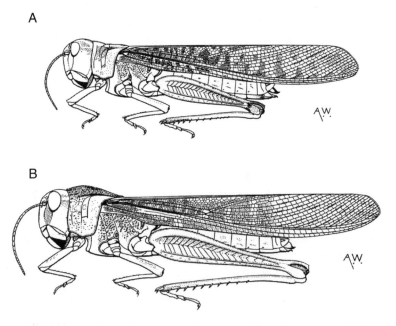

Figure 8.7. Migratory desert locusts can take on one of two forms as an adult. A. The *gregaria* form, which makes up the dense, massive, devastating swarms of locusts. B. The *solitaria* form, which does not swarm, and which typically lives at much lower densities than the gregaria form. Redrawn from V. M. Dirsh 1965 [20] with permission of Cambridge University Press. (S.T.)

Africa and the Middle East. High-flying swarms of gregaria locusts tend to move downwind. Within the swarm, locusts fly in random directions, but at the swarm's edges, locusts tend to fly back toward the center. This flight behavior maintains the cohesiveness of the swarm and allows it to drift downwind. The popular image of a swarm of locusts devouring everything in its path is accurate. Locust swarms actually tend to "roll" over the landscape like giant bulldozer treads. Flying members are blown downwind, eventually landing to eat. The swarm is blown overhead, and at the trailing edge, feeding locusts take off and rejoin the aerial part of the swarm.

For decades, biologists thought that these swarms were migratory mechanisms to move the swarms toward the Intertropical Convergence Zones, where trade winds bring the moisture that causes heavy rains [14]. Recently, biologists have discovered that locust migration is much more complex. A gregaria locust is not really behaving in true migratory fashion, because it actually spends more time on the ground feeding than in flight, and it does not make straight, undistracted flights. When biologists began using radar sensitive enough to detect individual insects, they discovered that solitaria locusts migrate, but they do it at night, and individually, rather than in swarms. In many cases, solitaria locusts move much farther than swarms of their gregaria brethren. Furthermore, solitaria locusts do not always migrate downwind; in some situations, they actually migrate upwind [15]. Even when moving largely downwind, solitaria locusts show some signs of orienting and navigating. Biologists now believe that the swarming behavior of the gregaria form evolved as a mechanism to exploit widespread but unpredictable favorable conditions rather than as true migratory behavior. In contrast, the solitaria form fits the criteria for migratory behavior quite well.

Some African armyworm moths also have gregaria and solitaria forms, much like the migratory locusts. In these moths, the gregaria forms seem to be doing most of their long-distance journeying away from crowded, overgrazed areas. The worse the crowding and competition among the armyworm caterpillars, the more likely the resulting adult moths are to fly long distances. These moths tend to spend the dry season near the coast or in high, moist areas. If the wet-season rains are abundant, the moths spread gradually inland without long migratory excursions because the rains produce plenty of new foliage to keep moth population densities from rising sharply. When the wet-season rains are sparse or sporadic, the armyworm

caterpillars become crowded into fewer, smaller patches of favorable vegetation. These conditions lead to the production of gregaria adults, which are
much more likely to fly long distances. At night, the moths take off and fly
downwind, carried inland by coastal winds. Scattered rain storms tend to
concentrate the moths and literally wash them out of the air at exactly the
location where new plant growth will soon appear. As the rainy season ends,
plants die back and armyworm populations plummet [1]. Biologists do not
yet know how or how many of the inland armyworm moths make it back to
their dry-season refuges. One thing is certain: a significant number must return or natural selection would eliminate the migratory behavior of the species. The armyworm pattern may be similar to that of the fall armyworms
and aphids in the United States, mentioned earlier, where a combination of
gigantic populations and some random (or oriented or wind-borne) flight ensures that some migrants or their descendents will make it back to the coastal
dry-season refuges.

Migration Distances

Most people tend to think of migration as a phenomenon of very long distance movement. When migration is defined as a particular set of behaviors,
however, migration can include journeys on a wide range of scales. As mentioned earlier, migrations can be as short as a fraction of a kilometer or as long
as halfway around the globe.

Short Distances

Insects, not surprisingly, are the main flying migrants at the short end of
the distance scale. Probably the shortest known migration by a flying animal
is made by Eurasian milkweed bugs on an island in the Baltic Sea (Fig. 8.8).
In the fall, these milkweed bugs fly from patches of milkweed to sheltered
sites in stone outcrops and old stone buildings. They wait out the winter in
these sheltered sites. In late spring, the bugs leave their shelters and fly back
to milkweed patches. At the milkweeds, the bugs mate, lay eggs, and produce
a new generation, which completes the cycle. These milkweed bugs migrate
from a few hundred meters to a kilometer or two [16], probably a typical distance for many insect migrations. For example, many aphid individuals migrate over similar distances, but a difference between them and the Eurasian

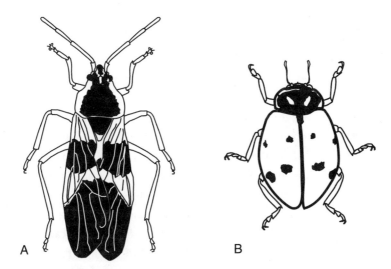

Figure 8.8. Two examples of migratory insects. A. A short-distance migrant, the small Eurasian milkweed bug. B. A medium-distance migrant, the convergent ladybird beetle. (S.T.)

milkweed bug is that the milkweed bugs only produce one generation per summer, while aphids may go through dozens. The aphids' rapid generation time means that if many aphid generations include migratory individuals, the species as a whole can migrate hundreds of kilometers even though any individual may only fly a few kilometers.

Moderate Distances

Migratory flight distances of approximately 100 kilometers seem to be rare, or at least rarely described. The California convergent ladybird beetle is one well-known example (Fig. 8.8). These beetles migrate from mountains to agricultural valleys in the spring and back in the fall, roughly 100 to 200 kilometers each way [17]. Even though the distance is modest, an inadvertent experiment just before World War I showed that this is a true migration. The beetles and their larvae are voracious predators of agricultural pests, so ladybird beetles are welcomed by farmers. Enterprising entomologists discovered that the beetles could easily be collected from their overwintering sites. The entomologists collected millions of adults, kept them cool until spring, and then released the beetles into orchards and fields. Rather than staying put and eating aphids, the beetles responded to their internal clocks and flew off

to the west, just as they would if they had started from their mountain re-
treats. No one knows whether they flew out to sea and drowned or dispersed
north and south upon reaching the Pacific coast. It took several attempts before the people involved realized that the beetles are programmed to fly west
upon being warmed up in the spring, and bringing them down from the
mountains to release them was actually counterproductive [17].

Among birds, a few small songbird species migrate similar distances to the
convergent ladybird beetles. In North America, some juncos migrate between high-elevation and low-elevation sites in mountainous regions, flying
distances of a few dozen to perhaps 100 kilometers. When the same species
of junco lives in lowland habitats, it follows typical north-south migratory
routes, covering much longer distances. Some North American species of
warblers do likewise, migrating short distances in mountainous areas and
long distances in lower areas. Long-tailed chickadees probably belong in the
short-distance category: they migrate two or three kilometers in elevation in
the Rocky Mountains, and perhaps four or five times that distance horizontally [18]. At least three Himalayan warbler species nest at elevations above
2100 meters, but overwinter in the foothills. Again, these migratory distances are on the order of 100 kilometers or less [19]. As these examples
demonstrate, medium-distance migrations from high elevations to low elevations can have the same effect as long-distance migrations over large
changes in latitude.

Though biologists have put less effort into studying bat migration than
bird migration, they have observed a few bat species migrating. For example,
many small New England cave-dwelling bats disperse during the summer but
migrate in the fall back to a small number of caves. Most of these bats fly well
under 100 kilometers, and 300 kilometers was the longest distance any of
them migrated [19].

Long Distances

Flight permits animals to migrate spectacular distances, and many animals take advantage of this ability. Recall such examples as the snow geese
that fly 2700 kilometers nonstop over eastern North America (Fig. 8.1), songbirds and shorebirds that fly more than 3000 kilometers between New England and eastern Canada and the northeast coast of South America (Fig.
8.2), and cuckoos that migrate 3500 kilometers between New Zealand and

Samoa. The Manx shearwater migrates about 10,000 kilometers between Great Britain and Brazil. It makes this journey nonstop, using dynamic soaring much of the way to keep energy costs as low as possible. Aside from birds, a species of solitary, forest-roosting bat summers in New York State and winters 1300 to 1400 kilometers away in South Carolina and Georgia. Some African bats migrate similar distances between the wet and dry seasons. Long-distance migration is not limited to birds and bats: aside from the 3000-kilometer migration of monarch butterflies, many other insects migrate in large numbers both in North America and Europe. In his book on insect migration, C. B. Williams describes mass movements of many species of butterflies and dragonflies, but no one has traced the exact routes and distances of these insect migrations as yet [17]. Some may just be moving a few dozen kilometers, but some of the others may make journeys every bit as spectacular as the monarch's.

The long nonstop bird migrations are, by any standards, amazing, but some birds that stop to eat and rest along the way migrate much longer total distances. Colin Pennycuick's white storks (Fig. 8.9) fly about 7000 kilometers from Europe around the Mediterranean to east and west Africa, and the same distance back. Two species of waders, the ruff and the curlew sandpiper, summer in the Siberian Arctic and winter in Africa (Fig. 8.9). The ruff winters

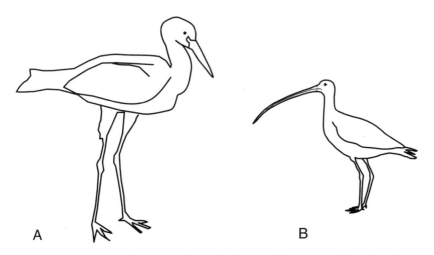

A   B

Figure 8.9. Examples of birds that migrate very long distances. A. Stork. B. Curlew, a type of sandpiper. (S.T.)

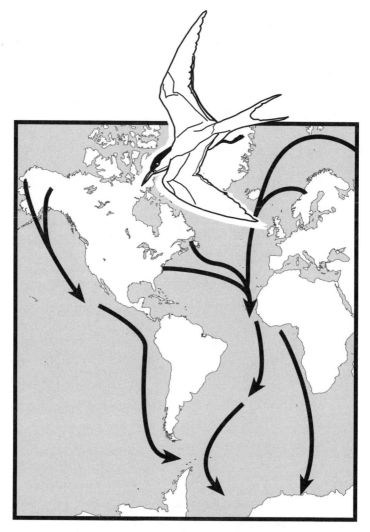

Figure 8.10. An Arctic tern and a map of its migration routes. (S.T. and J.P.)

in Senegal, West Africa, about 10,000 kilometers from its breeding area. The curlew sandpiper has two wintering grounds, in Mauritania (8500 kilometers one way) and near Cape Town, South Africa (over 13,000 kilometers from Siberia) [4]. Then there is the common buzzard, which makes a leisurely 10,000-kilometer migration between eastern Europe and southern Africa.

The Ultimate Long-Distance Migration

The unquestioned champion of long-distance migration is the Arctic tern, a small seabird (Fig. 8.10). Arctic terns breed along the far northern coasts of North America, Europe, and Asia and on Arctic islands. In the fall, terns fly south, generally following the western edges of continents (Fig. 8.10). Their destination is the ice off Antarctica, mostly south of Africa. In March, terns head back north. They follow approximately the same route as in the fall, but in reverse. Not all terns make the 20,000-kilometer journey each fall and each spring; many juvenile birds do not return north their first year. These juveniles spend over a year feeding off the southern continents or Antarctica before heading back north to their breeding areas. Thus the first round-trip is a multiyear journey for many, if not most, terns.

Arctic terns could not possibly migrate nonstop: they do not weigh much more than a pigeon, and the distance is simply too great. In order to minimize travel time, they could make the trip in several nonstop legs with feeding stops between each leg. Though their entire route is not known, terns do fly from Greenland and eastern Canada to Europe largely nonstop (about 3000 kilometers). Their migration routes along the western edges of the continents take them over some of the most nutrient-rich areas of the oceans, so they probably combine a few long nonstop flights with legs where they feed as they go. No matter how they do it, the ability of these small birds to make a 20,000-kilometer trip in the fall and another in the spring is almost beyond comprehension. In proportion to body length, for instance, each trip would be equivalent to a person bicycling all the way around the globe four or five times.

One Last Question

Flight clearly allows animals to travel great distances. Now the question remains: how does a migrating animal know where to go? A migratory animal must be able to find its way or navigate; that is the subject of the next chapter.

NINE

Finding the Way

*A*ny flying animal needs to have some way to figure out what direction to fly in order to reach its destination. Many features in the environment give directional information, and when an animal responds to one of these features by flying in a particular direction, biologists call this *orientation*. Animal flyers as well as nonflyers use a wide variety of orienting cues, including the position or movement of the sun or stars, the earth's magnetic field, odors, or remembered landmarks. Biologists have studied just about every directional cue, and one or more kinds of animal seem to respond to every cue they have studied. Moreover, most animals probably use more than one cue, depending on circumstances. Perhaps an animal might use the sun's position on clear days and the earth's magnetic field on cloudy days. Any given animal usually has one preferred cue, but falls back on one or two other cues if the preferred one is not available. For example, honey bees prefer to use the direction of the sun as their main orienting cue, but they resort to remembered landmarks on cloudy days [1].

When a bee travels from its hive in search of nectar, it may fly a couple kilometers or more, tens or hundreds of times farther than the food-gathering trips of a mouse on foot. Few people would be surprised by the mouse's ability to return to its nest over a few dozen meters, but the bee's ability to return home after such a long flight is remarkable. Even the bee's navigational ability pales in comparison with that of such long-distance migrants as monarch butterflies, North American songbirds, or Arctic terns (Chapter 8). The navigational feats performed by these migrants over thousands (or even tens of thousands) of kilometers are little short of astounding. How do these animals find their goals?

Orientation ability is critical, but it is only part of the story. Given that an animal can determine east using the sun or north using the magnetic field, how does the animal actually use this directional information to figure out the direction to its destination? Consider human navigation: if I am taken to some unfamiliar location and given a compass, I can easily determine which direction is north, but I cannot find my way home with the compass alone. I would need some sort of navigational system in addition, such as a map with my present location marked on it. *Navigation* in the broadest sense, then, means using various types of orientation to travel to some destination.*

Short-Range Navigation

Foraging—searching for food—is when flying animals most commonly use short-range navigation. The search for food may take a bee a kilometer or more from its hive. A heron might make foraging flights of several kilometers from its nest. Returning home requires short-range navigation: travel over distances too long to see, smell, or hear home directly, but short enough that the animal can make more than one round-trip per day. All mobile animals—walkers, swimmers, and flyers—must use short-range navigation to

---

* Specialists sometimes define *true navigation* to mean using indirect cues to travel over an unfamiliar route to a destination. While this narrow definition has some technical advantages, it excludes the use of remembered landmarks for orientation, which most flying animals surely use—at least to some degree—on almost every flight. I thus use *navigation* in the broad sense in this chapter.

some extent. But the great foraging distances covered by flyers makes the problem of finding home a much greater challenge for them.

Landmarks and Maps

Pilotage, or traveling from one remembered landmark to another, is an important component of short-range navigation. Indeed, the two favorite orientation subjects of scientists, homing pigeons and honeybees, have a keen ability to memorize terrain features in the vicinity of their loft or hive. They use these remembered landmarks to quickly locate home once they enter familiar territory [1, 2, 3]. Animals build up their "route maps" by memorizing significant landmarks either during specific flights for that purpose, during flights to and from feeding sites, or incidentally as they fly around the local area. Whether animals actually combine these route memories into a general mental or "cognitive" map of an area is controversial, and the question has triggered many studies attempting to demonstrate cognitive maps [4]. So far, biologists have found behavioral evidence for cognitive maps in birds, but the evidence is weak for such maps in insects. Instead, insects memorize a sequence of landmarks for a given route, but they do not combine knowledge from two different routes to figure out shortcuts [5].

Although animals often use landmarks for orientation in the immediate vicinity of their home (nest, hive, loft), landmark orientation becomes less and less practical the farther an animal flies from home. First, as the animal gets far from home, it is more likely to fly over unfamiliar territory—places it has never been before. Obviously, a honeybee cannot have any stored memories of landmarks for a patch of clover it is visiting for the first time. Second, there is a limit on how much landmark information animals can store in their memories. Biologists do not yet understand how long-term memories are stored, but as any student can confirm, repetition and familiarity are powerful memory tools. Thus, flying animals are most likely to remember landmarks that they see frequently—in other words, near home. The complexity of a given animal's nervous system and its lifespan also play roles. A worker honeybee only spends a few weeks in the adult (flying) stage, and it may spend this whole period within three or four kilometers of its hive. It will probably be familiar with many landmarks within 100 meters of its hive, but it will only recognize landmarks along specific routes when it flies more than a kilometer from the hive. In contrast, most birds live several years; some nest in the same place year after year, and some even live continuously in the same place for several years. A five-year-old crow who has lived in the same woodlot for several years may be intimately familiar with landmarks within several kilometers of its nest. Yet in the case of either the bee or the crow, the farther the animal gets from home, the more likely it is to fly over unfamiliar territory, and the less likely it is to use landmarks for orientation. When animals cannot see familiar landmarks, they must rely on other navigation methods; some are the same as long-range methods, but at least one—inertial navigation—is especially well suited to short-range navigation.

Inertial Navigation

Inertial navigation means remembering turns and speeds traveled in order to reconstruct the path from some starting point. In its simplest form, animals can use it to retrace the outbound course in reverse. Most animals that use inertial navigation, however, do something more sophisticated: they integrate turns and distances to keep track of the straight-line direction home at all times. This more sophisticated use of inertial navigation is called *path integration*. Most people are fairly good at path integration. For example, tell a person the direction of north, take him or her on a tour of a building, and then ask what direction north is. Most of the time, the person will be able to give a reasonably accurate answer (particularly if the turns are all sharp and right-angled). This ability is quite widespread throughout the animal kingdom, even in animals with simple nervous systems like ants.

Given its pervasiveness, surprisingly little research has been done on inertial navigation in flying animals. Homing pigeons appear to use it while they are being transported to unfamiliar release sites. Whether other flying animals use it to any great extent is as yet an open question. On the one hand, inertial navigation does not require any specific directional cue. In other words, animals do not need to see the sun or stars, or sense the earth's magnetic field, in order to use inertial navigation. On the other hand, it may be poorly suited to animal flight. Inertial navigation requires accurate speed or distance measurements, which may be difficult for animals to make as they fly at different heights. Moreover, inertial navigation in animals does not take crosswinds into account. These two sources of error may make iner-

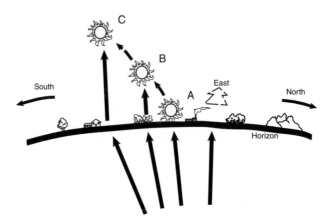

Figure 9.1. The sun's azimuth is the direction to the point on the horizon directly below the sun. The apparent path of the sun is not directly across the sky from east to west, but shifted to the north or south depending on the season and hemisphere. In the Northern Hemisphere winter, as shown here, the sun's azimuth shifts to the south as it rises. The sun is almost due east shortly after sunrise (A); an hour later is a bit south of due east (B), and an hour after that, it is even farther south (C). (S.T.)

tial navigation too unreliable for flying animals.* Note that neither of these factors would affect walkers: a terrestrial animal can get a very good distance measurement just from paying attention to how much effort it is putting into walking, and crosswinds rarely influence the paths of ground-bound animals.

Sun Compasses

Many birds and insects make use of a *sun compass*. The sun compass works by projecting the sun's location in the sky onto the horizon. An animal compares this location with the time of day to determine directions, such as east in the early morning and west in the late afternoon. The sun's direction projected onto the horizon—its *azimuth*—changes slowly near dawn (sun's azimuth roughly east) and dusk (sun's azimuth roughly west) but more rapidly in the middle of the day (Fig. 9.1). The details of this change depend on the

---

* Humans use inertial navigation systems in many types of vehicles—submarines and spacecraft, for example. Even though these systems use much more precise sensors than those of any animal, they still tend to drift after long periods of use. Thus, they must be reset periodically using external cues: typically stars, landmarks, or satellites.

season and the latitude at the animal's location. A sun compass is thus useful during much of the day only if an animal has some sense of the time of day.

Most animals (including humans) have a daily cycle of physiological processes, or *circadian rhythm*. Animals that use the sun compass have such an accurate circadian rhythm that they have what amounts to an internal clock. For instance, honeybees can learn the daily pattern of the sun's motion at their location and time of year.* They combine this knowledge with an uncannily accurate internal clock to give them a handy directional cue. In fact, the sun compass appears to be their preferred directional cue for middle-distance navigation [4, 6, 7]. Homing pigeons also use a sun compass; older, experienced pigeons rely heavily on sun compasses on clear days [8]. Day-flying migratory species certainly could use a sun compass, but so far, only starlings and one or two other species are known to use them [9, 10].

Magnetic Compasses

Many animals, including some that do not fly, can sense the earth's magnetic field; they literally have an internal magnetic compass. A magnetic compass is simpler than a sun compass because no time compensation is needed: the direction of the magnetic field is constant over the course of a day, and it does not change appreciably with seasons. Thus, animals using a magnetic compass do not need an internal clock. When the first experiments over thirty years ago suggested that some birds could sense the earth's magnetic field [11, 12], many scientists were skeptical. No one had imagined that animals might have a sensing ability completely unknown in humans. Over time, however, more and more studies confirmed the ability of birds to use a magnetic compass for orienting. Homing pigeons use it for homing, and some migrants use it for long-range navigation if the sun or stars are not visible. Among insects, some migrating moths and walking ants also use magnetic compasses [13, 14]. Surprisingly, honeybees seem to be able to sense the magnetic field, but they do not use it for navigation even when other cues

---

* Insects can detect the sun's direction even when the sun itself is not visible. Unlike humans, insects can detect the polarization of light. They use a pattern of polarization in the sky called the *e-vector,* which is linked to the sun's azimuth. Insects can see the e-vector even in a small patch of blue sky through clouds, and they can use this to determine the sun's direction.

are not available [1, 15]. Although biologists have overwhelming behavioral evidence for magnetic orientation in a variety of animals, no one has been able to find a receptor organ for this sense in any flying animal, despite years of searching. Magnetite, a magnetic mineral, has been found in the brains of some animals [16], so some biologists have suggested that the brain itself (or some part) senses the magnetic field in these animals. So far, however, few researchers have tested this idea with experiments.

Magnetic compasses can actually be used in a couple of different ways. The ordinary compass, familiar to most people as a navigation instrument, works horizontally. When the user holds the compass horizontally, the needle points toward the magnetic North Pole. However, the lines of the earth's magnetic field are not actually horizontal except near the equator, and surveyors and geologists sometimes use a special vertical compass, called an *inclination* or *dip compass,* which shows the angle of the magnetic field from the vertical. A dip compass points downward at an angle toward the magnetic North Pole in the Northern Hemisphere, but it reverses in the Southern Hemisphere, where it points down in the direction of the magnetic South Pole. Birds seem to use a dip compass rather than a horizontal compass. The dip compass seems less useful than a conventional compass, because it does not work near the equator, and it changes directions as an animal flies from the Northern to the Southern Hemisphere (as do many migrants). However, the dip compass tells a bird the direction of the equator in either hemisphere (Fig. 9.2). For birds that spend the winter in one hemisphere and the summer in the other, a dip compass could simplify orientation at the beginning of migration: in either case, they would start their journey in the "equatorward" direction.

Middle-Distance Navigation

In addition to their skills at short-range navigation, homing pigeons can also navigate over longer distances. Homing pigeons can be released dozens or hundreds of kilometers from their home loft and still manage to find their way home—hence, the term *homing.* Other flying animals may have this homing ability to some degree, but homing pigeons have been bred for this specific ability for centuries, so they may have the most highly developed middle-distance navigation ability among flying animals. Moreover, because

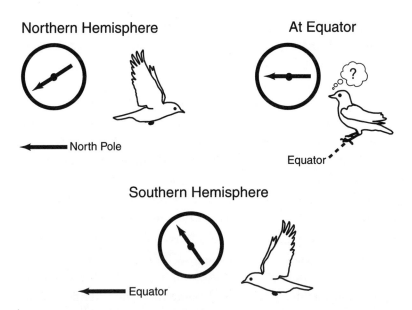

Figure 9.2. In the Northern Hemisphere, a dip compass points down toward the North Pole, and in the Southern Hemisphere, it points up toward the equator. Birds can only sense the direction in which one end of the dip compass needle points down; in other words, with bird's internal compasses, they can sense the direction to the nearest pole. Since the needle of the dip compass is horizontal close to the equator, birds' internal compasses give no directional information near the equator. (S.T.)

homing pigeons are domesticated, they are especially handy for studying navigational behavior.

Homing pigeons can be hauled hundreds of kilometers to some unfamiliar location and released, and they can find their way home. This ability almost defies belief. Pigeons appear to use as many cues as possible (sun and magnetic compasses, inertial navigation) to figure out their route on the outbound trip. They then use these same cues, along with some new ones, to find their way home. Pigeons rely mainly on their sun compass on clear days and their magnetic compass on cloudy days. If they can track their outbound path well enough, they can use path integration to figure out the direction to home. Then, using either a sun or magnetic compass, they fly toward home until they encounter familiar landmarks. They use these landmarks in the vicinity of their goal to guide the last stage of their trip.

What happens if pigeons are transported without being able to sense their outbound route? They can be carried in windowless cages, surrounded by electromagnetic coils to confuse their magnetic sense, and along a route with lots of turns and circles to confuse any inertial abilities. When researchers do this, they still find that a substantial number of birds arrive back at home. A number of theories have been proposed to explain this ability. Some are based on sensing two directional properties at angles to each other, which could provide unique position information analogous to latitude and longitude. These schemes have yet to be demonstrated in pigeons, and in any case, they do not always provide unambiguous position information. One rather controversial suggestion is that pigeons use *olfactory* navigation: they "smell" their way home. Some researchers have found that if they block pigeons' ability to smell, the birds have a harder time finding home. The behavioral evidence in favor of olfactory homing seems convincing. For example, birds with impaired senses of smell headed away from the release site in random directions, while normal pigeons started out in roughly the home direction, and fewer nonsmelling birds arrived home than their normal loft mates. This mechanism, however, remains controversial for two reasons. First, researchers in the United States and Germany have had difficulty replicating the results of the original studies in Italy. Second, and more fundamentally, meteorologists and atmospheric chemists vigorously deny that coherent, detectable odor gradients or trails can exist over dozens of kilometers in a normal atmosphere, let alone the hundreds of kilometers over which pigeons can home. More research is needed to reconcile the birds' apparent abilities with the physics of the atmosphere.

Long-Distance Navigation: Migration

The ability of migratory birds and insects to fly thousands of kilometers to relatively precise locations certainly ranks as among the more astonishing behaviors in the animal kingdom. Biologists have thus put a great deal of effort into studying migratory navigation. At least in the case of birds, the magnetic compass plays a significant role, but they use many other directional cues as well.

At the time a given type of bird normally begins its migration, it exhibits a behavior called "migratory restlessness." Biologists noticed long ago that individual caged birds huddle by the wall in the direction that they normally migrate. Several researchers have developed methods to measure this behav-

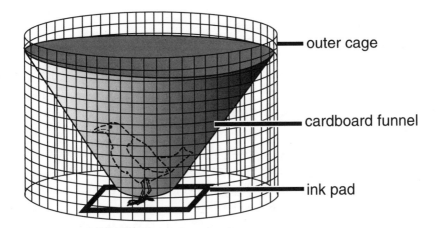

Figure 9.3. Biologists use Emlen funnels to measure birds' tendencies to orient in a particular direction. The Emlen funnel consists of an inverted, truncated cone of cardboard in a circular cage. The bird stands at the bottom of the cone on an ink pad. Every time the bird hops up onto the wall of the funnel, it leaves an ink mark. The researcher can count the number of ink marks in each direction and use these to calculate the bird's preferred direction of hopping. (S.T.)

ior. For example, F. W. Merkel and colleagues built octagonal cages with perches near each wall; the perches were attached to electrical switches connected to a counting device. Thus, the number of times a bird used a given perch was automatically counted. Stephen Emlen developed the simplest and most reliable cage of all: a circular cage lined with a shallow funnel of heavy paper, and with an inkpad on the floor (Fig. 9.3).* When the bird hops up on the funnel attempting to leave the cage, it leaves a mark on the funnel. The experimenter can count the marks on the funnel in different directions to determine the bird's desired direction [17]. Emlen used his cages to show that indigo buntings normally orient by the pattern of stars in the night sky. Other researchers, however, using other types of birds, found that even young, naïve birds reared in windowless rooms are still able to orient in the appropriate migratory direction. When biologists rotate the magnetic field around the birds 90 degrees with electromagnetic coils, the birds also change the direction of their orientation by 90 degrees [9]. Thus, at least some birds can use magnetic compasses for migratory orientation.

---

* Today, these are often called *Emlen cages.*

Magnetic versus Celestial Compasses

Birds—at least the songbirds studied so far—seem to use the magnetic field to learn the direction of north early in development. Later in development, birds often use their internal clocks to learn the pattern of star movement around the North Star (Polaris) throughout the night. Although some birds develop their magnetic sense before other orientation abilities, and this sense is vital during their development, the magnetic field has shortcomings as an orientation tool: it does not point exactly north at most places on the planet. In fact, birds might benefit if they could realign or recalibrate their magnetic compass with their celestial (sun or star) compass as they move through different regions (just as aviators and ship navigators do). Some birds appear to do just that, as they stop at staging areas on long migrations. Strangely, in one study, Savannah sparrows seemed to use their celestial compass to calibrate their magnetic compass under some conditions, but they calibrated their celestial compass with their magnetic compass under other conditions [9]. The age and rearing conditions of the bird, as well as the environmental conditions encountered by a migrating bird, all affect which compass calibrates which.

Many birds travel nonstop for several days at a time, and others migrate mainly at night. During daytime flights, these birds rely mostly on sun compasses, and at night they use star compasses. Even though they can sense the magnetic field, many birds use it only when they cannot see a clear sky. Some migratory birds actually stay grounded when the sky is overcast. If they can sense the magnetic field, why not use it for navigation? The answer may lie in the directional accuracy of the cues. Polaris (the "North Star") is exactly in the north, and a sun compass can be just as accurate when adjusted for season and latitude. The magnetic field is not as precise. Because the magnetic North Pole is hundreds of kilometers from the "true" (geographic) North Pole, in most places on the globe, the magnetic field does not point directly at true north. There are only very restricted longitudes where magnetic and true north are aligned; in other places, the magnetic variation tells how far off magnetic north is from true north. At many longitudes—western North America, for example—magnetic north can be 15° or 20° off from true north (Fig. 9.4). Moreover, local magnetic anomalies are common; a large deposit of iron ore can change the direction of the magnetic field by several degrees.

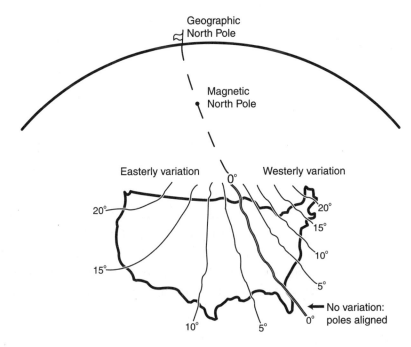

Figure 9.4. Magnetic variation across different regions of the United States. Note that the magnetic and geographic poles are aligned along the line of zero variation. West of that line, compasses point to the east of true north (easterly variation) and east of that line, compasses point to the west of true north (westerly variation). The number of degrees for each contour line gives the amount of variation. (S.T.)

Over short distances, the magnetic deviation is not a big problem. However, a 10° error over a journey of 1000 kilometers introduces a navigation error of almost 200 kilometers! This kind of error could mean the difference between making landfall on an island or peninsula, or missing it altogether, or arriving at a destination so far from the desired location that the bird could never find it. Clearly, if celestial cues are available, they are much more accurate for guiding long-distance travel.

Given that many birds—even those that are normally active during the day—migrate at night, and many others fly around the clock, most migratory birds need to be able to use the stars for navigation. How do they develop this ability? Young birds learn the patterns of sun and star movements at a very specific period during development. If a bird is prevented from seeing the sky during this critical period, it is never able to learn to use celestial navigation. Clever experiments with birds reared in planetariums showed that these birds have an innate ability to determine which star is the North Star—they watch the night sky and learn which star appears to stand still while other stars rotate around it. They then memorize a few general patterns of stars (not unlike our constellations), which allow them to easily locate the North Star. Thus birds do not need to memorize a detailed map of all the stars; they just need to learn enough of the general pattern of stars to be able to locate Polaris, and use that as a cue to indicate the direction of north. Once they learn to find the North Star, they no longer need to pay attention to the *movements* of the stars, so they do not need an internal clock to find north. If birds are reared in a planetarium where the stars appear to rotate about some other star, these birds treat this new star as the North Star, and in the autumn, they try to migrate away from it [18, 19]. Any given bird species has its own particular preferred migratory direction—relative to the North Star—and these directions seem to be genetically programmed into their nervous systems.

Vector Navigation (Dead Reckoning)

Compasses, whether based on the sun, stars, or the magnetic field, are useful tools for orientation. However, more is needed for navigation. At the very minimum, the traveler must know a time or distance as well as a direction. For short and medium-range trips, such *dead reckoning* or vector navigation may be sufficient: "If I fly east for 85 minutes at a comfortable cruising speed, I shall be able to see the lake adjacent to my nest site." Rather than using time, many animals estimate distance by watching objects on the ground as they appear to move past the flyer. Although this method does not require an accurate internal clock, variations in altitude affect the visual "flow" of objects on the ground and can thus add errors to distance estimates.

Crosswinds and Wind Drift

For long trips, crosswinds pose a serious problem. After an 85-minute flight, a bird with a crosswind of 10 kilometers per hour—a fairly gentle breeze—at right angles to its heading could easily be about 14 kilometers off course (Fig. 9.5). This amount of error would clearly be unacceptable on a flight of 1000 or 2000 kilometers. If the animal flies over land, it could compensate for crosswinds in a couple ways. It could monitor its ground track (by

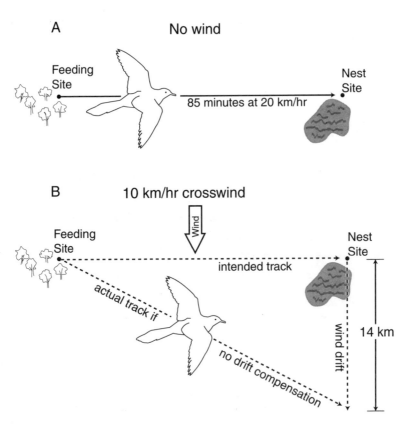

Figure 9.5. Effects of crosswinds. A. With no wind, a bird flies a known time and speed to return to its nest site from its feeding site. B. In a crosswind, if the bird maintains the same compass heading that it used with no crosswind, it will be blown off course. To compensate for the crosswind, the bird would need to fly a bit to the left of the direction back to the nest site. (S.T.)

paying attention to the direction of visual flow), and adjust its heading in the air so that its ground track is in the proper direction. A simpler approach is just to note whether objects on the ground appear to drift right or left as you fly over them, and change your heading slightly into the direction of drift. With either approach, the animal must be able to see the ground. Determining whether a flying bird is compensating for a crosswind (especially if the researcher does not know the animal's destination) is extremely challenging. For example, is the bird in Fig. 9.5B attempting to fly straight east, or does it actually *intend* to fly southeast? Flocks of migrating birds show up on some kinds of radar, but researchers usually cannot tell the kind of bird or the destination from this radar information. Statistical techniques comparing directions of the birds with wind directions can suggest whether birds are compensating for crosswinds, but these techniques do not work if birds seek out tailwinds (by flying at particular altitudes or times of day). When scientists can combine radar tracking data with other kinds of observations of the same birds, they find that some kinds of birds do have the ability to compensate partly or completely for crosswinds [20].

In contrast, some birds fly fixed compass headings, regardless of wind. When thrushes were fitted with tiny radio-tracking transmitters that signaled the heading of the birds (the direction they actually faced, as opposed to their ground track), the thrushes showed no sign of compensating for crosswinds. However, for long-distance migrants that encounter variable winds, the most economical strategy turns out to be to ignore the winds in the early part of the journey and then correct for winds late in the trip. The radio-tracked thrushes may not have been at the stage where correction was important. Even with its limits, radar data suggest that at least some birds do compensate for crosswinds, and some can even do it without being able to see the ground. Although researchers have several theories to explain how birds might tell they are being blown sideways when they cannot see the ground, testing these theories will be extremely challenging [20]. The easiest way to deal with crosswinds, however, is to avoid them, and some migrating birds appear to choose cruising altitudes or times of day with tailwinds or very light crosswinds [20, 21].

Complex Migration Routes

Long-distance navigation by most flying migrants is a good bit more complex than simply flying in a given direction for a given distance, and then doing the opposite to return. For example, most birds fly different routes in the fall and spring. Wheatears, members of the thrush family, spend the summer in Greenland and the winter in western Africa. They stop over in southern France and Spain on the way south in the fall, but they stop over in England and northern France on the northward trip in the spring. They seem to be taking advantage of tailwinds and typical movements of weather fronts during different seasons [22]. Even more impressive, albatrosses routinely fly thousands of kilometers over the open ocean, yet they find their way back to small, remote islands to nest. Moreover, we know from satellite tracking data that albatrosses often fly to these islands in a relatively straight line over many hundreds of kilometers of open ocean [23]. Again, how they locate these tiny islands and approach them while compensating for crosswinds over open water is largely unknown.

Biologists know a bit more about the navigation used by some long-distance migrants. Whooping cranes, for example, migrate from central Canada to the Texas coast each fall and back each spring. Because of their endangered status, many of these birds have been carefully radio-tracked and observed from airplanes [22, 24]. They follow a 100- to 200-kilometer-wide corridor that deviates noticeably from a direct compass course. Part of the northern corridor is west of the direct line, and over half of the southern corridor is east of the direct line (Fig. 9.6). Moreover, the fall and spring corridors are not entirely the same. The spring corridor is narrower and does not completely overlap the fall corridor at the southern end. The cranes' route keeps them over land less than 1000 meters in elevation. Also, the cranes clearly follow landmarks, especially those like ridges and rivers that may be almost (but not quite) aligned with the cranes' direct course. In such long-lived birds, remembered landmarks surely help them find their way over this long route.

Innate Programming or Learning from Experience?

Many migrating songbird species appear to be genetically programmed with some direction and time (vector) navigation information. Naïve birds tend to follow this genetic program, but experienced birds that have already flown the route can be more flexible. In a classic experiment, biologists captured thousands of European starlings part way through their southerly fall migration. The trapped birds were carried hundreds of kilometers to the east and then released. The naïve young birds continued on their original southerly heading, but the older, experienced birds adjusted their heading to the west to compensate for the displacement [25]. Other bird species have responded similarly in displacement experiments.* In some species, young

* Orley Taylor of the University of Kansas has seen similar compensation for displacement in migrating monarch butterflies, according to his preliminary, as yet unpublished results.

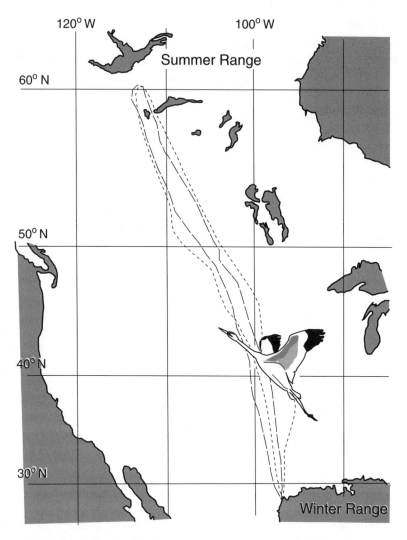

Figure 9.6. Map of the whooping crane migration corridor. The boundaries of the fall, southward migration corridor are dotted, and the boundaries of the spring, northward corridor are dashed. Redrawn from T. Alerstam 1996 [22], by permission of the Company of Biologists Ltd. (S.T.)

birds fly the first migration on their own and depend almost entirely on vector navigation. In other species, most young birds follow their parents or migrate in flocks and memorize some route information along the way; the genetic program acts as a general guide, emergency backup, and probably as a motivating drive. Learning the details of a route may be crucial for those species where their migration routes have turns or doglegs, or where they must deviate to avoid obstacles. Amazingly, however, even some of these complex routes seem to be genetically programmed in some species. Scientists tested one type of European warbler in orientation cages, and found that young birds oriented in one direction for a few days, and then shifted slightly later during the migratory period. The warblers' preferred directions matched the directions of a detour that their migration route takes around a mountain range [26].

Exactly how experienced birds compensate for displacement is not yet known, but some scientists speculate that the birds use large-scale landmarks—mountain ranges, coastlines—to correct large-scale course deviations. Also, the closer the birds get to their goal, the more detailed their memories of landmarks become. This landmark navigation near the destination is especially important for animals that return to specific nest sites year after year.

Navigation: An Incomplete Picture

What is our current state of knowledge of animal navigation over long distances? Animals can orient using a variety of cues: sun, stars, magnetic field, large- and small-scale landmarks. In many cases, they use genetically programmed vector navigation information (direction and distance) as either a primary or a backup system. Long-distance flyers often supplement vector navigation by adjusting for wind drift and detouring around obstacles. They often follow large landscape features such as coasts or rivers (even if they are only approximately in the right direction), and they use the same stopover places from year to year. Some migratory birds can return to the same patch of woods—or even the same tree—after round-trip journeys of thousands of kilometers. Many mysteries remain, however, and the fresh ideas and clever experiments of the next generation of researchers will be needed to solve some of them. How do migrating animals sense magnetic fields, and how do they know to fly south in the fall, rather than east or north? Just how do they correct for wind drift (crosswinds), especially over open oceans, and can they really sense the wind direction without being able to see the ground? How do they select their routes when they follow different paths in the fall and spring? How do first-time, naïve migrants know when they have reached their goal if they do not follow experienced leaders or fly in flocks? These are just a few of the major unanswered questions waiting to be tackled by students interested in studying animal navigation.

TEN

The Global Impact of Animal Flight

*F*lying animals have had profound effects on the living world. Flying is such a potent innovation that flying animals have become extremely successful, and in many ways dominate their nonflying relatives. Moreover, without flying animals, the vegetation on earth would look very different today. Thus both animal and plant communities owe much of their present form to the presence of flying animals. Humans are not immune to these effects, either. Flying animals have important and pervasive influences on humans, affecting everything from our health to the food we eat.

Flying Animals and the Natural World

Dispersal

Flight is both a fast and an energy-efficient way to travel (Chapter 6), so flying animals have an advantage for spreading into new territory. Consider the virgin land of islands, freshly formed by volcanoes or changes in sea level. The first animals to arrive are the flyers. When the volcanoes on Krakatau Island erupted in 1883, the blast was so powerful that it killed all living things on the remnant of Krakatau and on two nearby islands. Biologists surveyed the island group in 1908 and found that 92 percent of the animals that had recolonized the islands had got there by air. Most were insects, most of the rest were birds, and a couple were bats. Several spiders were there as well; although spiders cannot fly, they do travel aerially by *ballooning* on the wind with a long thread of silk, much like dandelion or thistle seeds [1]. At first, flyers may just stop to rest in a freshly denuded disaster area like Krakatau, but as vegetation becomes established, animals also soon become long-term or permanent residents. The first transient visits of flying animals usually help reestablish vegetation: their droppings are a source of nutrients for new soil and contain seeds that colonize the new land. On Krakatau, floating seeds or fruit gave rise to most of the new plants, but figs, the dominant large forest trees, arrived as seeds in the droppings of fruit-eating birds and bats. On another volcanic island, Motmot, ducks were present within a year of its eruption, and almost all the vegetation on the island is derived from seeds carried by animals [1]. Flying animals are also the first to arrive in continental areas newly denuded by forest fires, floods, volcanoes, or other disasters. As soon as food is available (if not before), flying animals will find and visit such regions. Within weeks after Mount St. Helens erupted in Washington State, researchers commented on the presence of abundant insects, including flies, wasps, and hordes of mosquitoes [2].

Of course, flying animals do not require virgin, empty territory for dispersal. Flight permits these animals to move into a new region, exploit a food source quickly, and then move on when the food is gone. The plagues of locusts mentioned in the Bible represent an extreme of this tactic (Chapter 8), but many other insects and some birds and bats use this same general method on a more scattered, less intense basis. The key ability to pick up and move long distances, as well as being able to search large areas quickly, gives flyers an enormous advantage over walkers. Moreover, generalist flyers—those that can prosper in a wide range of habitats—can spread rapidly over large areas, and specialists can find desirable habitats over wider regions than their ground-bound competitors. Thus the dispersal ability that flight confers on a species has many benefits that pay off in the game of evolution.

The dramatic spread of African honeybees in the Western Hemisphere provides a lesson in the effectiveness of the dispersing ability of a flying animal. In 1956, to stimulate beekeeping in hot areas, W. E. Kerr introduced a variety of honeybee from South Africa into Brazil that was distinct from the European bees kept by most beekeepers. Some African bees escaped into the wild and established feral populations. The African bees are more aggressive than their European cousins, and they spread throughout South America, replacing both feral and cultivated populations of honeybees.* They moved through Latin America at over 300 kilometers per year. By the 1970s, African bees had reached Central America, and by the early 1990s, they began to spread across the border into the United States. The African bees do not survive cold winters, so entomologists do not expect them to spread north of Texas.† Forty-five years is, however, a staggeringly brief time for an animal species to spread over a continent and a half. Biologists and paleontologists are accustomed to thinking of this kind of invasion taking centuries, or even millennia, but the African bees could spread much faster because they could fly [3].

Aside from dispersing themselves, flying animals play a role in helping other organisms disperse. Such "passengers" are surely unintentional, perhaps even undesirable, from the flyer's point of view. Flying animals do not voluntarily carry a different species around with the intent of turning the passenger loose unharmed at the end of the flight. Nevertheless, many flying animals do inadvertently or unwillingly carry passengers, enabling the latter species to disperse more widely. Passengers fit into two general categories, those that become attached to their flying hosts accidentally and those that

* African bees have been referred to as "killer bees" because of their aggressiveness: unlike European honeybees, when one African bee stings an intruder, its hive mates are stimulated to swarm and sting the intruder. Even though African bees are smaller than European bees, they can be deadly when dozens or even hundreds are aroused to attack an intruder. African bees' fierce reputation is exaggerated, however: they are kept commercially in South Africa and Brazil, and the term *killer bees* may actually have been coined by spokesmen of the military government of Brazil, which Kerr had outspokenly opposed.

† Originally, biologists assumed the African bees were hybridizing with local bees, and described them as "Africanized" bees. Genetic evidence, however, shows that the African variety completely replaces local bees, presumably by outcompeting them. Thus, genetically, they are almost pure descendents of the original African bees.

attach intentionally. Accidental passengers are those that somehow become attached to a flying animal fortuitously or by chance—fungal spores that stick to the feet of a fly walking on rotting fruit, or seeds trapped in the feathers of a bird after it takes a dust bath. Intentional passengers are those that have specific adaptations that help them find and attach to the correct flying host. Many intentional passengers are benign; they simply latch onto their host to get a free ride. Biologists call this hitchhiking process *phoresy* and describe the freeloaders as *phoretic*.

One of the more entertaining examples of benign hitchhikers concerns fish in farm ponds. Throughout rural parts of the midwestern United States, farmers build small ponds in almost every pasture and meadow to provide water for livestock. Even though these ponds are not connected to streams or rivers, if they contain water year-round, they will inevitably contain fish in a year or so. How do the fish get there? Every angler or biologist of whom I ask this question replies that fish arrive as eggs on the feet of ducks. I have not found any hard evidence supporting this assertion, but it does seem plausible. Minnows and sunfish cannot travel overland, but they reliably arrive in isolated ponds. Sticky eggs could surely be carried by any animal that swims in a pond with fish and then climbs out and proceeds to swim in a fishless pond. Walking animals could do the job, but fish eggs would probably not survive long periods out of the water, so they require a fast-moving carrier. What animal is more likely to fly quickly from one pond to another than a duck? Actually, any swimming or wading bird would suffice, and I suspect geese and herons may transport fish eggs just as well as ducks.

Many intentional passengers are benign hitchhikers, but others are unwelcome pests. For example, beekeepers in North America recently discovered that their hives were becoming infested with a mite in the genus *Varroa*. These mites suck the bees' body fluids and eventually kill the bees. The mites are probably spread when an infested bee visits a flower, dropping off some mites, and a bee from another hive subsequently visits the flower and contracts them. These mites spread very rapidly because of the large foraging areas covered by bees.*

---

* Controlling the mites is a challenge: some pesticides can kill the mites without seriously harming the bees, but may contaminate the honey, making it unfit for human consumption.

Not all passengers carried by flying animals are other animals. Most fruits (in the culinary, not the botanical sense) evolved as dispersal mechanisms. Plants evolved nutritious outer coverings to attract animals, which feed on the covering and then deposit the seed some distance from the parent plant. Small, brightly colored fruits, such as cherries and berries (blackberries, blueberries, raspberries, etc.) evolved to attract birds, which eat them seeds and all. The seeds pass through the birds' digestive systems, which takes time, ranging from about half an hour to a couple of hours. When the birds defecate, they sometimes deposit the seeds far from the parent plant. The seeds of some plants will not germinate until they have passed through the digestive tract of an animal. Plant breeders often have to treat wild plant seeds with acids to get them to germinate. Junipers, called "red cedars" in North America, show the effectiveness of this dispersal mechanism. In many rural areas in the United States, rows of juniper trees grow along fence lines around meadows, pastures, and farm fields. Junipers produce berries that some birds eat. The birds then fly to fences or telephone wires at the edges of meadows or fields where they perch, rest, and defecate, depositing seeds. I recently found two juniper seedlings under an oak tree in front of my house, and the nearest juniper is at least a couple of hundred meters away.

In the tropics, bats also disperse seeds, although bats tend more to eat the flesh of the fruit, sucking the juice and spitting out the pulp and seeds. Quite a few tropical fruits that people now grow agriculturally evolved originally as bat-dispersed fruits, including mangoes, papayas, guavas, bananas and avocados [4]. Bat-dispersed fruits tend to be green or yellow, rather than brightly colored, because bats, unlike birds, are color-blind [5]. Birds are only effective dispersers if they ingest the seeds, because they usually eat fruit where they find it; in contrast, bats tend to pick fruit, and then fly to perches to eat it, where they drop the seeds [4].

Pollination

Pollination is the most biologically and ecologically significant interaction between flying animals and plants. Plants use pollen grains containing sperm cells to transfer sperm from one plant to another for sexual reproduction, and many plants depend on animals to move their pollen among plants. Flying animals have obvious advantages for transferring pollen, and almost all pollinating animals are flyers. People tend to associate bees with flowers and pollination, and bees are largely specialized as pollinators. But many other animals are important pollinators, including a variety of insects and vertebrates, which we shall encounter shortly.

The flowering plants, or *angiosperms,* are by far the most common and successful plants today. If a plant lives on land and is not a conifer, fern, or moss, it is almost certainly an angiosperm. Many even live in the water. All angiosperms have flowers, although the flowers of an oak tree or a grass may only be obvious to a botanist. Their early fossil record is murky, but angiosperms apparently first appeared in the early Cretaceous (over 150 million years ago). They began to flourish during the Cretaceous, and by the end of the Cretaceous (65 million years ago), they were becoming the dominant land plants. We know from the abundant and varied flower fossils at the beginning of the Eocene (60 million years ago) that angiosperms were already quite diverse by that time.* These ancient flower fossils have characteristics that suggest they were insect-pollinated [6]. Their anatomy is strikingly similar to that of modern flowers specialized for particular pollinators, such as bilaterally symmetrical flowers like snapdragons and larkspurs, favored by large bees, and tubular flowers like morning glories, favored by lepidopterans (moths and butterflies). Moreover, the origins of bees—the preeminent pollination specialists—seem to overlap the very period when angiosperms began their great diversification (Chapter 7).

Scientists believe that the association between pollinators and plants probably started much earlier, when some insects evolved the habit of eating pollen or seeds. Like today's conifers, the ancestors of angiosperms were wind-pollinated. Insects that ate pollen or seeds would have inadvertently picked up pollen on their bodies and carried it to other plants, where it fertilized those flowers. For some plants, this additional pollen transfer was beneficial enough to compensate for the loss of pollen eaten by the insect. The benefit must have been at least partly due to the insects' ability to move long distances by flying. From the insects' perspective, their small size allowed them to visit small, wind-pollinated flowers and get significant nutritional benefits from tiny

---

* In everyday language, people often use *flower* to refer to the whole plant. I use *flower* in the strict sense, meaning the reproductive structures of a plant that produce pollen, ovules, or both. If pollinated, the flower develops into a fruit with seeds.

pollen grains. Some plants evolved to produce extra pollen as an attractant to insects, and nectar similarly evolved as a "reward" for visiting insects. Eventually, plants evolved obvious, showy flowers to guide insect pollinators to nectar and pollen sites [6]. The better the insects became at transferring pollen, the more they were rewarded by the plants, so ancestral bees evolved specializations like hairy bodies to carry more pollen. Getting insects to help in reproduction apparently gave early angiosperms a great competitive advantage: almost as soon as recognizable flowers appear in the fossil record, they are quite diverse. As they diversified, they acquired many new pollinators. All butterflies and most of their moth relatives are pollinators. Hummingbirds of the New World are the most familiar vertebrate pollinators, but ecologically similar birds have evolved the nectar-drinking, pollen-carrying habit in other regions, including honeyguides and sunbirds. At night, in addition to moths, quite a few bat species visit flowers for nectar, thus carrying pollen. Saguaro cactus and balsa trees are both pollinated by bats [7, 8].

Today, angiosperms have evolved a variety of reproductive tactics. Some can "self" or use their own pollen to fertilize their flowers, such as jimsonweed and violets. Others, like grasses and some trees, have secondarily evolved wind pollination, even though their ancestors were probably insect-pollinated. Nevertheless, the vast majority of angiosperms produce flowers that depend to some extent on animal pollinators, and, as noted earlier, almost all pollinators are flyers. Moreover, the early insect pollinators clearly helped the early angiosperms diversify. Insect pollination gave them such a competitive advantage that they replaced other formerly important plant groups, such as horsetails and ferns, and became the dominant type of land vegetation.

Success and Diversity

Biologists define *success* in a rather special way: successful groups of organisms are both diverse and abundant. Diverse groups include many species, and abundant groups have many individuals in some or most species. The extant groups of flying animals are unusually successful by these criteria.* We

* Pterosaurs may have been equally successful, but the fossil record is not complete enough to be sure.

cannot say for sure that flight is what makes these groups so successful. Perhaps they would have been just as successful if they had not evolved flight. However, the advantages of flight are obvious enough that most biologists believe that flight contributed significantly to the success of these groups, and may be the dominant factor.

Insects: By any measure, insects are the most successful organisms on earth. Scientists have named approximately one million insect species and estimate that as many as ten to fifty times that number remain undescribed. There are more scientifically named species of insects than all other named organisms—plants and animals—combined. (In contrast, biologists have named only 50,000 extant species of vertebrates, and because vertebrates are much larger and more apparent, most are surely known to science.) Biologists have named more species of true bugs alone (stink bugs, seed bugs, cicadas, aphids, leafhoppers and relatives) than all vertebrates combined. Some of the really large groups include the dipterans (true flies), with over 120,000 named species, and the ultimate diversity champions, the beetles, which include over 300,000 named species [9]. Moreover, scientists believe that these numbers represent a vast undercount, because insects are much more diverse in the tropics, which have been little surveyed at the level of detail necessary to find tiny animals like most insects.

Insects are numerous as well as diverse. Ants alone make up nearly one-third of the total weight of land animals in Amazon rain forests, and termites are not far behind. Although worker ants and termites are wingless, the reproductive individuals—queens and drones (ants) or kings (termites)—fly during the dispersal phase of their life cycles. Insects live in all terrestrial and freshwater habitats; few are marine, probably because crustaceans already filled the arthropod niche in the ocean before insects evolved. Because of their small size, insects can get into all sorts of places, and because of their short lives and rapid life cycles they can evolve rapidly if environmental conditions change. They are also hardy: for example, many insects are much less susceptible to ionizing radiation (radioactivity) than plants and other animals.

In earlier chapters (Chapters 3, 6, and 7), we looked at the advantages of flight. Is flight itself enough of an advantage to explain the success of insects? The primitively wingless insects, those that evolved from insect ancestors before flight evolved, provide at least a partial answer.* Proturans and diplurans are so tiny and rare that they have no common names, and neither group includes more than 200 species. Colembollans, or springtails, can be locally abundant, but they are tiny (one or two millimeters long) and not at all diverse, with only 2000 species. Silverfish and bristletails are bigger, and the former can be pests in buildings, so they may be a bit more familiar; even so, they include fewer than 1000 species. All other insects—the other 99.8 percent—are pterygote (winged) insects. This is compelling circumstantial evidence, although it alone cannot prove that insects are successful because of flight.

Vertebrates: Although we cannot say much about pterosaur success, birds and bats both stand out among vertebrates as unusually successful air-breathers. Scientists have named about 22,000 species of tetrapods (land vertebrates). With over 9000 species, birds make up nearly half of this total. Bats are the second largest group among the mammals, including about 1000 of the 4200 known mammal species (only rodents are more diverse, with about 1800 species) [10, 11].

If flight helped produce the diversity of birds and bats, it led to an almost contradictory pattern: although birds and bats are evolutionarily diverse, they show little variation structurally or anatomically. Do not be misled by the dazzling variation in colors, shapes, and patterns of bird plumage: beneath their plumage, and aside from plumage-related features, birds are all quite similar. Consider the anatomical difference between nonflying mammals such as a horse, a gorilla, and a seal. All are mammals, yet they show striking differences in limb and skull anatomy and overall body form. The mechanical requirements of flight have prevented vertebrate flyers from evolving such gross variations in body form. All birds have streamlined bodies, hindlimbs for walking or perching, forelimbs used as wings, a short, rigid vertebral column between the shoulders and hips, and a light-weight skull with a beak. Any major departure from this general "bird" arrangement would compromise the bird's ability to fly. The flight constraints are relaxed in those birds that have lost the ability to fly, such as ostriches and emus.

---

* Some insects are secondarily flightless, that is, they evolved from flying ancestors. The most numerous and best-known of these are parasitic groups such as fleas and lice.

They are not particularly streamlined, and their wings are vestigial, but they otherwise still show the imprint of their flying ancestors on much of their anatomy.

Likewise, bats have evolved a large number of species (for mammals, at any rate) while maintaining a great deal of structural similarity. Again, the structural requirements of the wing and flight muscles have constrained bats to a generally uniform body plan. Their proportions may differ somewhat, and facial adornments (used for echolocation) show wide and striking variations, but the underlying body structure does not change much from one bat species to the next.

If the anatomy of birds and bats is constrained so heavily, how is it that flying insects vary so dramatically in form? Part of the answer lies in the wing apparatus of insects: their wings are completely independent of their other appendages, thus allowing the other appendages (e.g., legs and antennae) to evolve separately. Another part of the answer is their small size. Insects operate in a Reynolds number range where streamlining is much less effective than for vertebrates (Chapter 2), so large variations in body form have little aerodynamic effect. Finally, insects have been evolving much longer than birds and bats. Insects have had three or four times as long to evolve new variations that still can be aerodynamically successful.

The Effect of Flying Animals on Humans

Flying animals have had enormous effects on human life, which continue today. Human agriculture has been influenced, some might even say controlled, in both helpful and harmful ways by flying animals. Other flying animals have dramatic effects on human health. Indeed, some of these flying animals limit the ability of people to prosper in particular regions of the world.

Agriculture: Beneficial Effects

Flying insects provide some of the most important beneficial services to farmers. Two crucial services they provide are pollination and regulation of pests. The three primary staple crops of humans—wheat, rice, and corn—are wind-pollinated and do not require animal pollinators. Almost all other fruits and vegetables, however, are animal-pollinated, and most of these are pollinated by bees. In some places, native wild bee species are abundant enough to provide the needed pollination for crops. Elsewhere, native bees may be rare or may not visit cultivated plants, so farmers arrange with bee-keepers to place hives of honeybees near their crops. Among the many crops that depend heavily on bees are some citrus fruits, blackcurrents, squash and pumpkins, blueberries, cranberries, strawberries, blackberries, apples, pears, plums, peaches, cherries, passionfruit, and forage crops like alfalfa and clover. Many crops can *self-pollinate,* or "self," where flowers use some of their own pollen to produce seeds, while others produce little or no fruit by selfing. Even those crops that can self, such as cotton, grapes, and coffee, usually improve their yields quite a bit when pollinated by bees. True flies are often minor pollinators of "bee-pollinated" crops, but they are the primary pollinators of cashews and cacao [12].* Hummingbirds and lepidopterans are important pollinators of many wild plants but not, apparently, of any crop plants. Finally, as noted earlier, bats appear to be the main pollinators of balsa trees [8].

Many predatory animals assist farmers by feeding on herbivorous insects, mites, and nematode worms that consume crop plants. Without predators, the only check on the numbers of many herbivores would be food: the herbivores would eat until no plants were left, wild or cultivated. Insect-eating birds and bats help to control larger herbivorous insects such as grasshoppers and moths; predatory insects help keep the smaller herbivores under control. Many insects eat other insects, including praying mantises, dragonflies, lacewings, and beetles. For example, both the larvae and adults of ladybird beetles are voracious predators of aphids. Since aphids are some of the most injurious crop pests, ladybird beetles are bred and sold commercially to combat aphid infestations. The beetles' ability to disperse themselves over an area by flying is one of the traits that makes them so valuable for controlling aphids.

Ladybird beetles are not particularly choosy about which aphids they eat, and they sometimes end up attacking insects other than the ones a farmer might want to control. For this reason, scientists developing biological con-

---

* The identity of the pollinators of cacao, from which we get chocolate, has been the source of much controversy. Recent studies point to tiny gnatlike flies called midges as the most important pollinators [8].

trol methods often turn to parasitoids. Parasitoids are insects whose larvae live inside the larvae of another species. The parasitoid larvae allow the host to live until the parasitoids are ready to emerge as adults, at which point they usually kill the host. Some of these parasitoids are very host-specific, meaning that they only attack one type of host, which is very useful for controlling pest species. Most parasitoids are tiny flies or wasps, and several are available to farmers to use against herbivorous insects. For example, *Encarsia formosa* is a tiny parasitoid wasp that growers often use in greenhouses to combat whiteflies, a type of aphid.

In some cases, the herbivorous insect, rather than the predator, is the one that helps the farmer. Weeds compete with crop plants and can reduce crop yields. When farmers use chemical insecticides, they sometimes discover that weeds become more of a problem because the insecticide kills the insects that eat the weeds. This can be even more of a problem where weeds have been introduced to new areas accidentally, so that they are no longer held in check by their natural enemies. In several such cases, agricultural researchers found herbivorous insects from the home region of the introduced plant, and introduced the herbivore in turn. For example, people accidentally introduced a prickly pear cactus into Australia, which became a major nuisance. In the 1920s, several insects were introduced to try to control the cactus, without much success until a cactus moth from South America was reared in great numbers and released in the late 1920s. By 1932, the cactus population was greatly reduced [13]. Again, the ability of the moth to fly for dispersal was key to its ability to quickly control the cactus.

Agriculture: Origins of Fruits

Most, if not all of the items in the fruit section of the grocery probably evolved as dispersal devices to attract birds or bats. In most cases, our cultivated versions are much larger than their wild ancestors. Some, such as blueberries or raspberries, are still relatively small, although cultivated forms tend to be a bit larger than wild berries. Cultivated cherries and strawberries are substantially larger than the wild fruit, and apples and related fruit (pears, plums, peaches) are positively huge compared to their wild relatives. A robin or sparrow could easily eat or carry off a crabapple, but a fruit of that small size is inefficient for humans to harvest. People have thus selected for larger fruit size over hundreds or thousands of years, to the point where a robin cannot eat or carry a modern apple. People have selected for large size in bat-dispersed fruits as well, although the wild progenitors may have been larger to start with. Fruit-eating bats tend to be among the largest of bats, but even a flying fox might have difficulty carrying off a modern cultivated mango, banana, or papaya.

Figs clearly evolved as a seed-dispersal mechanism, probably to attract bats, and they also have a wonderfully intricate pollination system. Fig flowers are tiny and contained on the inside of a hollow structure, the receptacle, that only opens to the outside through a tiny passage. Females of a special type of tiny wasp, fig-wasps, crawl into the hollow receptacle and lay eggs. The fig plant nourishes the larvae, which grow to adulthood entirely within the receptacle. Males are wingless and do not leave the receptacle. They mate with females that have grown up within the same receptacle, which then leave and fly to other receptacles to lay eggs, transferring pollen in the process. After the flowers are pollinated, the receptacle develops into the fig. Many cultivated varieties of figs form fruit without pollination, but some of the most highly prized types will not set fruit unless they have been visited by the fig-wasp. However, these cultivated figs do not contain the special modified flower needed to nourish the wasp larvae (probably because people would rather not eat wasp larvae in their figs). So fig growers must plant wild figs (*caprifigs*) in and around their orchards to provide a source for pollinating fig-wasps [14].

Agriculture: Harmful Effects

Some of the most harmful interactions between flying animals and humans are in agriculture. Insects play the major role, which is not surprising: humans are by far the most abundant large animals, and insects are the most abundant tiny animals, so competition and conflict between these groups is inevitable. In agricultural terms, a *pest* is any animal that interferes with crop production, and many of the most dreaded pests are flying animals.

The damage that herbivorous insects do to crops is hard to overestimate. Insect pests reduce harvests by roughly one-third; in other words, humans could probably harvest about 50 percent more if all insect pests suddenly disappeared [9]. The estimated loss to insects of vegetable crops alone in the United States was about $200 million in the early 1960s, and the situation has improved little since then. Today, total worldwide agricultural losses to insects may be as much as $300 billion, in spite of the application of $20 billion worth of insecticides [15]. Even after harvest, we lose 5 to 10 percent of our stored food products to insects.

Every crop species has more than one major insect pest, and many important crops have numerous problem insects. Usually, one pest is dominant in a field at one time, but two or more pests may attack simultaneously or in sequence. Moreover, insects can attack all parts of a plant, and in many ways. As a final insult, many insect pests also transmit crop diseases.

Wheat is the most important food crop in the world, and wheat has a staggering array of pests. For example, wire worms (moth caterpillars) and white grubs (beetle larvae) eat the roots and underground parts of the stems; Hessian fly maggots and greenbugs (a type of aphid) suck fluids from the stems; cereal leaf beetles eat the leaves, and weevils eat the ripening seeds. Grasshoppers and armyworms (caterpillars) eat just about all parts of the plant above ground [16]. Usually, insect larvae do most of the damage to plants, but adult insects find plants and lay eggs on them. The boll weevil demonstrates the devastating speed that flight confers on the dispersal ability of pests. Boll weevils are tiny beetles that attack cotton plants. Both larvae and adults feed on flower buds and prevent them from forming cotton fibers (or ruin the fibers that do form). Boll weevils entered the United States from Mexico in about 1892. Spreading from 65 to 250 kilometers a year, by 1922, they had infested most of the southeastern United States. This region was then essentially a one-crop agricultural economy, which the weevils devastated [17]. From 1909 to 1949, farmers lost an average of over $200 million annually to boll weevil damage, which bankrupted suppliers, shopkeepers, and bankers as well as farmers.* In this way, one insect species dramatically changed the economy of a large region of the United States.

For sheer dispersal and rapid colonizing ability, aphids are the champions. These insects are too small to fly against the wind, so they essentially drift along with weather fronts. Many species overwinter along the Gulf

---

* In an interesting twist, the citizens of Enterprise, Alabama, actually built a monument of appreciation to boll weevils. The near-collapse of cotton farming forced farmers to diversify into a variety of crops and livestock. This stabilized the economy and allowed small farmers to prosper in a region formerly dominated by large plantation owners.

Coast, the far southwest of the United States, or Mexico. They reinvade the agricultural regions of the Midwest and Middle Atlantic states every summer (Chapter 8). Aphid populations can grow and disperse so rapidly because of their peculiar life history. They only reproduce sexually in the fall, and the offspring of the sexual generation are all female. During the spring and summer, the females reproduce parthenogenically: they produce genetically identical copies of themselves without mating. Moreover, they do so via live birth rather than by laying eggs [18]. Since the females have no need to copulate for reproduction, their embryos develop very quickly. In some species of aphids, even before the young are born they have developing embryos in *their* reproductive tracts. This reproductive arrangement means that females can generate offspring at a stunning rate, who in turn begin reproducing immediately. Early in the spring, the offspring are wingless, but as the aphids become more crowded, new individuals are born with wings. The winged individuals disperse to new areas, often many kilometers from the parent's location. Moreover, a single individual can start a new population. When she finds a good food source, a dispersing female can immediately begin parthenogenic reproduction with no need to find a male first. This combination of fast, parthenogenic reproduction and flying dispersal allows aphids to quickly spread over many hundreds of kilometers every growing season.

Insects also transmit crop diseases, often with devastating effect. Aphids, with their piercing-sucking mouthparts, resemble tiny animated hypodermic needles, and they transmit numerous viral diseases. These include the viruses that cause squash mosaic disease, bean mosaic disease, and potato leafroll. Thrips, even smaller sucking insects, transmit tomato spotted wilt virus. Striped cucumber beetles are the vectors for a wilt disease of cucumbers (and their relatives—gourds, pumpkins, etc.) caused by bacteria. Many fungal diseases are transmitted by wind-blown spores, but even some of these are vectored by insects, such as yeast spot of lima beans, which is transmitted by stink bugs [19].

Food crops are not the only important plants affected by insects. For example, Dutch elm disease is a fungal disease of elms, carried by a bark beetle that tunnels under the bark. The disease was introduced into the United States along with some Old World elm trees in the 1930s, and within 20 years, it had wiped out most of the native elms [20]. Before the disease was introduced, American elms were by far the most common trees planted in U.S. cities. These trees were prized for their stately appearance and rapid growth, so the elms in most cities were essentially monocultures. Bark beetles spread the disease by flying from infected to healthy trees, and the disease spread rapidly across cities and throughout forests. Over time, many cities replaced American elms with more resistant Asian and European elms. Unfortunately, in the 1950s, a more virulent strain of the fungus appeared that proceeded not only to wipe out the few remaining native elms but to devastate the imported ones as well.

The Impact of Flying Animals on Human Health

Quite a few types of insects feed on vertebrate blood. Most of the time, the bites of these insects are merely annoying or irritating, but not medically important. In rare situations, however, biting insects can become so numerous that their bites pose actual medical or veterinary problems. But biting insects in small numbers can become medically significant when they act as carriers, or "vectors," of disease-causing organisms. Indeed, flying animals can have staggering effects on humans when they transmit diseases. Malaria is the top public health problem worldwide (although AIDS has surpassed malaria's impact in parts of Africa). Malaria is caused by microscopic protozoans that are transmitted by mosquitoes. In the 1980s, public health officials estimated that about 800 million people suffered from malaria worldwide, and about 1,500,000 people died from malaria each year [21]. The parasite has a complex life cycle, including asexual and sexual reproduction phases, and is just as much a disease for the mosquitoes as for humans (because it interferes with feeding, an infected mosquito bites more victims than normal, further spreading the disease). Malaria has several forms, caused by different species of the parasite. Most are "benign," causing a chronic, nonfatal disease, but one "malignant" form causes a more acute disease, which is sometimes fatal. Even the "benign" forms can be fatal for children, and sufferers often remain listless and drained of energy. Moreover, fatigue can trigger relapses, so people fearful of relapses often become lethargic and idle [22].

At the beginning of the twentieth century, malaria was common in the southeastern United States, particularly along the Atlantic coast and the Mississippi River valley.* It even occurred sporadically north of the Ohio River. Public health experts believe that as many as three or four million people in the United States contracted malaria annually during the Depression [23]. During World War II, military officials realized that malaria was nearly as great a threat as enemy soldiers, so they initiated major efforts to control the disease. (The laboratory that eventually grew into the world-famous Centers for Disease Control and Prevention, or CDC, in Atlanta, Georgia, started out as a World War II malaria research center.) By the end of the war, researchers had developed an effective control strategy, which they used successfully overseas. After the war, federal and state health authorities decided to apply this strategy at home in the United States too. Part of the strategy required using persistent pesticides like DDT to eliminate mosquitoes in and around dwellings. This eventually led to other problems, which we shall look at later in this chapter. Other components of the control strategy ranged from basic measures such as screening windows to sophisticated new anti-malarial drugs. By the late 1950s, the strategy was largely successful. Almost all malaria in the United States since that time has occurred in, or been spread by, returning military personnel (particularly from Korea and Vietnam) and immigrants and travelers from places where the disease is endemic [24].

Mosquitoes carry quite a few other diseases, and some are or were major health problems in particular areas. These include the yellow fever and dengue viruses and one of the nematodes that causes filariasis. Many epidemics of yellow fever have occurred throughout the subtropics and tropics of the New World, and it was such a scourge that it almost defeated the construction of the Panama Canal. People fear yellow fever more than malaria, because the former is more frequently fatal: up to 50 percent of severely infected yellow fever patients die from a combination of liver, kidney, and heart failure [25].

Many other insects besides mosquitoes suck blood, and several carry diseases [26]. In Central and South America, cone-nosed bugs—true bugs, relatives of the predatory bugs called assassin bugs—carry a trypanosome parasite that causes Chagas' disease. Cone-nosed bugs are nocturnal, so their

---

* My grandfather contracted malaria while in army training in the southern United States during World War I.

victims are normally asleep. They are also called "kissing bugs" because they bite sleepers where they are uncovered, usually on the face, most often around the lips and eyes. Chagas' disease is particularly insidious because its initial symptoms (skin swellings) are quite mild. It lies dormant for years and then attacks the heart and certain nerves. Some people never come down with the later symptoms, but many people die of heart or digestive system failure brought on by the disease [27]. Although much less notorious than malaria, Chagas' disease is a major problem: roughly 20 million people contract it each year and as many as 60,000 die from it per year. In Bolivia alone, 2500 children die from the disease every year [21, 27].

Although tsetse flies and the sleeping sickness they transmit are probably better known, a disease called "river blindness" (*onchocerciasis*) is a much greater public health problem than sleeping sickness in Africa. River blindness is caused by a nematode parasite (a microscopic worm) transmitted by the bites of black flies, small, midgelike insects whose larvae develop in swiftly flowing water (hence the *river* in the name *river blindness*). River blindness is the third leading cause of blindness in Africa. In 1987, about 18 million people had the parasite, and a million of them were partially or totally blind [28]. Following an approach similar to the malaria eradication campaign in the United States, international health and aid agencies teamed up with several governments in west African countries to fight river blindness. These efforts have been successful in some areas, but they require a long-term commitment: the parasites survive up to fourteen years in humans, so local populations of black flies must be essentially eliminated (and other control measures maintained) for at least that long. In many poverty-stricken areas or places where civil wars rage, these measures are not practical, and river blindness continues to be a major scourge.

The Effect of Humans on Flying Animals

Clearly, flying animals like insects have had a tremendous impact on human society, both positive and negative. Conversely, humans have had enormous effects on many flying animals. Of course, human activities have probably had local effects on flying animals ever since humans first evolved. When humans mostly lived in villages and towns and farmed small areas near their settlements, few species were endangered. However, since the industrial revolution—and accelerating over the past century—the activities of people began to affect whole populations of animals. When land-hungry farmers and resource-hungry corporations destroy an area of tropical forest the size of Denmark every year, and when people use many *tons* of pesticides on golf courses alone every year, many species are endangered and some go extinct.*

Direct Human Effects

As groups of humans move into new areas, they bring along many species of plants and animals. Some are intentional imports, such as honeybees and apple trees, while others, such as rats and cockroaches, are not. When new species are introduced into new areas, they often flourish due to lack of natural enemies. If they can fly, so much the better for them. European immigrants brought honeybees to the New World, where the bees established themselves in the wild in all the warmer regions. People may have accidentally introduced birds to new places, but some of the most successful invaders were deliberately introduced. People introduced house sparrows into New England in 1899 to help control insects, and now they are among the most common birds in North America. They range from Canada to Mexico and have displaced many native species. Starlings were also deliberately introduced. Hard as it may be for birdwatchers to believe, introducing starlings did not work the first time. Eugene Schieffelin tried unsuccessfully to introduce them in New York in 1890. In 1891, he released 40 birds into Central Park, where they bred successfully [31]. Within eight years, starlings had reached Connecticut, and by the 1930s, they were breeding in Wisconsin and Georgia [32]. They reached San Francisco in 1959, and they now live in Alaska and Mexico too. Biologists estimate that about 200 million live in North America, making them one of the commonest birds on the continent.

Humans have also spread species unintentionally. For example, some mosquitoes (*Aedes aegypti*) have a nearly worldwide distribution thanks to humans, and the ideal breeding sites offered by discarded tires—both on the

---

* The area of conservation biology and biodiversity is a large and complex field. To keep my discussion to a manageable size, I shall describe samples of key issues and a limited number of examples where flying animals are significant components. Many textbooks are available that provide a more comprehensive treatment of the subject [29, 30].

ground and in transit to disposal sites—have allowed other mosquito species to invade new areas as well.

Not all of our activities, however, are beneficial to flying animals. Humans have caused the extinction of several abundant bird species in recent history. Subsistence hunting rarely causes problems, but market or commercial hunting often seals the fate of animals. The Great Auk, a flightless bird of the Arctic (Chapter 4), was hunted to extinction for its feathers, as was the Carolina parakeet in the southeastern United States. The Labrador duck was hunted to extinction for the urban restaurant market in the mid 1800s, as was the heath hen shortly afterward. These birds were all so common and abundant that people did not even consider until it was too late that they could all be exterminated. The most astounding example is the passenger pigeon,* the commonest single type of bird in North America by a wide margin in the early 1800s, which was nonetheless extinct by the beginning of the twentieth century. Descriptions of migrating flocks of passenger pigeons beggar the imagination: flocks taking several days to pass a given point on the ground; flocks filling forests to the point where tree limbs and whole trees collapsed under their weight; roosting flocks so noisy that they drowned out the sound of hunters' gunshots [33]. Estimates put their population in the *billions.*

In the 1800s, people killed passenger pigeons en masse, largely for shipment to urban restaurants. Moreover, natural mortality, wastage, and leftovers from dressing carcasses were so great that farmers were known to drive large herds of hogs dozens of miles to fatten them at passenger pigeon roosts [33]. When the pigeon populations started to decline rapidly at the end of the nineteenth century, people took it for granted that hunting was the cause. However, the overall passenger pigeon population was so large and its decline so precipitous that population biologists have recently become suspicious. Some biologists have concluded that hunting alone could not have exterminated the pigeons so quickly. Habitat destruction (commercial logging and conversion of breeding forests to croplands) probably had as much or more effect than hunting in reducing the population at first, and then the pigeons' unusual foraging system broke down. These birds were seed-eaters, and they took advantage of a process called *masting,* or mast production, in

---

* The *passenger* in *passenger pigeon* refers to these pigeons' migratory habits, rather than to their having been passengers or carried passengers in the typical modern sense.

oaks and some other trees. Masting is an enormous, synchronized increase in seed production over a significant area by one or more types of tree, and it occurs irregularly and unpredictably. Passenger pigeons apparently needed to forage in gigantic flocks so that they could cover lots of ground when searching for masting stands of trees. When the population of pigeons dropped below some critical threshold, their flocks were no longer large enough to effectively hunt for masting trees. At that point, the population began its precipitous decline, because these gregarious birds could not find enough food to rear their young without masts. Finally, by 1900, no passenger pigeons were left in the wild [34].

Unintentional Human Effects

Many times in history, when people have tried to control one kind of animal, they have produced catastrophic effects on other animals. Nowhere is this more common than with pesticides. DDT, the first of a long line of chlorinated-hydrocarbon (CH) insecticides, was discovered at the beginning of World War II. As we saw earlier, DDT proved to be a valuable tool in the eradication of malaria. But the chemical has a decidedly mixed legacy. In a vain attempt to stem the spread of dutch elm disease, city officials sprayed elms with DDT to kill the bark beetles. Soon afterward, huge numbers of songbirds died from eating insects full of DDT. Moreover, the spraying did little to halt the beetles, probably because elms outside the cities provided a refuge. One of DDT's (and other CH's) advantages was also its downfall: its persistence. DDT could be applied to the walls of a house, and mosquitoes landing there would be killed for many months, even years. Unfortunately, however, once DDT enters the food chain, it persists there for years as well. DDT and other CH's also tend to accumulate in fatty tissues, rather than being excreted or broken down, leading to *biomagnification;* insects and worms accumulate a little DDT, bug- and worm-eating birds accumulate more, and predators of such birds accumulate high levels. This accumulated DDT does not usually kill vertebrates, but its effects can be just as devastating. Rachel Carson warned about this phenomenon in her 1962 book *Silent Spring* [35], but it took the near-extinction of some very popular and symbolic animals for governments to limit the use of DDT.

In the 1960s, British scientists noticed a decline in peregrine falcon populations, and discovered that they were rarely able to rear their young successfully, because DDT caused them to lay eggs with such thin shells that they usually broke before they hatched [36]. Within a few years, biologists discovered that the same thing was leading to alarming declines in other bird- and fish-eating birds, like bald eagles, ospreys, and pelicans. DDT and other CH insecticides were then banned in North America and Europe. Conservation organizations started captive breeding programs to help the affected birds, and they also received legal protection. Today these predatory birds have come back remarkably, to the point where they are no longer rare in some places. Peregrine falcons have been released and are thriving even in big cities. They like to nest on high ledges and eat birds, and what better habitat than city skyscrapers for nesting, with plenty of pigeons for food?

Just as habitat loss may have led to the demise of the passenger pigeon, changes that humans are making today are having dramatic effects on many animal species. In addition to outright loss of habitat, populations of songbirds (in addition to other plants and animals) have declined in North America due to habitat fragmentation. Fragmentation occurs when large tracts of natural habitat become broken up into many small "islands" surrounded by a "sea" of agricultural and urban land. North American songbirds are probably the most noticeable and treasured victims of these changes.

In the 1980s, recreational birdwatchers and scientists alike began to notice a decline in breeding populations of woodland songbirds that migrate long distances. Several lines of evidence, from annual counts to the numbers of flocks tracked on radar, show decreases in the number of songbirds that breed in North America and spend the winter in the tropics [36]. Biologists believe that these declines are caused by habitat changes in both their summer and winter ranges.

Numerous studies show that when extensive forests are fragmented into many small woodlots surrounded by farm fields, these birds fare poorly. The main problem seems to be edge effects: deep in a huge forest, birds are far from the edge, but in a woodlot of a few hectares, everywhere is close to the edge. Borders between wooded and open areas attract both competitors— birds that inhabit both open areas and edges of forests—and predators like raccoons and opossums, which thrive in disturbed areas. Moreover, although North America's premier nest parasite, the brown-headed cowbird, does not penetrate deep into forests, it does move into forest margins. Cowbirds lay large, fast-hatching eggs in the nests of other birds. Because they are larger, cowbird nestlings tend to muscle aside the nest owners' own young, even to the point of kicking them out of the nest entirely. Birds parasitized by cowbirds raise, at best, fewer young, and, at worst, none of their own offspring. Birds that normally nest in open areas are better adapted to coping with cowbirds than forest-nesting birds [32]. As people cut down more and more of the once-continuous eastern forest, many of these songbirds face an uncertain future.

What happens at the winter end of the migration route is just as important. Many of these birds, including thrushes, warblers, and flycatchers, pass the winter in the mature tropical forests of Central and South America and the Caribbean islands. There, deforestation is a dire problem, driven largely by pressures of rapidly expanding human populations. In the early 1990s, scientists estimated that about 150,000 square kilometers of tropical forest were being cut down annually: that works out to about 29 hectares (70 acres) every minute of every day, around the clock [29]. These estimates are notoriously difficult to make, and published estimates range from as little as 0.6 percent to as much as 2 percent of tropical forests lost per year [29, 36]. Two percent may not sound like much, but if this continues, it means that after 20 years, 40 percent of tropical forests would be gone. Moreover, as these forests are cut down, some of the remnants become fragmented, just as in North America. If a bird species has a limited wintering range, tropical deforestation can be devastating. Bachman's warbler formerly bred in the southern United States and wintered in Cuban rain forests. When the Cuban forests were cut down to make way for sugarcane plantations, Bachman's warbler went extinct.

Of course, North American songbirds are merely a conspicuous indicator. As the tropical forests are destroyed, local plants and animals are destroyed too. The tropics are well known to harbor enormous biological diversity. Tropical rain forests cover just a few percent of the earth's surface, but may contain almost half of all species [36]. A single tree in the tropics may harbor more types of insects than whole counties in the United States. Deforestation in the tropics affects all plants and animals, including most of the world's insects, birds, and bats. Some of these animals have limited ranges and many have already gone extinct.

Stemming the Tide

Conservationists are working hard to try to halt some of the damage that humans are doing to other species, and even to reverse it when possible. When a number of bird species were threatened with extinction by the effects of DDT or habitat loss, biologists established captive breeding programs to increase the birds' numbers. The goal was to build up a sustained population in captivity and then gradually reintroduce some of the captive birds back into the wild. These programs were highly successful for peregrine falcons and eagles. Similar programs have kept whooping cranes and California condors in existence, but reintroducing them into the wild has been only partly successful. Of course, these kinds of programs only work for "charismatic" animals: those large enough and interesting enough for many people to notice. Trying to save every warbler being outcompeted by house sparrows, or every native bee being outcompeted by honeybees is simply not realistic.

Rather than trying to save individual species, conservation biologists recommend preserving habitats. They have put much effort into producing guidelines for game reserves and habitat preserves that save as many species as possible. In the past decade or so, they have accumulated a large body of theory on the best preservation arrangements and have begun testing their theories with real reserves. Guidelines include recommendations such as preferring a single large reserve to several smaller ones (to reduce habitat fragmentation) and leaving "corridors" of undeveloped land connecting reserves wherever possible. Setting aside reserves and parks at all may be difficult for many developing countries, let alone following such guidelines. After all, population pressures are driving the habitat destruction in the first place, and those same pressures work against governments setting aside much untouched land. Some countries, however, have been quite successful. In Central America, for example, Costa Rica has put over 10 percent of its land area into parks and nature preserves [36], and the country derives a substantial income from "ecotourists" who visit to see this natural wealth. Kenya, another developing country with a large system of parks and reserves, received over $200 million annually in income from tourists visiting to see the wildlife in the 1990s.

The United Nations Educational, Scientific and Cultural Organization

(UNESCO) has developed guidelines for establishing preserves to include all the major types of ecosystems on earth. Its "Biosphere" preserves include an untouched and undevelopable core area, surrounded by a zone of minimal impact where experimental research and traditional subsistence exploitation are allowed. Outside of these zones is a buffer area of controlled and limited development. Quite a few preexisting parks and reserves have been included in this program, such as Grand Canyon National Park, and new ones have been established around the world. One of the problems that governments and conservation biologists often face is a local population that opposes establishing new parks for economic and cultural reasons. If the local people do not support the park, poaching and illegal exploitation will be a major problem. Many of the famous African parks and reserves have to contend with this problem.

One way to avoid such problems is to involve the people living nearby in the process of establishing and operating the park or reserve. A fascinating example of this resulted in a park and a set of reserves that may prevent the world's largest bat from going extinct. On the Pacific islands of American Samoa (administered as a protectorate by the United States), the Samoan government began requiring villages to build schools and medical clinics, but provided no money. The villagers had to find the money on their own. Many villages are quite poor, and the only resources they have are the surrounding rain forests. The forests are essentially the common property of all villagers, and the village chiefs control access to them. In 1988, the village of Falealupo had to find money to build a school, so they reluctantly agreed to allow logging in their forest. Paul Cox, a biologist doing research on bats nearby, saw the anguish of the people of Falealupo as they saw their forest being destroyed. He took out a personal loan to cover the cost of building the school and began discussions with the villagers to try to come up with a solution that would preserve their forest [37]. Cox had a professional, as well as a humanitarian interest, because Samoa is home to two species of flying foxes (megabats) including the world's largest bat, *Pteropus samoensis*. He knew that these bats were seriously threatened by human activities.

The seeds of the decline in flying foxes were sown much earlier, in yet another demonstration of the destructiveness of market hunting. In the 1970s, the island of Guam in the Western Pacific developed a thriving market for flying foxes as a luxury food item. The high prices offered for these bats induced many Pacific islanders, including some Samoans, to hunt flying foxes and sell them for export [38]. In the 1980s and 1990s, researchers discovered an alarming decline in flying foxes on many Pacific islands, resulting from a combination of deforestation and market hunting. Cox and several of his colleagues, particularly the Swedish biologist Thomas Elmqvist, began seeking ways to protect Samoa's flying foxes. The biologists realized that a plan that helped the flying foxes could also help the local people, and vice versa. They brought charitable organizations together with the chiefs of Falealupo village and developed a unique covenant. Donors provided money to pay for the school, and in return, the villagers agreed to preserve and protect the rain forest for 50 years. Moreover, the covenant allows the villagers to continue to use the forest in traditional ways. In Cox and Elmqvist's words, "[t]he covenant encourages the villagers to continue to harvest the forest for medicinal plants and other cultural purposes on a sustainable basis, but prohibits commercial logging or any other activity that may significantly damage the forest. Under the covenant, complete control and management authority for the preserve is maintained by the Falealupo chiefs council" [37]. The Falealupo Preserve has been in operation for a decade now, and audits have always shown that the chiefs have used all the donated funds for their intended purposes.

At about the same time that Falealupo was forming its preserve, other Samoans began discussions with the U.S. government to develop a national park in Samoa. The National Park Service and the Samoan government studied the issue and came up with a plan remarkably similar to the Falealupo covenant. The park would be established on land with a 50-year lease, local people would continue to use the forest in traditional ways, and the Park Service would be forbidden to build roads or overnight accommodations and was required to run the park in close consultation with village chiefs. In 1988, Congress passed, and the president signed, legislation creating the American Samoa National Park [39].

Since the creation of the Falealupo Preserve, a number of other villages in Samoa have set up similar covenants to preserve their rain forests. Similar, if less sweeping, programs to involve local people in management and traditional use of reserves in Africa have had some success in reducing poaching and other illegal uses. Not surprisingly, if people on the edge of a preserve believe that they benefit from the preserve's presence, many will become protectors rather than illegal exploiters of the preserve.

Sometimes these conservation efforts have substantial unexpected benefits. For example, in Samoa, researchers discovered that the flying foxes are major pollinators of many trees and shrubs [40]. Without the flying foxes, many plants that people use heavily, like the kapok tree, set no fruit at all. Had the flying foxes gone extinct, they would have taken significant parts of the vegetation with them, and the villagers might well have been unable to continue their traditional lifestyle and culture.

Although such conservation measures benefit all animals and plants, flying animals—especially birds—are often the most obvious and highly prized beneficiaries. Birds are up in the air, where people can see them, many are gaudy or loudly vocal, and some are important symbols—the robin as a herald of spring, the eagle representing power and majesty. Moreover, from the casual backyard observer to the dedicated birdwatcher, people also notice when birds become rare or disappear. Conservationists can use the plight of birds to raise the awareness of the public and at the same time apply measures that benefit other organisms. When people stopped using DDT in the United States to save birds of prey, many other species, from ants to salmon, benefited as well. Similarly, regardless of whether or not the northern spotted owl is ecologically important, if preserving the owl also preserves the old-growth forests of the Pacific Northwest, the ecosystem and ultimately the people benefit as well. If people support efforts to preserve animals, like birds, that they care about, conservationists can take steps to save the birds. In doing so, they will also save habitats, along with more obscure but equally threatened animals and plants.

ELEVEN

Have the Birds and Bees Taught Us Anything Useful?

*I*n Greek mythology, Daedalus fashioned wings from bird feathers, string, and wax so that he and his son Icarus could escape from the island of Crete, where they were imprisoned. As they flew over the sea, however, Icarus ignored his father's warning and flew too close to the sun. The wax holding his wings together melted, and he fell into the sea and drowned, leaving his father to fly on alone. The myth of Icarus and Daedalus makes a fine parable but is poor engineering: a person simply cannot fly by flapping wings attached to his or her arms. From dreamers leaping off castle walls with their arms strapped to wood and canvas contraptions to careful scientists such as Leonardo da Vinci, many would-be aeronauts have made the mistake of assuming that if they could just do what the birds do, they would be able to fly. This approach has several fatal flaws, however. First and foremost, humans are simply underpowered. Our arm muscles cannot produce nearly enough power to carry our body weight.* Our *legs* are just barely powerful enough, but no one considered using leg power for flight until well into the twentieth century. And even then, pilots used their legs to pedal, not to flap wings. Even if our arms were as strong as our legs, until recently, we still would not have been able to fly by flapping mechanical wings, because the materials available a century or more ago could not produce light enough structures with sufficient strength. Wood, canvas, leather, and iron can make strong structures, but they will be very heavy for their strength. Finally, the early would-be aeronauts understood neither the movements nor the control ramifications of flapping. Birds (and bats) do not simply wave their forelimbs up and down: in most cases they move their wings down and forward while tilted leading-edge-down—the downstroke—and up and back while tilted leading-edge-up—the upstroke. The tip often follows a figure-eight path or one even more complex (Chapter 4). Moreover, flapping animals adjust their wing-beat patterns for maneuvering and stability (Chapter 5). None of these aspects of flapping flight could have been known to people before the invention of high-speed photography. Even if people were strong enough and could have built light-enough wings, who could have instructed the trainees in the crucial wing movements? How would they have dealt with the need for active-stabilization reflexes?

To us, it is self-evident that wings are required for flight, because we have the examples of birds, bats, and insects before us; would we even have discovered wings and flight if there were no flying animals? Yet flying animals have provided humans with much more inspiration than instruction. Indeed, beyond the need for wings, and the basic principles for turning flight, flying animals have contributed surprisingly little to the technology of aviation.

Direct Copying: What We Have Learned from Birds

Wings and Wing Designs

The example of flying animals so clearly demonstrates that wings are necessary for flight that for humans, wings have become a powerful symbol for

---

* Humans probably are strong enough to fly on arm power alone in the reduced gravity on the moon, which has inspired more than one science-fiction story. Of course, these flights would have to be indoors where there is air, and I am not entirely convinced that people could make the proper wing movements without a lot of training.

flight. Consider that military services almost universally indicate aviator status with badges or pins showing stylized bird wings. And flying animals have given would-be airplane designers some strong hints about wing design. For instance, animal wings tend to have much longer spans than chords—their aspect ratios are much greater than 1.0 (Chapter 2). Animal wings are usually thick in front, taper to a thin trailing edge, and are cambered. These last three characteristics form a good starting point for airfoil design. Because of size effects and functional constraints—birds fly at much lower Reynolds numbers than airplanes and must be able to flap and fold their wings—bird airfoils do not make particularly good airplane airfoils. However, as Horatio Phillips confirmed experimentally in the 1880s [1], bird airfoils are more aerodynamically effective than flat plates, and they provide a useful starting point for designing more efficient airfoils for airplanes.

Aerial Turns

The story of how inventors developed airplane turn mechanisms illustrates a common result of borrowing from nature: nature provides the general principle, but engineers must apply it in practice. The resulting devices almost always differ a great deal from the animal or plant that provided the inspiration [2].

The definition of a successful airplane is surprisingly detailed: a heavier-than-air, person-carrying machine that takes off under its own power and lands at an elevation as high or higher than the takeoff point. Orville and Wilbur Wright were the first people to fly a successful airplane. Several people built machines that "flew" (or at least hopped) before the Wright brothers, but these earlier efforts were not successful, because their builders did not understand how to *control* a flying machine. The few machines that actually got off the ground typically had too much built-in stability, and their inventors intended to turn them with rudders. The key innovation of the Wrights' design was inspired by birds: the Wrights observed that gliding birds turned by twisting their wings lengthwise [1, 2]. A bird twists the trailing edge of one wing down (increasing lift) and the trailing edge of the opposite wing up (decreasing lift) to roll into a bank (Chapter 5). The Wrights were the first inventors to realize that aerial banking is not a byproduct of a turn (as in ships) but rather what *causes* the turn [1]. The Wrights developed essentially the same mechanism for their gliders and powered craft as used by birds. Their

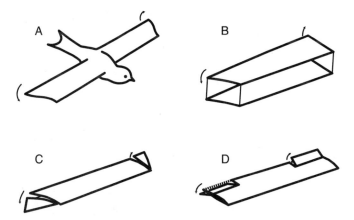

Figure 11.1. Development of airplane turning mechanisms from the mechanism used by birds. A. A gliding bird turns by differentially twisting its wings: it twists the wing to the outside of the turn trailing-edge-down (increasing its lift) and the inner wing trailing-edge-up. This rolls the bird into a bank, which produces the turn (Chapter 5). B. The Wright brothers used essentially the same mechanism, which they called *wing-warping.* This schematic view of the biplane wings of their airplane shows how they twisted the part of the wings to the outside of the turn to a higher angle of attack, and the part of the wings to the inside of the turn to a lower angle of attack to roll into a bank. C. Glen Curtiss built airplanes similar to the Wrights', but Curtiss put moveable extensions on his wing tips that could be pivoted to raise or lower the extension's leading edge; the sketch shows a wing with early triangular extensions. He called these moveable surfaces *ailerons.* D. Within less than ten years, ailerons became moveable flaps on the trailing edge of the wing, and this has been the most common roll-control arrangement on airplanes ever since. (S.T.)

control mechanism actually bent the trailing edge of the outer wing down, which increased its angle of attack and camber, and bent the trailing edge of the inner wing up, which decreased its angle of attack and camber (Fig. 11.1). This so-called wing-warping system (actually wing twisting) was the first effective roll-control system on a person-carrying flying machine. Some engineers and historians view the 1902 Wright glider as more significant than the 1903 powered airplane, because Orville and Wilbur solved their control problems and taught themselves to fly on the 1902 glider [3].

The Wright brothers discovered the general principle of aerial turns, but the mechanism that they used to produce turns was not especially practical. Wing twisting is fine for birds and bats with their flexible wings, and for the Wright airplane with its springy, wood-framed, wire-braced wings. However, building a wing that is rigid enough to carry a load yet twistable enough for wing-warping control is a serious design challenge. In fact, some of the Wrights' competitors sought to avoid this challenge altogether. Glen Curtiss built airplanes very similar to the Wrights' but with more rigid wings. Rather than twisting the wings, he had small, moveable extensions on the end of each wing. The pilot could pivot the trailing edges of these extensions up or down, with the extensions on opposite sides working in opposite directions. These extensions were dubbed *ailerons,* which comes from a French word for "pinion," the outer part, or flight feathers, of a bird's wing. Eventually, ailerons evolved into hinged sections of the trailing edge of an airplane's wings near the tip: these are what most airplanes use for turns today (Fig. 11.1). Although a jumbo jet does not turn by warping its wings, it does have ailerons. It turns by banking with asymmetrical lift, using exactly the same principle as a gliding bird or the Wrights' first airplane.*

Blind Alleys?

When early aircraft designers turned to nature for inspiration in other areas, however, sometimes they were led seriously astray. Some features of Otto Lilienthal's gliders fall into this category. Lilienthal (1848–1896) was a brilliant researcher, and his pioneering studies on wing properties were a major influence on later inventors, including the Wright brothers. Although he made great strides in understanding the aerodynamics of wings, he was convinced that successful flight depended on emulating nature. Even the title of his classic book *Der Vogelflug als Grundlage der Fliegekunst* ("bird flight as the basis for aviation") reflects his prejudice [1]. For example, he rejected the use of propellers for thrust and planned eventually to add a wing-flapping mechanism to one of his gliders to try to achieve powered flight [4]. In the 1890s, Lilienthal built and flew a series of gliders that we would today call hang gliders, designed so that the pilot guided them by shifting his weight. His gliders were quite successful for their day, and his longest glides of over 360 meters were not surpassed until many years after his death. He thought that flying machines should be extremely stable in order to make piloting easier. Reasoning that the samaras of *Alsomitra* (the "flying cucumbers" of Chapter 3)

---

* Although some modern airplanes turn using devices other than standard ailerons, all fixed-wing airplanes still turn by generating asymmetrical lift to roll into a bank.

glide perfectly well without a pilot, Lilienthal borrowed some of the stabilizing tricks of *Alsomitra*. His gliders were so stable that by modern standards,
the pilot had little control authority, and, sadly, Lilienthal was killed in one
of his gliders in 1896. A gust of wind apparently caused the craft to stall, and
Lilienthal could not shift his weight enough to recover before crashing. This
is a modern interpretation of the accident; at the time, scientists had little
understanding of the stall phenomenon and how to avoid or recover from it.

Several later designers went even further in copying *Alsomitra* (known at
the time as *Zanonia*). The Austrians Igo Etrich and Franz Wels built a series of
gliders from about 1900 to 1907 that were closely modeled on *Alsomitra*,
down to the wing outline. They flew their 1907 glider successfully, but it
could not have been very maneuverable, and they never followed through
on plans to add an engine. In England, J. W. Dunne also used *Alsomitra* as the
basis for the design of a successful glider and a marginally flyable powered
craft [5].

The record of flying machines with wing planforms borrowed directly from
nature is decidedly mixed. In the late 1800s, Clement Ader built a craft powered with a steam engine, with bat-shaped wings. In 1890, it hopped off the
ground briefly, but it had no provision for steering, and Ader never attempted
to fly it again. Before World War I, many inventors designed flying machines
with wings copied from birds (and more than one was intended to flap its
wings). Few of these early birdlike designs were particularly successful [6].

A few airplanes with birdlike wing designs were more successful. The best
example is Igo Etrich's "Taube" (dove), a successful Austrian design of about
1909 that grew out of Etrich's work on *Alsomitra*-shaped gliders. The Taube's
wing outline was clearly inspired by bird wings (Fig. 11.2), and Etrich also
used a wing-warping system reminiscent of birds (and the Wright brothers)
to turn.

Curiously, the Taube retained some of the stabilizing features of Etrich's
earlier tailless gliders. Although these graceful airplanes were underpowered,
they were reputedly easy to fly by the standards of the day, perhaps because
of their stability. Etrich licensed his design to a German company, Rumpler,
which built a series of reconnaissance airplanes based on the Taube design
that were widely used in World War I. Other European firms built airplanes
with birdlike wings, including Albatrosse in Germany and Handley-Page in
England [5].

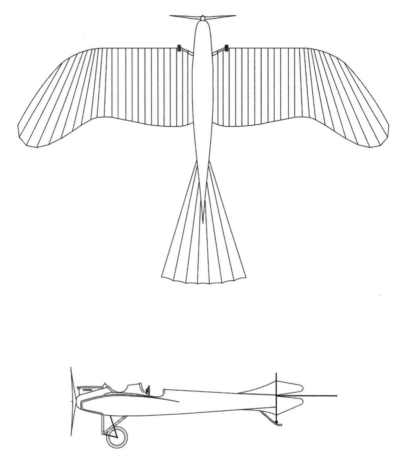

Figure 11.2. The Etrich "Taube" (dove) was the most successful of many early airplane designs based on birdlike wing planforms. Redrawn by permission, National Air and Space Museum, Smithsonian Institution (Acc. XXXX-0002: Etrich Taube 7/1/13). (O.H.)

Parallel Achievements

Convergent evolution is a common phenomenon in the history of life. When two organisms evolve to perform similar tasks under similar conditions, they often evolve adaptations that are strikingly similar. Bird, bat, insect, and pterosaur wings illustrate convergent evolution. Other examples include the webbed feet of ducks and beavers, and the eyes of fish and squid. These animals did not inherit their similar structures from a common ancestor; the common ancestor of the first two was a terrestrial stem reptile, and the common ancestor of the latter pair was probably a wormlike creature with no eyes at all. The phenomenon of convergence is quite widespread. Trout and porpoises have similar body shapes because natural selection favored animals having shapes with low drag, so trout and porpoises converged on similar, streamlined, low-drag shapes. Although such convergent features may make two animals appear quite similar, the adaptations are only superficially similar and have fundamental differences. Fish are cold-blooded, scaly animals with gills, but porpoises are warm-blooded, smooth-skinned breathers of air. Hummingbirds and bumblebees have almost identical wing-beat patterns, but hummingbirds' wings are made of bone, muscle, and feathers; bee wings are made of a pleated membrane supported by stiff, hollow veins.

Just as selection in nature leads to convergent adaptations, the much more rapid process of technological evolution has produced several areas of convergence between flying animals and flying machines. These convergences were not intentional copies of mechanisms used by animals, but technological solutions to common challenges faced by all flyers.

Gyroscopic Instruments

One challenge of flying is the need to sense orientation or movement when out of sight of the ground, such as in or above clouds or in darkness. Flies evolved halteres, which work on the same physical principle as the solution employed by engineers: gyroscopes (Chapter 5). True flies (Diptera) are the only major group of flying animals with gyro sensors. In contrast, humans use a variety of gyroscopic instruments in airplanes, and these instruments were crucial to the development of safe, versatile aircraft over the last half of the twentieth century. Today, even a small, single-engined, four-seat airplane typically has at least three gyroscopic instruments. The most venerable of these is the turn and bank indicator, which is a single-axis instrument: a gyroscope's resistance to turning is only monitored in one direction. This instrument has two parts: the turn indicator part uses a gyroscope to detect yaw (nose right or nose left) movements, connected to a needle that shows the direction and speed of the turn. The other part is a nongyro instrument, simply a ball in a curved tube that tells the pilot if the airplane is sideslipping. The turn and bank indicator is a simple, reliable instrument, but it requires training and practice to be used effectively. Moreover, the instrument's name is a bit misleading, because it only measures banking indirectly.

Another common gyro instrument is the gyrocompass, or directional gyro. The pilot manually adjusts the gyroscope's axis in this instrument so that it points north, and then the gyroscope is set free in its gimbals and continues pointing north, regardless of how the airplane moves. Unfortunately, friction causes gyroscopes to *precess,* or gradually drift away from their original orientation. So gyrocompasses must be periodically reset, which the pilot usually does by referring to a magnetic compass and a chart of magnetic deviation.*

The most sophisticated gyro instrument is the *attitude indicator,* or "artificial horizon." Attitude indicators represent a particularly strong functional convergence with fly halteres: both primarily detect pitch and roll movements. Standard attitude indicators do not detect yaw, and flies do not seem to use their halteres to detect yaw changes. The attitude indicator is a two-axis instrument designed to tell the pilot of pitch changes (nose up or nose down) as well as roll changes. The artificial horizon is quite intuitive to use. Its display has a reference line that moves in the same direction that the horizon appears to move as the airplane changes orientation. When the airplane's nose drops, the reference line rises, just as the real horizon appears to the pilot to rise; when the airplane rolls right, the reference line rolls to the left, just as the horizon appears to roll left (Fig. 11.3). The attitude indicators of some high-performance aerobatic and military airplanes allow—and can display—full 360° rotations, which is important for airplanes that include loops and rolls as normal maneuvers. The attitude indicators of most aircraft have limited operating ranges, so that if the airplane maneuvers too drastically, the attitude indicator loses its fixed orientation, or "tumbles." These attitude indicators are perfectly adequate for normal flight, but the pilot must lock them in place, or "cage" them, before doing any aerobatic maneuvers.

In one specialized application, humans have developed an instrument eerily similar to halteres. Helicopters are notoriously difficult to fly, partly be-

* Why have a gryocompass at all if it must be reset using a magnetic compass? Gyrocompasses are easier to read, and they respond more quickly and precisely during turns. Magnetic compasses lead or lag during turns, depending on the direction of the turn, which makes precise maneuvering difficult.

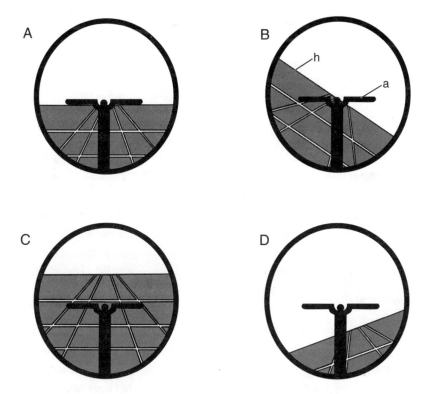

Figure 11.3. Instrument display of an airplane's *attitude indicator*, or "artificial horizon." The horizon reference line of the instrument, h, moves in the same way that the real horizon appears to the pilot to move as the airplane maneuvers. The airplane symbol, a, is fixed to the face of the instrument and does not move. A. Straight and level flight. B. Level left turn. C. Straight descent. D. Climbing right turn. (S.T.)

cause they have little passive stability, particularly when hovering, and most need to be hovered to take off and land. Perhaps the most demanding piloting task of all is piloting a helicopter by remote control. A pilot must fly a remotely controlled helicopter entirely by watching its movements, with none of the visual, tactile, or instrument feedback available to an on-board pilot. This piloting task is so demanding that very few people actually achieved controlled flight with them when radio-controlled helicopters first became available to hobbyists in the 1970s. Then someone tried putting a miniature gyroscope in a radio-controlled helicopter. This gyro instrument was connected to the controls so that it automatically corrected unintentional yaw (nose left or nose right) movements. With the gyro controlling yaw, the pilot "only" has to control pitch, roll, and total lift. With gyros, pilots were soon able to master hovering without unrealistically intensive efforts.

The early radio-controlled gyros were a mixed blessing, however. They were heavy, delicate, power-hungry, and expensive. Later, "solid state" gyros, based on technologies developed for other purposes, became available to hobbyists. Solid state gyros are based on a vibrating mass rather than a spinning wheel, so they function in the exact same way as a fly's haltere: they use oscillation rather than rotation to provide a reference that resists turning. These gyros are engineering marvels, only a centimeter or two on a side and weighing a few dozen grams. They are smaller, lighter, cheaper, more rugged and reliable, and use less power than conventional rotating gyros. Clearly, a similar requirement—automatic, or reflex, course correction—led to the evolution of surprisingly similar mechanisms in flies and radio-controlled helicopters.

Active Stabilization

The use of active stabilizing mechanisms by flying animals is a major topic of Chapter 5. Humans developed active stabilizing systems for airplanes—at least initially—for the same reasons as animals. Flying animals, being either predators or prey (or both), gain a clear advantage if they are more maneuverable than other individuals or species, so they evolved active stabilization with little passive stability. Fighter aircraft face almost identical selection pressures: if a fighter pilot cannot outmaneuver his or her opponent in a dogfight, the consequences may be just as fatal as for the pigeon that cannot outmaneuver the stoop of a falcon. Fighter aircraft have always been the most maneuverable airplanes, but for most of aviation's history, they had to have enough passive stability to allow the pilot to control them without excessive effort in noncombat flight. That changed with the availability of electronic computers in the 1970s and 1980s. Engineers began to design airplanes that were barely stable. These airplanes use computers that sense deviations from the intended heading and constantly adjust the control surfaces automatically, dozens or hundreds of times per second. The pilot has conventional-looking controls, but they are actually connected to a computer, and the pilot has no direct, physical link to the control surfaces. Thus, the pilot uses the controls to signal the computer where he or she wants to go, and the computer actually moves the control surfaces to get there. By using such a computer control system in a marginally stable (or even unstable) craft, the pilot has the illusion of a rock-steady airplane in level flight, but one that can change directions with whip-cracking speed when desired.

These computer control systems are part of what is termed a *fly-by-wire* system, where the computer sends an electrical signal to actuators near the control surface hinge, which actually move the control surfaces. Fly-by-wire systems normally have two or more sets of wires from the computer to the actuator, each routed along a different path. That way, battle damage is less likely to cut off control to a given control surface, which clearly adds survival value. In some of the earlier fly-by-wire designs, if the computer failed, the pilot could revert to a manual backup system and still control the airplane. Many of today's modern frontline fighter jets are reputedly uncontrollable without the computer, which is why there are backup computer systems. If the computers all fail in one of these craft, the pilot has no choice but to eject.

Maneuverability and survivability are not the only advantages of fly-by-wire systems. When the Space Shuttle was being developed, NASA engineers programmed a computerized control system built into a small business jet to fly it so that it mimicked the feel of flying the Space Shuttle. This gave the astronauts realistic flight training before they ever climbed into the real Space Shuttle's cockpit. The engineers realized that by changing the computer's software, they could change the jet's flight characteristics, which in turn allowed them to experiment with shuttle flight characteristics without actually modifying and flying the shuttle. At about the same time, engineers at companies that build airliners began to realize that fly-by-wire systems might have advantages for them as well. For example, an airliner's handling characteristics could be changed by changing the computer software, rather than modifying the craft's structure. The computer can be programmed to give the airliner a different "feel" to the pilot if needed to make the airplane easier to fly. Moreover, the redundancy aspect was not lost on the builders of airliners. Multiple backup signal routes are much lighter and easier to install, so they can be used where multiple mechanical or hydraulic systems would be prohibitively heavy and complex. Just as they can improve the survivability of combat aircraft, redundant fly-by-wire control systems improve the chances of an airliner surviving accidents, mechanical failures, or sabotage.

Collisions with birds, for instance, are a major concern for designers of airliners, according to a friend who is a retired engineer from a major aircraft

manufacturer. Even at relatively low landing approach speeds in the neighborhood of 220 kilometers per hour (120 knots), hitting a goose can cause
damage not unlike that done by a cannon shell. At typical airliner cruising
speeds of well over 950 kilometers per hour (over 500 knots), even a small
bird becomes a lethal projectile. By incorporating redundant fly-by-wire systems into modern airliners, designers have made them significantly better
able to survive such nasty encounters.

Airplane fly-by-wire systems are only partly convergent with animal
flight-control mechanisms. Both involve rapid, automatic adjustments—reflexes in animals, computers in airplanes—to keep an otherwise unstable
flyer moving in the desired direction. Animals, however, do not have the redundant signal pathways of a fly-by-wire system. Any given muscle is connected to the brain by a single nerve, and if that nerve is cut, the muscle (and
all others controlled by that nerve) will be paralyzed. Instead of redundancy,
animals rely on their ability to heal: if a nerve gets cut and the animal sur-
vives long enough, the nerve will regenerate.*

Adjustable Wings

Flapping and maneuvering require an animal's wing to change shape. Some
of these shape changes are passive, such as the bending of the flexible trailing
edge of an insect or the tip feathers of a bird at the top and bottom of the stroke.
Other changes are active, such as when birds flex their wings during the upstroke or bats change their camber for maneuvering. Today, aeronautical engineers are actively studying ways to cause airplane wings to make analogous
changes. In the passive realm, engineers have been studying *aeroelasticity*—distortions of wing shape by aerodynamic loads—for many years. Until recently,
their aim was mostly to avoid distortions as much as possible. With the advent
of high-tech, lightweight composite materials, researchers have begun to consider building wings with controlled flexibility, so that they flex under load
predictably and in ways that improve their aerodynamic performance. Composite structures made of materials like graphite-epoxy are constructed from
several layers of fabric-like material that can easily be cut into complex shapes

* Under ideal conditions, a severed nerve can regrow all its original connections, although in practice, tissue damage can limit this process. The brain and spinal cord are
fundamentally different, and they cannot regrow damaged parts.

and hardened by heating; this building method gives designers much greater ability to tailor the structural properties of a wing than they have with metal.

With the advent of active stabilization systems—computerized fly-by-wire systems—some engineers began to consider actively adjusting the shape of the wing itself. Large airplanes already have some ability to adjust their wing area and camber with the use of various movable control surfaces, but these adjustments are quite limited compared to the routine shape changes used by animals. These engineers envisioned a much more adjustable wing, often referred to as a *mission-adaptive* wing. If such a wing could twist, bend its trailing edge up or down, or change its camber, all while maintaining a smooth surface, it could perform better under different conditions. A thick, high-camber wing might be best at slow speeds or for landing, or a thin, low-camber wing might be optimal for high-speed cruising. Moreover, if such a wing could be deflected enough, the changes could be used for maneuvering; the smooth wing surface would produce less drag than conventional control surfaces that have an abrupt bend at their hinge line when deflected. The U.S. Air Force and NASA began experimenting with mission-adaptive wings in the late 1980s. They used pneumatic actuators to bend the leading and trailing edges of wings up or down to change the camber. Although these wings had aerodynamic advantages over conventional wings, the weight and complexity of their pneumatic actuators made the systems impractical. More recently, designers have turned to "smart" materials that literally change shape in response to electrical currents or similar signals. Researchers have actually built working wind tunnel models of wings incorporating smart materials as actuators [7]. In its most ambitious form, designers envision a wing that can change its camber, thickness, degree of twist, and perhaps even its area or span, under control of a computer that adjusts the wing to an optimal form for the given flight conditions. Larger adjustments or deflections would be used for maneuvering. At least one design being developed uses smart materials to twist the wing, thus returning full circle to the technique developed by the Wright brothers and borrowed from birds [7].

Modern aircraft, particularly large ones, already have camber-control devices: moveable control surfaces on the trailing edge (and sometimes leading edge) called *flaps,* which hinge downward to increase the wing's camber. Normally, these are controlled directly by the pilot, who deploys them manually for takeoff and landing. If a computer could constantly adjust the flaps to optimize the camber for any given altitude and air speed, some researchers think that airliners could reduce their fuel consumption or increase their range. Fly-by-wire systems are already performing very similar tasks, and it is just a matter of time until automatic camber adjustments are incorporated into the functions of the stabilizing computers of new airplanes.

Human (Muscle) Powered Aircraft

Long after the Wright brothers had ushered in the age of engine-powered flight, a few people still dreamed of building aircraft that could fly using muscle power alone. In 1959, Henry Kremer, a wealthy inventor and philanthropist, offered a prize for a successful flight by a human-powered aircraft (HPA). He put up £5,000 for the first aircraft to fly a figure-eight around two markers half a mile (805 meters) apart. At first, the Kremer prize was only open to British Commonwealth citizens, but when no British teams had come close to winning the prize by 1967, Kremer doubled the prize amount and opened it to all nationalities. In the early 1970s, several teams in Japan, the United States, and the United Kingdom built and flew HPAs, but none had the endurance or controllability to complete the Kremer course. So Henry Kremer once again increased the prize, this time to £50,000 [8].

One of the problems is that humans are marginal airplane engines. Physiologists had measured the maximum power output of humans pedaling on stationary bicycles, and engineers had calculated that trained athletes could produce just barely enough power to fly for a significant distance. With ever lighter materials—new aluminum alloys, composites like graphite-epoxy— becoming available during the decades after World War II, a few engineers were confident that someone would eventually win Kremer's prize.

A couple of the later machines achieved nearly the needed endurance, but even those could not turn enough to fly the Kremer course. Their enormously long wings—from 20 to more than 35 meters in span—caused several problems. These airplanes were designed to fly fairly low, at altitudes of just a few meters, to take advantage of the ground effect.* Because of their

* The *ground effect* is a decrease in drag on wings that fly close to the ground. As a wing gets close to the surface, its downwash angle decreases (the ground deflects the downwash), so its induced drag also decreases [10]). This phenomenon reduces power requirements and increases gliding distances when the wing is less than about one half of its span in height above the surface.

long wings, they could barely bank a few degrees before they risked dragging a wing tip on the ground. In addition, at the relatively low speeds at which these craft flew, when they turn, the inside wing moves much slower than the outside wing. When such a long-winged craft makes a turn sharper than a very gentle course correction, the inner wing may slow down enough to stall. In these light, fragile craft, stalling the wing on one side inevitably led to a hard landing at best, or a destructive crash at worst. Because they were built so light, even a hard landing could cause serious damage. If they could overcome these obstacles, HPAs faced yet a third problem with turns. When an airplane banks in a turn, the lift is tilted into the turn, so that the vertical component is reduced (Chapter 5, Fig. 5.3). Some teams managed to begin turns with their HPAs without stalling the inner wing, only to find that they sank to the ground anyway: the reduction in the vertical component of lift was so great that the craft could not maintain its altitude. In other words, the airplane was so underpowered that it could barely stay aloft in level flight, and it had no safety margin available for increasing lift in turns.

By 1976, the Kremer prize still had not been won, and an independent aeronautical engineer named Paul MacCready decided to try for it. Mac-Cready was a hang-glider designer and pilot, as well as a champion sailplane pilot. He decided he ought to be able to build a very large craft that would be light enough if he used the basic structure of a hang glider as his starting point. The wing would be braced with wires to a vertical tube called a king-post, extending above and below the wing. MacCready's success hinged on one creative insight and one practical principle, as well as an extremely innovative control arrangement. His insight was that if his craft flew slow enough, he could use external bracing wires and not pay a large drag penalty. His practical principle was that all structures should be easy to build, and especially, easy to repair [9]. His team developed ways to quickly splice broken structural tubes, and they used thousands of meters of plastic packing tape for repairs. The other HPAs that had been most successful up to that time had mostly been rather conventional-looking, except for some with tail-mounted propellers. MacCready's design was radically different. The enormous, wire-braced wing was set high above the pilot, whose pedals were linked by a light-weight wire and plastic chain to a large propeller on the back of the wing. It had no tail, but a small wing called a *canard* was suspended on a pole out in front of the pilot. The craft was built, rebuilt, and

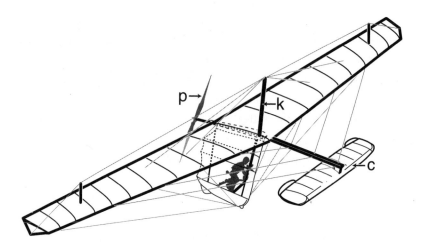

Figure 11.4. The Gossamer Condor, which won the original Kremer prize and became the first successful human-powered airplane. (S.T.)

modified by a dedicated team consisting mostly of MacCready's friends and
relatives. It was called the Gossamer Condor (Fig. 11.4) [8].

The canard was MacCready's first step in solving the turn problem. His design used the canard for both pitch and yaw control. Although a bit unusual,
quite a few airplanes had been built with canards for pitch control, including most of the Wright brothers' airplanes. On the Gossamer Condor, when
the canard tilts nose-up, its lift increases and pulls up on the main wing; the
main wing's increased angle of attack in turn causes the craft to climb. In the
same way, tilting the canard nose-down causes a descent. The truly innovative aspect of MacCready's canard was that it was slung from its pole so that
it could also roll right or left. Rolling the canard to the right, for example,
caused the canard to pull the rest of the craft to the right. In a sense, the canard towed the rest of the craft in whatever direction the canard flew [9]. The
canard worked well for course corrections, but it was not sufficient to turn
the Gossamer Condor in the sharp turns needed for the Kremer prize course.
Turns using the canard started well enough but often resulted in a sideslip
into the ground. The craft needed a more effective turn control.

The wire-braced structure of the Gossamer Condor made wing-warping an
obvious solution to roll control, so the pilot was given a lever to twist the
wings, just as the Wright brothers did on their airplanes more than a half century earlier. When MacCready's team first flew their craft, however, they ran

into one of the problems that plagued the other HPAs: when they tried to warp the wings to bank into the turn, the Gossamer Condor sometimes stalled and was forced to land. Moreover, wing-warping did not seem to produce turns predictably. After the aerodynamic consultant Henry Jex made an intense, detailed analysis of the Gossamer Condor's aerodynamic characteristics, the team learned that wing warping would not be effective for roll control, but could be quite effective for yaw (nose-left or nose-right) control [8]. They discovered that they could turn the Gossamer Condor successfully by first rolling the canard in the desired direction and then warping the wings in the opposite direction from conventional ailerons [8, 9]. Reverse wing-warping apparently worked because when the trailing edge on one side was twisted down, that side produced more drag. The drag slowed the wing on that side and pulled the wing around, in contrast to the conventional, expected effect of an increased angle of attack producing more lift and raising the wing into a bank. Any increase in lift that may have been produced by the higher angle of attack of the wing to the inside of the turn apparently just compensated for the loss of lift due to the wing slowing on the inside of the turn. The final piece of the turn puzzle was in place.

After several months of experimentation and modification (including completely replacing the main wing), and several failed attempts to fly the Kremer course, the Gossamer Condor made several test flights that included flying most or all of the Kremer course. MacCready's team felt sure enough of success to make another official attempt at the Kremer course (which required the presence of certified observers, and generally included lots of publicity). This particular attempt was made on such short notice that a couple of the key team members were not notified in time to get to the airport to witness it. On August 23, 1977, with the bicycle racer (and hang-glider enthusiast) Bryan Allen as both pilot and motor, the Gossamer Condor successfully flew the Kremer course, clearing a 10-foot (3-meter) obstacle at the beginning and end of the flight [8].

The Gossamer Condor looks distinctly old-fashioned and inelegant, compared to the sleek designs of the Japanese, European, and other U.S. teams. The Gossamer Condor's external bracing wires, canard, and wing warping strongly echo the earliest airplanes, and its many patches and taped-up repairs made it look jury-rigged and hastily thrown together. (The Gossamer Condor today hangs in the Smithsonian's National Air and Space Museum in

Washington, D.C., for all to see.) Misled by the Gossamer Condor's appearance, some European writers described it as "unsophisticated" and attributed its success to luck [8]. Luck may have played a small role, but thousands of hours of work, the ability to make frequent test flights at their desert airport, plus MacCready's insistence on easily repaired structures, gave the Gossamer Condor team much more flight time in a few months than other teams achieved in years [8, 9].

Within less than two years, the Gossamer Condor's critics were well and truly silenced. In 1978, Henry Kremer put up money for a new prize: he offered £100,000 for the first HPA to fly across the English Channel. MacCready rose to the challenge, and his team built the Gossamer Albatross. This new craft was a lighter, safer, and more high-tech version of the Gossamer Condor. On June 12, 1979, with Bryan Allen again at the controls, the Gossamer Albatross flew more than 30 kilometers and successfully crossed the English Channel from England to France [8].

Why So Little "Technology Transfer"?

Given that evolution has been refining and improving the flight ability of animals for millions of years, why have animal flight mechanisms had so little direct influence on airplane design? Why don't fighter jets look more like falcons, or airliners look more like storks? The most important factor that led airplane design along a different path from that taken by flying animals' is scale differences—that is, differences in size and speed. In order to carry one or more people, an airplane must be big: even small airplanes are bigger than any flying animals.* The pilot and passengers are essentially deadweight, contributing nothing to lift or thrust. Moreover, airplanes must carry one or more engines for thrust, so they must be far heavier than any flying animals. Since they are so heavy, airplanes must fly much faster than animals just to stay airborne. This minimum flight speed is just the beginning; higher speeds often have practical advantages, so airplane builders often design their craft to cruise at much higher speeds than the minimum practical speed. Scaled-

* Quetzalcoatlus northropi, the giant pterosaur, did have a wingspan about the same as a small airplane, but such an airplane is much longer and taller than the pterosaur, not to mention perhaps eight or nine times heavier.

up animal wings simply would not work well at these speeds. Animal wings operate over a Reynolds number range of about 50 to 100,000, while a Reynolds number of 1,000,000 is relatively low for an airplane (Chapter 2). Lift coefficients, drag coefficients, and lift-to-drag ratios for any given wing can be quite different at such different Reynolds numbers. A dragonfly wing scaled up to fit a jumbo jet would have a terrible lift-to-drag ratio compared to the airplane's actual wing; conversely, a jumbo jet wing scaled down to dragonfly-size might be aerodynamically effective, but would probably be no better than the insect's actual wing, and a miniaturized airplane wing would be too heavy and fragile for the insect to flap.

A second area where animals and airplanes face different design constraints is in building materials. Steven Vogel recently noted that nature tends to build with wet, flexible, nonmetallic, self-repairing materials, while humans build with dry, rigid materials, including metals, which do not heal from damage [2]. Because of these material limitations, animals face size limits. A bee's wing and a sparrow's feather are miracles of structural engineering, amazingly strong for their weight and length. But they would lose this advantage when scaled up, because weight increases with the length cubed. Doubling a feather's length but keeping its proportions the same increases its weight eight times. Long before they got big enough to use on airplanes, scaled-up insect wings or bird feathers would actually become too heavy for their strength, which is clearly undesirable for a wing. Thus, differences in materials lead us back to differences in scale: because of the material they are made of, animal wings cannot be scaled up past certain limits, so flying animals cannot get big enough to form the basis for useful airplane designs.

Animals use a completely different method of producing thrust, and this also limits their usefulness as patterns for airplane wing designs. Because animals get power from their muscles, and animals cannot use wheels or axles for rotary motion, they must flap their wings for thrust (Chapter 4). We humans use rotating motors of one kind or another for power in our machines, and converting rotary motion to the linear up-and-down motion of flapping is distinctly inefficient. Instead, we use rotating propellers, or rotating compressors and turbines in jet engines, which are much more efficient at producing thrust. As we saw in Chapter 4, the flexibility and adjustability of an animal wing is important in flapping. In contrast, airplanes do not use their wings for thrust, only for lift. Unintended or unexpected wing flexing can play havoc with an airplane's control and stability. Thus, flexibility is generally something that aeronautical engineers try to avoid, and even mission-adaptive "smart" wings will be much more rigid and less flexible than a typical animal wing.

A more subtle reason why the design of flying animals is not a suitable pattern for flying machines is that the functional constraints on the wings of animals are vastly different and more complex than those on airplane wings. Not only must animal wings be flapped for propulsion, they must usually be folded into a compact package when the animal is walking, swimming, or climbing. Moreover, animal wings serve an amazing variety of secondary functions in different animal groups. They can be used for solar collectors (many insects, some birds), visual or sound signals for recognition or courtship (many birds and insects), climbing and catching prey (many bats, some birds, probably pterosaurs), deception (the "broken wing" display of killdeers), or camouflage (moths). And this is far from an exhaustive list. Because airplane wings generally do not move, and can only change shape slightly (with flaps and other control surfaces), their aerodynamic properties are the overriding concern of their designer. Secondary functions, such as carrying fuel or landing gear, are essentially incidental compared to the multiple functions performed by animal wings.

Tiny Flying Machines

In the past few years, extremely lightweight materials and ultraminiaturized motors and actuators have become available. These tiny components allow engineers to build complex machines on the same scale as flying animals, and a number of researchers have begun to design tiny, autonomous flying machines. This field may finally get practical benefits from borrowing directly from nature.

For many years, military establishments have worked to develop small, pilotless surveillance aircraft. Dubbed "UAVs" (unmanned aerial vehicles), their advantages are obvious: they can fly over places too dangerous to risk a human pilot, and being small, they are harder to detect than a manned aircraft. With small, powerful computers now available, UAV's can be made "smart" enough to operate completely independently of human control (except for takeoff and landing). Continuing advances in miniaturization have prompted researchers to study the logical extension of this approach: if a UAV half the size of a Piper Cub can approach to within a kilometer of its surveillance target without being detected, think how close a UAV the size of a crow, or even a dragonfly, could get. Since the mid 1990s, the U.S. Department of Defense has been supporting research on "micro-UAVs," or MAVs, rather openly, with the goal of eventually producing insect-sized autonomous aircraft. In addition to their military uses, MAVs offer great advantages for civilian search and rescue operations. A hummingbird-sized flying robot could quickly and safely search for survivors of disasters and survey damage in places where human rescuers could only move slowly and at great risk to themselves and the survivors. MAVs might be of great help searching through collapsed tunnels or mines, or buildings damaged by earthquakes. They could also gather data in places where humans cannot go, such as inside damaged nuclear reactors or at the sites of dangerous chemical spills.

Although little specific information on this research is publicly available, most researchers are apparently taking the approach of scaling down conventional aircraft, if rumors of thumb-sized turbojet engines are accurate. However, at least one research group is taking the "reverse-engineering" approach and looking to nature for hard data, not just inspiration. This group's goal is to build a tiny ornithopter patterned closely on the shape and movements of a large dragonfly. If they can produce a wing-flapping mechanism and a control system small and light enough, they may find that this is one area where following nature's example is more effective than using the approach of conventional aeronautical engineering.

More "Borrowing" in the Future?

Because of the limitations of scale effects, MAVs are probably the only area where engineers will borrow major design features directly from flying animals. Full-sized aircraft will continue to develop along a separate path from flying animals; feathered airplanes or person-carrying ornithopters will not replace conventional aircraft in the foreseeable future. However, one area of convergence between airplanes and animals may benefit from techniques borrowed from animals: control systems. Biologists have long known that insects can generate quite complex behavior with anatomically simple brains, and engineers and robot designers have begun to show interest in insect control systems. In fact, this is currently a thriving area of research. Engineers may find that by emulating animal control systems, they can make aircraft computer control systems simpler and more adaptable, leading to safer and more reliable airplanes.

ONE: Introduction

1. J. H. McMasters 1986, *Perspectives in Biology and Medicine* 29.
2. M. H. Dickinson, F.-O. Lehmann, and S. P. Sane 1999, *Science* 284.
3. C. van den Berg and C. P. Ellington 1997a, *Philosophical Transactions of the Royal Society of London*, B, 352.
4. C. van den Berg and C. P. Ellington 1997b, *Philosophical Transactions of the Royal Society of London*, B, 352.
5. A. K. Brodsky 1994, *The Evolution of Insect Flight*.
6. M. S. Vest and J. Katz 1995, *AIAA Journal* 34.
7. M. J. C. Smith, P. J. Wilkin, and M. H. Williams 1996, *Journal of Experimental Biology* 199.
8. M. J. C. Smith 1996, *AIAA Journal* 34.

TWO: How Wings Work

1. S. Vogel 1994, *Life in Moving Fluids*.
2. D. E. Alexander and T. Chen 1990, *Journal of Crustacean Biology* 10.
3. D. E. Alexander 1990, *Biological Bulletin* 179.
4. S. Vogel 1981, *Life in Moving Fluids*.
5. G. Rüppell 1977, *Bird Flight*.
6. C. J. C. Rees 1975a, *Nature* 258.
7. B. G. Newman, S. B. Savage, and D. Schouella 1977, in *Scale Effects in Animal Locomotion,* ed. T. J. Pedley.
8. C. W. Burkett 1989, *Aeronautical Journal* 93.
9. S. F. Hoerner and H. V. Borst 1975, *Fluid-Dynamic Lift*.
10. S. Vogel 1967, *Journal of Experimental Biology* 46.
11. C. J. C. Rees 1975b, *Nature* 256.
12. S. Vogel 1988, *Life's Devices*.

THREE: Gliding and Soaring

1. J. M. V. Rayner 1981, in *Vertebrate Locomotion,* ed. M. H. Day.
2. J. A. Oliver 1951, *American Naturalist* 85.
3. H. E. Edgerton and C. M. Breder 1941, *Zoologica* 26.
4. J. Davenport 1994, *Reviews in Fish Biology and Fisheries* 4.

5. E. H. Colbert 1967, *American Museum Novitates* 2283.

6. R. W. J. Thorington and L. R. Heaney 1981, *Journal of Mammology* 62.

7. N. Wells-Gosling 1985, *Flying Squirrels*.

8. S. Vogel 1994, *Life in Moving Fluids*.

9. H. Tennekes 1996, *The Simple Science of Flight*.

10. G. Rüppell 1977, *Bird Flight*.

11. C. H. Zimmerman 1932, *National Advisory Committee on Aeronautics* 431.

12. A. Azuma and Y. Okuno 1987, *Journal of Theoretical Biology* 129.

13. F. Galé 1991, *Tailless Tale*.

14. R. Å. Norberg 1973, *Biological Reviews* 48.

15. A. Rosen and D. Seter 1991, *Journal of Applied Mechanics* 58.

16. D. Seter and A. Rosen 1992a, *Journal of Applied Mechanics* 59.

17. C. W. McCutchen 1977, *Science* 197.

18. R. Dudley and P. DeVries 1990, *Biotropica* 22.

19. G. A. Hazlehurst and J. M. V. Rayner 1992, *Paleobiology* 18.

20. C. J. Pennycuick 1975, in *Avian Biology,* ed. D. S. Farner and J. R. King.

21. U. M. Norberg 1990, *Vertebrate Flight*.

22. C. J. Pennycuick 1982, *Philosophical Transactions of the Royal Society of London,* B, 300.

23. V. A. Tucker 1993, *Journal of Experimental Biology* 180.

24. V. A. Tucker 1995, *Journal of Experimental Biology* 198.

25. V. A. Tucker and G. C. Parrott 1970, *Journal of Experimental Biology* 52.

26. V. A. Tucker 1987, *Journal of Experimental Biology* 133.

27. V. A. Tucker and C. Heine 1990, *Journal of Experimental Biology* 149.

28. U. M. Norberg and J. M. V. Rayner 1987, *Philosophical Transactions of the Royal Society of London,* B, 316.

29. P. A. Cox 1983, *Mammalia* 47, and unpublished observations of E. D. Pierson and W. E. Rainey.

30. M. J. Benton 1999, *Philosophical Transactions of the Royal Society of London,* B, 354.

31. W. Langston Jr. 1981, *Scientific American* 224.

32. J. H. McMasters 1976, *Science* 191.

33. D. L. Gibo 1981, *Journal of the New York Entomological Society* 89.

34. S. Vogel 1967, *Journal of Experimental Biology* 46.

35. R. Dudley and C. P. Ellington 1990a, *Journal of Experimental Biology* 148.

36. R. Dudley and C. P. Ellington 1990b, *Journal of Experimental Biology* 148.

37. W. Nachtigall 1977, *Fortschritte der Zoologie* 24 2/3.

38. C. J. Pennycuick 1971a, *Journal of Experimental Biology* 55.

39. J. Roskam 1979, *Airplane Flight Dynamics and Automatic Flight Controls*.

40. J. Davenport 1992, *Journal of the Marine Biological Association of the United Kingdom* 72.

41. S. Vogel 1981, *Life in Moving Fluids*.

42. C. K. Augspurger 1986, *American Journal of Botany* 73.

43. A. R. Ennos 1989, *Journal of Zoology* 219.

44. D. Seter and A. Rosen 1992b, *Biological Reviews* 67.

45. D. S. Green 1980, *American Journal of Botany* 67.

FOUR: Flapping and Hovering

1. U. M. Norberg 1976, *Journal of Experimental Biology* 65.

2. C. J. Pennycuick 1971b, *Journal of Experimental Biology* 55.

3. V. A. Tucker 1968, *Journal of Experimental Biology* 48.

4. V. A. Tucker and G. C. Parrott 1970, *Journal of Experimental Biology* 52.

5. D. E. Alexander 1986, *Journal of Experimental Biology* 122.

6. J. M. V. Rayner 1995a, *Symposium of the Society for Experimental Biology* 49.

7. G. R. Spedding 1992, in *Mechanics of Animal Locomotion,* ed. R. McN. Alexander.

8. R. J. H. Brown 1953, *Journal of Experimental Biology* 30.

9. R. J. H. Brown 1948, *Journal of Experimental Biology* 25.

10. G. R. Spedding, J. M. V. Rayner, and C. J. Pennycuick 1984, *Journal of Experimental Biology* 111.

11. G. R. Spedding 1986, *Journal of Experimental Biology* 125.

12. W. Nachtigall 1980, in *Aspects of Animal Movements,* ed. H. Y. Elder and E. R. Trueman.

13. J. M. V. Rayner, G. Jones, and A. Thomas 1986, *Nature* 321.

14. J. M. V. Rayner 1987, in *Recent Advances in the Study of Bats,* ed. M. B. Fenton, P. Racey, and J. M. V. Rayner.

15. K. Schmidt-Nielsen 1990, *Animal Physiology.*

16. T. A. McMahon 1984, *Muscles, Reflexes, and Locomotion.*

17. A. K. Brodsky 1994, *The Evolution of Insect Flight.*

18. J. M. V. Rayner 1991b, in *Acta XX Congressus Internationalis Ornithologici,* ed. B. D. Bell et al.

19. T. Weis-Fogh 1973, *Journal of Experimental Biology* 59.

20. R. F. Chapman 1982, *The Insects.*

21. T. Weis-Fogh 1956, *Philosophical Transactions of the Royal Society of London,* B, 239.

22. C. J. Pennycuick 1972a, *Animal Flight.*

23. J. M. V. Rayner 1995b, *Israel Journal of Zoology* 41.

24. R. H. J. Brown 1963, *Biological Reviews* 38.

25. J. McGahan 1973, *Journal of Experimental Biology* 58.

26. J. H. McMasters 1989, *American Scientist* 77.

27. T. Weis-Fogh and M. Jensen 1956, *Philosophical Transactions of the Royal Society of London,* B, 239.

28. M. Jensen 1956, *Philosophical Transactions of the Royal Society of London,* B, 239.

29. U. M. Norberg 1990, *Vertebrate Flight.*

30. J. M. V. Rayner 1980, in *Aspects of Animal Movement,* ed. H. Y. Elder and E. R. Trueman.

31. U. M. Norberg 1975, in *Swimming and Flying in Nature,* ed. T. Y. Wu, C. J. Brokaw, and C. Brennen.

32. H. H. Dathe 1982, *Zoologische Jahrbücher* 86.

33. R. Å. Norberg 1975, in *Swimming and Flying in Nature,* ed. T. Y. Wu, C. J. Brokaw, and C. Brennan.

34. R. Dudley and C. P. Ellington 1990b, *Journal of Experimental Biology* 148.

35. M. Cloupeau, J.F. Devillier, and D. Devezeau 1979, *Journal of Experimental Biology* 80.

36. T. Weis-Fogh 1975, in *Swimming and Flying in Nature,* ed. T. Y. Wu, C. J. Brokaw, and C. Brennan.

37. C. P. Ellington 1984c, *Philosophical Transactions of the Royal Society of London,* B, 305.

38. T. Maxworthy 1979, *Journal of Fluid Mechanics* 93.

39. M. J. Lighthill 1973, *Journal of Fluid Mechanics* 60.

40. H. J. Haussling 1979, *Journal of Comparative Physiology* 30.

41. R. H. Edwards and H. K. Cheng 1982, *Journal of Fluid Mechanics* 120.

42. G. R. Spedding and T. Maxworthy 1986, *Journal of Fluid Mechanics* 165.

43. W. S. Farren 1935, *Aeronautical Research Committee Reports and Memoranda* 1648.

44. M. H. Dickinson, F.-O. Lehmann, and S. P. Sane 1999, *Science* 284.

45. J. M. V. Rayner 1979a, *Journal of Experimental Biology* 80.

46. J. M. V. Rayner 1979b, *Journal of Fluid Mechanics* 91.

47. C. P. Ellington 1984a, *Philosophical Transactions of the Royal Society of London,* B, 305.

48. C. P. Ellington 1984d, *Philosophical Transactions of the Royal Society of London,* B, 305.

49. G. R. Spedding 1987, *Journal of Experimental Biology* 127.

50. A. Azuma and T. Watanabe 1988, *Journal of Experimental Biology* 137.

51. M. A. Reavis and M. W. Luttges 1988, *AIAA* 88-0330.

52. C. Somps and M. W. Luttges 1985, *Science* 228.

53. C. P. Ellington et al. 1996, *Nature* 384.

54. C. van den Berg and C. P. Ellington 1997a, *Philosophical Transactions of the Royal Society of London,* B, 352.

55. C. van den Berg and C. P. Ellington 1997b, *Philosophical Transactions of the Royal Society of London,* B, 352.

56. A. P. Willmot, C. P. Ellington, and A. L. R. Thomas 1997, *Philosophical Transactions of the Royal Society of London,* B, 352.

57. M. S. Vest and J. Katz 1995, *AIAA Journal* 34.

58. C. P. Ellington 1984b, *Philosophical Transactions of the Royal Society of London,* B, 305.

59. M. J. Lighthill 1977, in *Scale Effects in Animal Locomotion,* ed. T. J. Pedley.

60. Y. Hirashima, M. Inokuchi, and K. Yamagishi 1999, *Esakia* 39.

61. D. Boag and M. Alexander 1986, *The Atlantic Puffin.*

62. D. N. Nettleship and P. G. H. Evans 1985, in *The Atlantic Alcidae,* ed. D. N. Nettleship and T. R. Birkhead.

63. F. C. Wiest 1995, *Journal of Zoology* 236, 4.

64. J. D. DeLaurier 1993a, *Aeronautical Journal* 97.

65. J. D. DeLaurier and J. M. Harris 1993, *Aeronautical Journal* 97.

66. J. D. DeLaurier 1993b, *Aeronautical Journal* 97.

67. J. M. V. Rayner 1991a, in *Biomechanics in Evolution,* ed. J. M. V. Rayner and R. J. Wootton.

68. H. Aldridge 1986, in *Bat Flight—Fledermausflug,* ed. W. Nachtigall.

69. W. Nachtigall 1966, *Zeitschrift für Vergleichende Physiologie* 52.

70. J. M. Zanker 1990, *Philosophical Transactions of the Royal Society of London, B,* 237.

71. A. Azuma et al. 1985, *Journal of Experimental Biology* 116.

72. M. F. M. Osborne 1951, *Journal of Experimental Biology* 28.

73. W. Nachtigall 1974, *Insects in Flight.*

74. J. M. V. Rayner 1986, in *Bat Flight—Fledermausflug,* ed. W. Nachtigall.

FIVE: Staying on Course and Changing Direction

1. J. Roskam 1979, *Airplane Flight Dynamics and Automatic Flight Controls.*

2. C. P. Perkins and R. E. Hage 1949, *Airplane Performance Stability and Control.*

3. J. Maynard Smith 1952, *Evolution* 6.

4. A. L. R. Thomas 1996a, *Journal of Theoretical Biology* 183.

5. A. L. R. Thomas 1996b, *Journal of Theoretical Biology* 183.

6. P. S. Baker 1979, *Journal of Comparative Physiology A* 131.

7. A. J. Burton 1964, *Nature* 204.

8. P. Schneider and B. Krämer 1974, *Journal of Comparative Physiology* 91.

9. D. E. Alexander 1986, *Journal of Experimental Biology* 122.

10. H. D. J. N. Aldridge 1987, *Journal of Experimental Biology* 128.

11. D. R. Warrick and K. P. Dial 1998, *Journal of Experimental Biology* 201.

12. E. Boettiger and E. Furshpan 1952, *Biological Bulletin* 102.

13. W. Nachtigall and D. M. Wilson 1967, *Journal of Experimental Biology* 47.

14. J. J. Duggard 1967, *Journal of Insect Physiology* 13.

15. L. J. Goodman 1965, *Journal of Experimental Biology* 42.

16. G. Stange and J. Howard 1979, *Journal of Experimental Biology* 83.

17. C. P. Taylor 1981, *Journal of Experimental Biology* 93.

18. K. G. Götz 1968, *Kybernetic* 4.

19. R. D. DeVoe, W. Kaiser, J. Ohm and L. S. Stone 1982, *Journal of Comparative Physiology A* 147.

20. W. H. Warren and D. J. Hannon 1988, *Nature* 336.

21. M. R. Iddotson 1991, *Journal of Experimental Biology* 157.

22. A. Baarder, M. Schaefer, and C. H. F. Rowell 1992, *Journal of Experimental Biology* 165.

23. M. Gewecke 1974, in *Experimental Analysis of Insect Behavior,* ed. L. Barton-Browne.

24. J. Baker and M. Tyler 1979, *Physiological Entomology* 4.

25. M. Gewecke and M. Niehaus 1981, *Journal of Comparative Physiology* 145.

26. H. Mittelstaedt 1950, *Zeitschrift für Vergleichende Physiologie* 32.

27. J. W. S. Pringle 1948, *Philosophical Transactions of the Royal Society,* B, 233.

28. J. W. S. Pringle 1957, *Insect Flight.*

29. U. M. Norberg and J. M. V. Rayner 1987, *Philosophical Transactions of the Royal Society of London,* B, 316.

30. U. M. Norberg 1994, in *Ecological Morphology,* ed. P. C. Wainwright and S. M. Reilly.

SIX: Fueling Flight

1. A. F. Huxley 1974, *Journal of Physiology* 243.

2. D. S. Smith 1972, *Muscle.*

3. T. A. McMahon 1984, *Muscles, Reflexes, and Locomotion.*

4. T. Weis-Fogh and R. McN. Alexander 1977, in *Scale Effects in Animal Locomotion,* ed. T. J. Pedley.

5. K. Schmidt-Nielsen 1972, *Science* 177.

6. K. Schmidt-Nielsen 1990, *Animal Physiology.*

7. P. C. Withers 1992, *Comparative Animal Physiology.*

8. J. M. V. Rayner 1990, in *Bird Migration,* ed. E. Gwinner.

9. J. M. V. Rayner 1995b, *Israel Journal of Zoology* 41.

10. T. Alerstam 1981, in *Animal Migration,* ed. D. J. Aidley.

11. C. J. Pennycuick 1972b, *Ibis* 114.

12. C. J. Pennycuick, T. Alerstam, and B. Larsson 1979, *Ornis Scandinavica* 10.

13. A. Kvist, M. Klaasen, and Å. Lindström 1998, *Journal of Avian Biology* 29.

14. V. A. Tucker 1972, *American Journal of Physiology* 222.

15. V. A. Tucker 1968, *Journal of Experimental Biology* 48.

16. J. R. Torre-Bueno and J. Larochelle 1978, *Journal of Experimental Biology* 75.

17. K. P. Dial et al. 1997, *Nature* 390.

18. C. P. Ellington 1991b, *Journal of Experimental Biology* 160.

19. J. M. V. Rayner 1994, *Journal of Zoology* 234.

20. J. M. V. Rayner 1999, *Journal of Experimental Biology* 202.

SEVEN: Evolving Flyers

1. A. R. Ennos 1988, *Journal of Experimental Biology* 140.

2. D. Pomeroy 1990, *Biological Journal of the Linnean Society* 40.

3. S. J. Gould 1985, *Natural History* 94.

4. J. G. Kingsolver and M. A. R. Koehl 1994, *Annual Review of Entomology* 39.

5. R. J. Wootton 1990, in *Major Evolutionary Radiations,* ed. P. D. Taylor and G. P. Larwood.

6. C. P. Ellington 1991a, *Advances in Insect Physiology* 23.

7. J. H. Marden and M. G. Kramer 1994, *Science* 266.

8. R. D. Alexander and W. L. Brown 1963, *Occasional Papers of the Museum of Zoology of the University of Michigan* 628.

9. J. G. Kingsolver and M. A. R. Koehl 1985, *Evolution* 39.

10. R. J. Wootton and C. P. Ellington 1991, in *Biomechanics in Evolution,* ed. J. M. V. Rayner and R. J. Wootton.

11. A. K. Brodsky 1994, *The Evolution of Insect Flight.*

12. J. B. Graham et al. 1995, *Nature* 375.

13. R. Dudley 1998, *Journal of Experimental Biology* 201.

14. G. R. Spedding 1986, *Journal of Experimental Biology* 125.

15. P. J. Currie 1991, *The Flying Dinosaurs.*

16. M. J. Benton 1999, *Philosophical Transactions of the Royal Society,* B, 354.

17. C. J. Pennycuick 1988, *Biological Reviews* 63.

18. K. Padian 1983, *Paleobiology* 9.

19. K. Padian 1991, in *Biomechanics in Evolution,* ed. J. M. V. Rayner and R. J. Wootton.

20. K. Padian and J. M. V. Rayner 1993, *American Journal of Science* 293A.

21. S. C. Bennett 1997b, *Journal of Vertebrate Paleontology* 17.

22. D. M. Unwin 1997, *Lethaia* 29.

23. P. Wellnhofer 1975, *Palaeontographica* A148.

24. Ibid., A149.

25. S. C. Bennett 1997a, *Historical Biology* 12.

26. D. M. Unwin and N. N. Bakhurina 1994, *Nature* 371.

27. W. Langston Jr. 1981, *Scientific American* 224.

28. K. Padian 1985, *Palaeontology* 28.

29. D. A. Lawson 1975, *Science* 187.

30. J. M. V. Rayner 1988b, *Current Ornithology* 5.

31. G. A. Hazlehurst and J. M. V. Rayner 1992, *Paleobiology* 18.

32. G. Heilmann 1927, 1972, *The Origin of Birds.*

33. L. Dingus and T. Rowe 1998, *The Mistaken Extinction.*

34. A. D. Walker 1972, *Nature* 237.

35. J. H. Ostrom 1973, *Nature* 242.

36. J. H. Ostrom 1975, *Annual Review of Earth and Planetary Sciences* 3.

37. J. A. Gauthier 1986, in *The Origin of Birds and the Evolution of Flight,* ed. K. Padian.

38. K. Padian and L. M. Chiappe 1988, *Scientific American* 278, 2.

39. L. D. Martin 1991, in *Origins of the Higher Groups of Tetrapods,* ed. H.-P. Schultze and L. Trueb.

40. A. C. Burke and A. Feduccia 1997, *Science* 278, 5335.

41. G. P. Wagner and J. A. Gauthier 1999, *Proceedings of the National Academy of Sciences* 96.

42. R. D. Dahn and J. F. Fallon 2000, *Science* 289.

43. A. Feduccia 1996, *The Origin and Evolution of Birds.*

44. J. Welman 1995, *South African Journal of Science* 91, 10.

45. J. Ackerman 1998, *National Geographic* 194, 1.

46. F. E. Novas and P. F. Puerta 1997, *Nature* 387, 6631.

47. Q. Ji et al. 1998, *Nature* 393.

48. L. M. Chiappe, M. A. Norell, and J. M. Clark 1998, *Nature* 392.

49. J. H. Ostrom 1974, *Quarterly Review of Biology* 49.

50. J. H. Ostrom 1979, *American Scientist* 67.

51. R. Å. Norberg 1985, in *The Beginnings of Birds,* ed. M. K. Hecht et al.

52. A. Feduccia and H. G. Tordoff 1979, *Science* 203.

53. J. M. V. Rayner 1991a, in *Biomechanics in Evolution,* ed. id. and R. J. Wootton.

54. A. Feduccia 1980, *The Age of Birds.*

55. R. T. Bakker 1975, *Scientific American* 232.

56. R. T. Bakker 1986, *The Dinosaur Heresies.*

57. P. J. Chen, Z. M. Dong, and S. N. Zhen 1998, *Nature* 391, 6663.

58. J. Ruben 1996, *Society for Experimental Biology Seminar Series* 59.

59. J. A. Ruben et al. 1996, *Science* 273, 5279.

60. J. A. Ruben et al. 1997, *Science* 278, 5341.

61. J. A. Ruben 1991, *Evolution* 45.

62. J. A. Ruben 1993, *Evolution* 47.

63. G. R. Caple, R. T. Balda, and W. R. Willis 1983, *American Naturalist* 121.

64. J. M. V. Rayner 1988a, *Biological Journal of the Linnean Society* 34.

65. J. M. V. Rayner 1985, in *The Beginnings of Birds,* ed. M. K. Hecht et al.

66. U. M. Norberg 1985, in *The Beginnings of Birds,* ed. M. K. Hecht et al.

67. U. M. Norberg 1990, *Vertebrate Flight.*

68. W. J. Bock 1983, *The Sciences* 23.

69. S. L. Olson and A. Feduccia 1979, *Nature* 278.

70. M. K. Hecht et al. 1985, *The Beginnings of Birds.*

71. P. Dodson 1985, *Journal of Vertebrate Paleontology* 5.

72. K. Padian and L. M. Chiappe 1998, *Biological Reviews* 73.

73. J. P. Garner, G. K. Taylor, and A. L. R. Thomas 1999, *Proceedings of the Royal Society of London,* B, 266.

74. A. J. Charig et al. 1986, *Science* 232, 4750.

75. S. Rietschel 1985, in *The Beginnings of Birds,* ed. M. K. Hecht et al.

76. C. C. Swisher et al. 1999, *Nature* 400.

77. L. M. Chiappe et al. 1999, *Bulletin of the American Museum of Natural History* 242.

78. L. D. Martin et al. 1998, *Naturwissenschaften* 85.

79. K. Scholey 1986, in *Bat Flight—Fledermausflug,* ed. W. Nachtigall.

80. N. B. Simmons and J. H. Geisler 1998, *Bulletin of the American Museum of Natural History* 235.

81. U. M. Norberg 1986, in *Bat Flight—Fledermausflug,* ed. W. Nachtigall.

82. J. Rydell and J. R. Speakman 1995, *Biological Journal of the Linnean Society* 54.

83. N. B. Simmons 1995, *Symposium of the Zoological Society of London* 67.

84. J. D. Altringham 1996, *Bats.*

85. G. L. Jepsen 1970, in *Biology of Bats,* ed. W. A. Wimsatt.

86. P. Pirlot 1977, in *Major Patterns in Vertebrate Evolution,* ed. M. K. Hecht, P. C. Goody, and B. M. Hecht.

87. J. M. V. Rayner 1986, in *Bat Flight—Fledermausflug,* ed. W. Nachtigall.

88. J. D. Pettigrew 1986, *Science* 231.

89. J. A.W. Kirsch et al. 1995, *Australian Journal of Zoology* 43.

90. J. Kukalova 1970, *Psyche* 77.

91. S. C. Bennett 1995, *Pterosaurs.*

92. J. D. Smith 1977, in *Major Patterns in Vertebrate Evolution,* ed. M. K. Hecht, P. C. Goody, and B. M. Hecht.

93. J. H. Ostrom 1985, in *The Beginnings of Birds,* ed. M. K. Hecht et al.

EIGHT: Migrating

1. H. Dingle 1996, *Migration.*

2. H. Dingle 1985, in *Comprehensive Insect Physiology, Biochemistry and Pharmacology,* ed. G. A. Kerkut and L. I. Gilbert.

3. R. R. Baker 1978, *The Evolutionary Ecology of Animal Migration.*

4. T. Alerstam 1981, in *Animal Migration,* ed. D. J. Aidley.

5. R. Bainbridge 1958, *Journal of Experimental Biology* 35.

6. J. M. V. Rayner 1999, *Journal of Experimental Biology* 202.

7. H. Tennekes 1996, *The Simple Science of Flight.*

8. D. E. Alexander 1986, *Journal of Experimental Biology* 122.

9. V. R. Dolnik 1990, in *Bird Migration,* ed. E. Gwinner.

10. C. J. Pennycuick 1975, in *Avian Biology,* ed. D. S. Farner and J. R. King.

11. C. J. Pennycuick, T. Alerstam, and B. Larsson 1979, *Ornis Scandinavica* 10.

12. C. J. Pennycuick 1972b, *Ibis* 114.

13. D.W. Thomas 1983, *Canadian Journal of Zoology* 61.

14. R.C. Rainey 1951, *Nature* 168.

15. R. A. Farrow 1990, in *Biology of Grasshoppers,* ed. R. F. Chapman and A. Joern.

16. C. Solbreck 1985, *Contributions in Marine Science* 27.

17. C. B. Williams 1958, *Insect Migration.*

18. G. C. Aymar 1938, *Bird Flight.*

19. R. T. Orr 1970, *Animals in Migration.*

20. V. M. Dirsh 1965, *The African Genera of Acridoidea.*

NINE: Finding the Way

1. F. C. Dyer and J. L. Gould 1981, *Science* 214.

2. R. Wiltschko 1991, in *Orientation in Birds,* ed. P. Berthold.

3. T. S. Collett and J. Baron 1994, *Nature* 368.

4. R. Wehner, B. Michel, and P. Antonsen 1996, *Journal of Experimental Biology* 199.

5. F. C. Dyer 1996, *Journal of Experimental Biology* 199.

6. J. A. Dickinson 1994, *Naturwissenschaften* 81.

7. R. Wehner 1996, *Journal of Experimental Biology* 199.

8. R. Wiltschko and W. Wiltschko 1981, *Behavioral Ecology and Sociobiology* 9.

9. K. P. Able and M. A. Able 1996, *Journal of Experimental Biology* 199.

10. K. Schmidt-Koenig, J. U. Ganzhorn, and R. Ranvand 1991, in *Orientation in Birds,* ed. P. Berthold.

11. W. Wiltschko and R. Wiltschko 1972, *Science* 176.

12. W. Wiltschko and F. W. Merkel 1966, *Deutsche Zoologische Gesellschaft Verhandlungen* 59.

13. R. R. Baker and J. G. Mather 1982, *Animal Behaviour* 30.

14. R. Jander and U. Jander 1998, *Ethology* 104.

15. M. L. Winston 1987, *The Biology of the Honey Bee.*

16. J. L. Kirschvink, D. S. Jones, and B. J. MacFadden, eds., 1985, *Magnetite Biomineralization and Magnetoreception in Organisms.*

17. A. J. Helbig 1991, in *Orientation in Birds,* ed. P. Berthold.

18. S. T. Emlen 1967, *Auk* 84.

19. W. Wiltschko et al. 1987, *Ethology* 74.

20. W. J. Richardson 1991, in *Orientation in Birds,* ed. P. Berthold.

21. P. Berthold 1996, *Control of Bird Migration.*

22. T. Alerstam 1996, *Journal of Experimental Biology* 199.

23. F. Papi and P. Luschi 1996, *Journal of Experimental Biology* 199.

24. E. Kuyt 1992, *Canadian Wildlife Service Occasional Papers* 74.

25. A. C. Perdeck 1958, *Ardea* 46.

26. A. J. Helbig 1996, *Journal of Experimental Biology* 199.

TEN: The Global Impact of Animal Flight

1. I. W. B. Thornton 1996, in *The Origin and Evolution of Pacific Island Biotas, New Guinea to Eastern Polynesia,* ed. A. Keast and S. E. Miller.

2. C. Rosenfeld and R. Cooke 1982, *Earthfire.*

3. R. A. Morse 1992, in *Insect Potpourri,* ed. J. Adams.

4. D.C. Constantine 1970, in *Biology of Bats,* ed. W. A. Wimsatt.

5. M. B. Fenton 1983, *Just Bats.*

6. W. L. Crepet 1983, in *Pollination Biology,* ed. L. Real.

7. S. L. Buchmann and G. P. Nabhan 1996, *The Forgotten Pollinators.*

8. E. Crane and P. Walker 1984, *Pollination Directory for World Crops.*

9. R. C. Brusca and G. J. Brusca 1990, *Invertebrates.*

10. R. S. K. Barnes, ed., 1998, *The Diversity of Living Organisms.*

11. H. F. Pough, C. M. Janis, and J. B. Heiser 1999, *Vertebrate Life.*

12. J. B. Free 1970, *Insect Pollinations of Crops.*

13. M. J. Crawley 1983, *Herbivory.*

14. J. H. Comstock 1940, *An Introduction to Entomology.*

15. D. S. Hill 1997, *The Economic Importance of Insects.*

16. R. E. Pfadt 1971b, *Fundamentals of Applied Entomology.*

17. R. E. Pfadt 1971a, in *Fundamentals of Applied Entomology,* ed. id.

18. T. Kono and C. S. Papp 1977, *Handbook of Agricultural Pests.*

19. W. D. Fronk 1971, in *Fundamentals of Applied Entomology,* ed. R. E. Pfadt.

20. R. P. Scheffer 1997, *The Nature of Disease in Plants.*

21. J. A. Walsh 1984, in *Tropical and Geographical Medicine,* ed. K. S. Warren and A. A. F. Mahmoud.

22. J. L. Cloudsley-Thompson 1976, *Insects and History.*

23. G. C. Shattuck 1951, *Diseases of the Tropics.*

24. H. D. Pratt 1992, in *Insect Potpourri,* ed. J. Adams.

25. T. P. Monath 1984, in *Tropical and Geographical Medicine,* ed. K. S. Warren and A. A. F. Mahmoud.

26. K. G. V. Smith 1973, *Insects and Other Arthropods of Medical Importance.*

27. J. W. Bastien 1998, *The Kiss of Death.*

28. D. Wigg 1993, *And Then Forgot to Tell Us Why.*

29. G. K. Meffe and C. R. Carroll 1997, *Principles of Conservation Biology.*

30. O. S. Owen, D. D. Chiras, and J. P. Reganold 1998, *Natural Resource Conservation.*

31. J. Page 1990, *Smithsonian* 21, 6.

32. P. W. Hedrick 1984, *Population Biology.*

33. C. Howes 1997, *The Spice of Life.*

34. J. D. Williams and R. M. Nowak 1993, in *The Last Extinction,* ed. L. Kaufman and K. Mallory.

35. R. Carson 1962, *Silent Spring.*

36. G. W. Cox 1997, *Conservation Biology.*

37. P. A. Cox and T. Elmqvist 1993, *Pacific Conservation* 1.

38. E. D. Pierson et al. 1996, *Conservation Biology* 10.

39. P. A. Cox and T. Elmqvist 1991, *Ambio* 20.

40. T. Elmqvist et al. 1992, *Biotropica* 24.

ELEVEN: Have the Birds and Bees Taught Us Anything Useful? _____

1. C. H. Gibbs-Smith 1966, *The Invention of the Aeroplane (1799–1909).*

2. S. Vogel 1998, *Cats' Paws and Catapults.*

3. T. Crouch 1998, *Air & Space / Smithsonian* 13, 1.

4. J. D. Anderson 1997, *A History of Aerodynamics.*

5. W. T. Bonney 1962, *The Heritage of Kitty Hawk.*

6. H. S. Villard 1968, *Contact!*

7. J. M. Sater 1997, *Smart Structures and Materials, 1997.*

8. M. Grosser 1981, *Gossamer Odyssey.*

9. J. D. Burke 1980, *The Gossamer Condor and Albatross.*

10. S. F. Hoerner and H. V. Borst 1975, *Fluid-Dynamic Lift.*

aerodynamics   The study of air motions and the forces produced by objects moving through air or air flowing over objects

airfoil   The cross-sectional shape of a wing, viewed as if sliced from leading edge to trailing edge.

angle of attack   The angle a wing's chord makes with its direction of movement through the air (or the direction of the relative wind approaching the wing); a measure of whether the wing is tilted leading edge up or down. Formerly called "angle of incidence," the latter term is now used to mean the angle of the chord with respect to some arbitrary reference on an airplane, such as the fuselage centerline.

anterior   Anatomical direction meaning toward the front. (Opposite to *posterior.*)

Archaeopteryx   Technically *Archaeopteryx lithographica,* a species of feathered, flying animal with both birdlike and reptilian traits. Spectacularly well-preserved fossils of this extinct Jurassic species that even show details of feathers have been discovered in the Solnhofen limestone of Bavaria.

aspect ratio   The value of (wingspan)^2 ÷ planform area, a measure of how long and narrow a wing is. For rectangular wings, it reduces to wingspan ÷ wing chord.

autogyration   Lift production by a passively rotating wing, such as a maple samara, a windmill, or an autogyro (a craft with an unpowered rotor and a separate engine-driven propeller, now mostly obsolete). Essentially, gliding in a tight turn.

autorotation   In biology, refers to rotation of a wing about its long axis to generate lift using the same process as a Flettner rotor. (In aeronautics, refers to an emergency landing procedure for a helicopter with a failed engine; the process is actually identical with autogyration.)

body fat index   Proportion of an animal's weight made up of fat—an animal whose body weight is made up of 20 percent fat has an index of 0.2. Migratory animals can have body fat indices of up to 0.5.

boundary layer   A layer of air slowed by friction near a solid surface. When air flows over a surface, the flow speed is zero at the surface; moving away from the surface, up through the boundary layer, the flow speed increases with distance from the surface until the speed reaches the free stream speed at the top of the boundary layer.

bound vortex   A circulatory pattern of air movement about a wing, rearward over the top and forward under the bottom; it is a component of the flow pattern that must be added to a symmetrical flow over the top and bottom in order for the wing to produce lift.

camber   Upward convexity of an airfoil, usually measured as the maximum height of the airfoil centerline above the chord, divided by the chord length.

chord   A line connecting the forwardmost point of a wing (or airfoil) with the farthest rearward point, in a plane parallel to the wing's direction of movement or the animal's or airplane's long axis. Also, the length of such a line.

delayed stall   A transient increase in lift well above the normal maximum lift, produced either by rapidly increasing the angle of attack above the stall angle, or accelerating from rest with the wing held at an angle well above the stall angle. In either case, the wing will eventually stall if the angle of attack is not reduced.

dorsal   Anatomical direction that usually means toward the top side (opposite to *ventral*); in humans and other bipedal vertebrates, dorsal refers to the side of the body with the spine, thus the back.

downwash   A slight downward motion imparted to air after flowing over a wing.

drag   A force parallel with the direction of movement (or air flow) but in the opposite direction; a retarding force.

dynamic stall   See *delayed stall.*

echolocation   The process of producing sounds and listening for echoes to determine distance and direction to solid objects. Used by bats and porpoises; also called sonar.

ectotherm   An animal whose body is generally the same temperature as its surroundings; ectothermic animals do not produce enough metabolic heat to warm their bodies.

endotherm   An animal that produces extra metabolic heat in order to warm its body above the temperature of its surroundings. Birds and mammals are generally endothermic; a few other animals can be temporarily endothermic under certain conditions, such as bumblebees in flight.

flapping   Up-and-down wing movements used by flying animals to generate thrust.

Flettner rotor   A cylinder (or elongated flat plate) rotating about its long axis that uses the Magnus effect to produce lift in an airflow.

force   Any influence or phenomenon that can cause a change in speed or direction of movement of an object. (From Newton's Second Law of mechanics: force = mass × acceleration, where accelerations can be changes in speed or direction or both.)

gait   One of two or more qualitatively different patterns of wing flapping movements; usually one gait is more efficient over one speed range, and another gait is more efficient over a different speed range. (From its analogy with terrestrial gaits, such as walking, trotting and running.)

glide   Unpowered flight, in which the glider must descend (relative to the air) in order to maintain a steady speed.

glide ratio   The horizontal distance a glider travels for every unit distance of vertical descent. Equivalent in value to the lift-to-drag ratio.

halteres   Gyroscopic sense organs found only in true flies (Diptera); evolved from modified hindwings, they are tiny structures used to detect pitch, roll, and to a lesser extent, yaw. (A couple other groups of rare, poorly studied insects also have reduced, modified wings called halteres; scientists do not yet know if these structures perform the same function as the halteres of dipterans.)

hovering   Flight with little or no forward speed; probably the most energetically costly form of locomotion.

leading edge   The front edge or margin of a wing, or the frontmost point of an airfoil.

lift   A force perpendicular to a wing's motion through the air (or equivalently, perpendicular to the direction of oncoming air flowing over the wing). Not necessarily perfectly vertical, but usually with an upward component to offset a flyer's weight.

lift-to-drag ratio (L/D or C_L/C_D)   The ratio of the forces of lift and drag on a wing. The ratio is dimensionless and is an important figure of merit for the effectiveness of a wing or a flyer.

metabolic rate   The amount of energy an organism uses per unit time to maintain life and carry out activities.

model   Used scientifically in at least two ways. A *physical model* is a physical object that is a simplified representation of the natural object used to test physical properties of the natural object in a controlled setting, such as in a wind tunnel. A *theoretical model* is a conceptual or mathematical representation, usually simplified or idealized, of some process used to explore fundamental properties of the process, and to predict changes in the process under new conditions.

no-slip condition   A condition of normal air or water flows in which the fluid directly in contact with a solid surface does not move with respect to the surface; an infinitesimally thin layer of fluid sticks to the surface.

ornithopter   A flying machine that produces thrust for flight by flapping its wings.

pitch   Rotation about a horizontal axis parallel to the wings; nose up or nose down rotation.

planform area   The area defined by a wing's outline viewed from above or below.

posterior   Anatomical direction meaning toward the back. (Opposite to *anterior.*)

pressure   A force distributed over an area, divided by the area over which it acts; force per unit area.

pronate (pronation)   An anatomical term for a particular type of rotation of a limb. When applied to wings, it means a rotation of the wing about its span that tilts the leading edge down (opposite of *supinate*).

pterosaurs   Extinct flying vertebrates closely related to dinosaurs; they were the first vertebrates to evolve powered flight and they include the largest animals that have ever flown.

resultant force   The vector sum of the lift and drag, which must be equal and opposite to the thrust and weight for any flyer to maintain level flight.

Reynolds number (Re)   Ratio of pressure (inertial) and viscous contributions to drag, formally defined as (density × speed × characteristic length)/dynamic viscosity. A dimensionless index important in many areas of fluid mechanics, the Reynolds number is important aerodynamically because at low Reynolds numbers (small size and/or low speeds), wings tend to have lower lift-to-drag ratios.

roll   Rotation about a longitudinal axis; tilting to the right or left.

samara   A seed or fruit with a winglike structure that uses gliding to increase dispersal from the parent plant.

soaring   Using rising air (or other atmospheric energy) to remain aloft while gliding.

span   The length of a wing from tip to tip (perpendicular to the body's long axis). Biologists sometimes call the distance from a wing tip to the wing base the "span" but this distance is more appropriately called the "semispan."

stability   A tendency to return to one's original heading after a disturbance causes a deviation from that heading.

stall   When a wing is tilted to an angle of attack above some critical angle, the flow separates from the upper surface of the wing and leaves a large, turbulent wake. When a wing at a high Reynolds number stalls (such as those of birds and larger flyers), lift drops and drag rises abruptly and dramatically; at lower Reynolds numbers, such as those of insects, the lift decrease and drag increase are more gradual.

supinate (supination)   Rotating a wing about its span so that the leading edge tilts up. (Opposite of *pronate*.)

thermal   Air that rises because it is warmer than surrounding air.

thorax   Middle body region of an insect, to which legs and (usually) wings are attached. (Also the region defined by the rib cage in terrestrial vertebrates.)

thrust   A force parallel to the direction of movement and in the same direction; a force tending to maintain forward movement.

trailing edge   The rear edge or margin of a wing, or the rearmost point of an airfoil.

vector   Any quantity with both a magnitude and direction, as opposed to "scalars," which have only magnitude. Forces, velocities, and accelerations are vectors; mass, length, speed, pressure, and energy are scalars.

ventral   Anatomical direction that means "down or toward the lower or

belly side" (opposite to *dorsal*); in humans and other bipedal vertebrates, the belly is forward, so ventral means "toward the front."

viscosity   The property of a gas or liquid in which any given mass or parcel
of fluid resists sliding past other parcels (shearing), and this resistance
increases as the speed of shearing increases. It is a form of friction.

vortex (plural, vortices)   A pattern of air (or water) movement having a
cylindrical or conical mass of air rotating about its long axis. Torna-
does and whirlpools are vortices.

wing   A somewhat flattened, three-dimensional object intended to produce lift in an airflow; its cross section in the direction of flow is usually an airfoil.

wing loading   Total weight divided by planform area; weight per unit
planform area.

yaw   Rotation about a vertical axis; nose-left or nose-right rotation.

BIBLIOGRAPHY

Able, K. P., and M. A. Able. 1996. The flexible migratory orientation system of the Savannah sparrow (*Passerculus sandwichensis*). *Journal of Experimental Biology* 199: 3–8.

Ackerman, J. 1998. Dinosaurs take wing. *National Geographic* 194, 1: 73–99.

Aldridge, H. D. J. N. 1986. The flight kinematics of the greater horseshoe bat, *Rhinolophus ferrumequinum*. In *Bat Flight—Fledermausflug,* ed. W. Nachtigall, 127–38. Stuttgart: Gustav Fischer.

———. 1987. Turning flight of bats. *Journal of Experimental Biology* 128: 419–26.

Alerstam, T. 1981. The course and timing of bird migration. In *Animal Migration,* ed. D. J. Aidley, 9–54. Cambridge: Cambridge University Press.

———. 1996. The geographic scale factor in orientation of migrating birds. *Journal of Experimental Biology* 199: 9–19.

Alexander, D. E. 1986. Wind tunnel studies of turns by flying dragonflies. *Journal of Experimental Biology* 122: 81–98.

———. 1990. Drag coefficients of swimming animals: Effects of using different reference areas. *Biological Bulletin* 179: 186–90.

Alexander, D. E., and T. Chen. 1990. Comparison of swimming speed and hydrodynamic drag in two species of *Idotea* (Isopoda). *Journal of Crustacean Biology* 10: 406–12.

Alexander, R. D., and W. L. Brown. 1963. Mating behavior and the origin of insect wings. *Occasional Papers of the Museum of Zoology of the University of Michigan* 628: 1–19.

Altringham, J. D. 1996. *Bats: Biology and Behaviour.* Oxford: Oxford University Press. 262 pp.

Anderson, J. D. 1997. *A History of Aerodynamics.* Cambridge: Cambridge University Press. 478 pp.

Augspurger, C. K. 1986. Morphology and dispersal potential of wind-dispersed diaspores of neotropical trees. *American Journal of Botany* 73: 353–63.

Aymar, G. C. 1938. *Bird Flight.* Garden City, N.Y.: Garden City Publishing Co. 234 pp.

Azuma, A., S. Azuma, I. Watanabe, and T. Furuta. 1985. Flight mechanics of a dragonfly. *Journal of Experimental Biology* 116: 79–107.

Azuma, A., and Y. Okuno. 1987. Flight of a samara, *Alsomitra macrocarpa. Journal of Theoretical Biology* 129: 263–74.

Azuma, A., and T. Watanabe. 1988. Flight performance of a dragonfly. *Journal of Experimental Biology* 137: 221–52.

Baarder, A., M. Schaefer, and C. H. F. Rowell. 1992. The perception of the visual flow field by flying locusts: A behavioural and neuronal analysis. *Journal of Experimental Biology* 165: 137–60.

Bainbridge, R. 1958. The speed of swimming of fish as related to size and to the frequency and amplitude of the tail beat. *Journal of Experimental Biology* 35: 109–33.

Baker, J., and M. Tyler. 1979. The innervation of the wind-sensitive head hairs of the locust *Schistocerca gregaria*. *Physiological Entomology* 4: 301–9.

Baker, P. S. 1979. The wing movements of locusts during steering behaviour. *Journal of Comparative Physiology A* 131: 49–58.

Baker, R. R. 1978. *The Evolutionary Ecology of Animal Migration*. New York: Holmes & Meier. 1012 pp.

Baker, R. R., and J. G. Mather. 1982. Magnetic compass sense in the large yellow underwing moth, *Noctua pronuba* L. *Animal Behaviour* 30: 543–48.

Bakker, R. T. 1975. The dinosaur renaissance. *Scientific American* 232: 58–78.

———. 1986. *The Dinosaur Heresies*. 1st ed. New York: William Morrow. 481 pp.

Barnes, R. S. K., ed. 1998. *The Diversity of Living Organisms*. London: Blackwell Sciences. 345 pp.

Bastien, J. W. 1998. *The Kiss of Death: Chagas' Disease in the Americas*. Salt Lake City: University of Utah Press. 301 pp.

Bennett, S. C. 1995. *Pterosaurs: The Flying Reptiles*. New York: Franklin Watts. 64 pp.

———. 1997a. The arboreal leaping theory of the origin of pterosaur flight. *Historical Biology* 12: 265–90.

———. 1997b. Terrestrial locomotion of pterosaurs: A reconstruction based on *Pteraichnus* trackways. *Journal of Vertebrate Paleontology* 17: 104–13.

Benton, M. J. 1999. *Scleromochlus taylori* and the origin of dinosaurs and pterosaurs. *Philosophical Transactions of the Royal Society of London*, B, 354: 1423–46.

Berthold, P. 1996. *Control of Bird Migration*. London: Chapman & Hall. 355 pp.

Boag, D., and M. Alexander. 1986. *The Atlantic Puffin*. New York: Poole (for Blandford Press, Australia). 128 pp.

Bock, W. J. 1983. On extended wings: Another view of flight. *The Sciences* 23: 16–20.

Boettiger, E., and E. Furshpan. 1952. The mechanics of flight movements in Diptera. *Biological Bulletin* 102: 200–211.

Bonney, W. T. 1962. *The Heritage of Kitty Hawk*. New York: Norton. 211 pp.

Brodsky, A. K. 1994. *The Evolution of Insect Flight*. Oxford: Oxford University Press. 229 pp.

Brown, R. J. H. 1948. The flight of birds. I. The flapping cycle of the pigeon. *Journal of Experimental Biology* 25: 322–33.

———. 1953. The flight of birds. II. Wing function in relation to flight speed. *Journal of Experimental Biology* 30: 90–103.

———. 1963. The flight of birds. *Biological Reviews* 38: 460–89.

Brusca, R. C., and G. J. Brusca. 1990. *Invertebrates*. Sunderland, Mass.: Sinauer Associates. 922 pp.

Buchmann, S. L., and G. P. Nabhan. 1996. *The Forgotten Pollinators*. Washington, D.C.: Island Press. 183 pp.

Burke, A. C., and A. Feduccia. 1997. Developmental patterns and the identification of homologies in the avian hand. *Science* 278, 5335: 666–68.

Burke, J. D. 1980. *The Gossamer Condor and Albatross: A Case Study in Aircraft Design.* New York: American Institute of Aeronautics and Astronautics. 49 pp.

Burkett, C. W. 1989. Reductions in induced drag by the use of aft swept tips. *Aeronautical Journal* 93: 400–405.

Burton, A. J. 1964. Nervous control of flight orientation in a beetle. *Nature* 204: 1333.

Caple, G. R., R. T. Balda, and W. R. Willis. 1983. The physics of leaping animals and the evolution of pre-flight. *American Naturalist* 121: 455–67.

Carson, R. 1962. *Silent Spring.* Boston: Houghton Mifflin. 368 pp.

Chapman, R. F. 1982. *The Insects: Structure and Function.* 3d ed. Cambridge, Mass.: Harvard University Press. 919 pp.

Charig, A. J., F. Greenaway, A. C. Milner, C. A. Walker, and P. J. Whybrow. 1986. *Archaeopteryx* is not a forgery. *Science* 232, 4750: 622–26.

Chen, P. J., Z. M. Dong, and S. N. Zhen. 1998. An exceptionally well-preserved theropod dinosaur from the Yixian formation of China. *Nature* 391, 6663: 147–52.

Chiappe, L. M., S.-A. Ji, Q. Ji, and M. A. Norell. 1999. Anatomy and systematics of the Confuciusornithidae (Theropoda: Aves) from the late Mesozoic of northeastern China. *Bulletin of the American Museum of Natural History* 242: 1–86.

Chiappe, L. M., M. A. Norell, and J. M. Clark. 1998. The skull of a relative of the stem-group bird *Mononykus. Nature* 392: 275–78.

Cloudsley-Thompson, J. L. 1976. *Insects and History.* New York: St. Martin's Press. 242 pp.

Cloupeau, M., J. F. Devillier, and D. Devezeau. 1979. Direct measurements of instantaneous lift in desert locust: Comparison with Jensen's experiments on detached wings. *Journal of Experimental Biology* 80: 1–15.

Colbert, E. H. 1967. Adaptations for gliding in the lizard *Draco. American Museum Novitates* 2283: 1–20.

Collett, T. S., and J. Baron. 1994. Biological compasses and the coordinate frame of landmark memories in honeybees. *Nature* 368: 137–40.

Comstock, J. H. 1940. *An Introduction to Entomology.* 9th ed. Ithaca, N.Y.: Comstock Publishing Co. 1064 pp.

Constantine, D.C. 1970. Bats in relation to the health, welfare and economy of man. In *Biology of Bats,* ed. W. A. Wimsatt, 2: 320–419. New York: Academic Press.

Cox, G. W. 1997. *Conservation Biology.* 2d ed. Dubuque, Iowa: William C. Brown. 382 pp.

Cox, P. A. 1983. Observations on the natural history of Samoan bats. *Mammalia* 47: 519–23.

Cox, P. A., and T. Elmqvist. 1991. Indigenous control of tropical rainforest reserves: An alternative strategy for preservation. *Ambio* 20: 317–21.

———. 1993. Ecocolonialism and indigenous knowledge systems: Village-controlled rainforest preserves in Samoa. *Pacific Conservation* 1: 6–13.

Crane, E., and P. Walker. 1984. *Pollination Directory for World Crops.* London: International Bee Research Association. 183 pp.

Crawley, M. J. 1983. *Herbivory: The Dynamics of Animal-Plant Interactions.* Oxford: Blackwell Scientific Publications. 437 pp.

Crepet, W. L. 1983. The role of insect pollination in the evolution of angiosperms. In *Pollination Biology,* ed. L. Real, 29–50. New York: Academic Press.

Crouch, T. 1998. The thrill of invention. *Air & Space / Smithsonian* 13, 1: 22–30.

Currie, P. J. 1991. *The Flying Dinosaurs.* Red Deer, Alberta, Canada: Red Deer College Press. 160 pp.

Dahn, R. D., and J. F. Fallon. 2000. Interdigital regulation of digit identity and homeotic transformation by modulated BMP signalling. *Science* 289: 438–41.

Dathe, H. H. 1982. Efficiency calculations on the hovering flight of the pigeon (*Columba livia*) and the black-headed gull (*Larus ridibandus*). *Zoologische Jahrbücher: Abteilung für Allgemeine Zoologie und Physiologie der Tiere* 86: 209–42.

Davenport, J. 1992. Wing-loading, stability and morphometric relationships in flying fish (Exocoetidae) from the north-eastern Atlantic. *Journal of the Marine Biological Association of the United Kingdom* 72: 25–39.

———. 1994. How and why do flying fish fly? *Reviews in Fish Biology and Fisheries* 4: 184–214.

DeLaurier, J. D. 1993a. An aerodynamic model for flapping-wing flight. *Aeronautical Journal* 97: 125–30.

———. 1993b. The development of an efficient ornithopter wing. *Aeronautical Journal* 97: 153–62.

DeLaurier, J. D., and J. M. Harris. 1993. A study of mechanical flapping-wing flight. *Aeronautical Journal* 97: 277–86.

DeVoe, R. D., W. Kaiser, J. Ohm, and L. S. Stone. 1982. Horizontal movement detectors of honeybees: Directionally selective visual neurones in the lobula and brain. *Journal of Comparative Physiology A* 147: 155–70.

Dial, K. P., A. A. Biewener, B. W. Tobalske, and D. R. Warrick. 1997. Mechanical power output of bird flight. *Nature* 390: 67–70.

Dickinson, J. A. 1994. Bees link local landmarks with celestial compass cues. *Naturwissenschaften* 81: 465–67.

Dickinson, M. H., F.-O. Lehmann, and S. P. Sane. 1999. Wing rotation and the aerodynamic basis of insect flight. *Science* 284: 1954–60.

Dingle, H. 1985. Migration. In *Comprehensive Insect Physiology, Biochemistry and Pharmacology,* ed. G. A. Kerkut and L. I. Gilbert, 9: 375–415. Oxford: Pergamon Press.

———. 1996. *Migration: The Biology of Life on the Move.* Oxford: Oxford University Press. 474 pp.

Dingus, L., and T. Rowe. 1998. *The Mistaken Extinction: Dinosaur Evolution and the Origin of Birds.* New York: W. H. Freeman. 332 pp.

Dirsh, V. M. 1965. *The African Genera of Acridoidea.* Cambridge: Cambridge University Press. 578 pp.

Dodson, P. 1985. Conference report: International *Archaeopteryx* conference. *Journal of Vertebrate Paleontology* 5: 177–79.

Dolnik, V. R. 1990. Bird migration across arid and mountainous regions of Middle Asia and Kasakhstan. In *Bird Migration:Physiology and Ecophysiology,* ed. E. Gwinner, 368–86. Berlin: Springer-Verlag.

Dudley, R. 1998. Atmospheric oxygen, giant paleozoic insects and the evolution of aerial locomotor performance. *Journal of Experimental Biology* 201: 1043–50.

Dudley, R., and P. DeVries. 1990. Tropical rain forest structure and the geographical distribution of gliding vertebrates. *Biotropica* 22: 432–34.

Dudley, R., and C. P. Ellington. 1990a. Mechanics of forward flight in bumblebees. I. Kinematics and morphology. *Journal of Experimental Biology* 148: 19–25.

———. 1990b. Mechanics of forward flight in bumblebees. II. Quasisteady lift and power requirements. *Journal of Experimental Biology* 148: 53–88.

Duggard, J. J. 1967. Directional change in flying locusts. *Journal of Insect Physiology* 13: 1055–63.

Dyer, F. C. 1996. Spatial memory and navigation by honeybees on the scale of the foraging range. *Journal of Experimental Biology* 199: 147–54.

Dyer, F. C., and J. L. Gould. 1981. Honey bee orientation: A backup system for cloudy days. *Science* 214: 1041–42.

Edgerton, H. E., and C. M. Breder. 1941. High speed photographs of flying fish in flight. *Zoologica* 26: 311–13.

Edwards, R. H., and H. K. Cheng. 1982. The separation vortex in the Weis-Fogh circulation-generation mechanism. *Journal of Fluid Mechanics* 120: 463–73.

Ellington, C. P. 1984a. The aerodynamics of hovering insect flight. I. The quasisteady analysis. *Philosophical Transactions of the Royal Society of London,* B, 305: 1–15.

———. 1984b. The aerodynamics of hovering insect flight. III. Kinematics. *Philosophical Transactions of the Royal Society of London,* B, 305: 41–78.

———. 1984c. The aerodynamics of hovering insect flight. IV. Aerodynamic mechanisms. *Philosophical Transactions of the Royal Society of London,* B, 305: 79–113.

———. 1984d. The aerodynamics of hovering insect flight. V. A vortex theory. *Philosophical Transactions of the Royal Society of London,* B, 305: 115–44.

———. 1991a. Aerodynamics and the origin of insect flight. *Advances in Insect Physiology* 23: 171–210.

———. 1991b. Limitations on animal flight performance. *Journal of Experimental Biology* 160: 71–91.

Ellington, C. P., C. van den Berg, A. P. Willmot, and A. L. R. Thomas. 1996. Leading-edge vortices in insect flight. *Nature* 384: 626–30.

Elmqvist, T., P. A. Cox, W. E. Rainey, and E. D. Pierson. 1992. Restricted pollination on oceanic islands: Pollination of *Ceiba pentandra* by flying foxes in Samoa. *Biotropica* 24: 15–23.

Emlen, S. T. 1967. Migratory orientation in the indigo bunting, *Passerina cyanea.* Part I: Evidence for use of celestial cues. *Auk* 84: 309–42.

Ennos, A. R. 1988. The importance of torsion in the design of insect wings. *Journal of Experimental Biology* 140: 137–60.

———. 1989. The effect of size on the optimal shapes of gliding insects and seeds. *Journal of Zoology* 219: 61–69.

Farren, W. S. 1935. The reaction on a wing whose angle of incidence is changing rapidly: Wind tunnel experiments with a short period recording balance. *Aeronautical Research Committee Reports and Memoranda* 1648: 1–24.

Farrow, R. A. 1990. Flight and migration in acridoids. In *Biology of Grasshoppers,* ed. R. F. Chapman and A. Joern, 227–314. New York: John Wiley & Sons.

Feduccia, A. 1980. *The Age of Birds.* Cambridge, Mass.: Harvard University Press. 196 pp.

———. 1996. *The Origin and Evolution of Birds.* New Haven, Conn.: Yale University Press. 420 pp.

Feduccia, A., and H. G. Tordoff. 1979. Feathers of *Archaeopteryx:* Asymmetric vanes indicate aerodynamic function. *Science* 203: 1021.

Fenton, M. B. 1983. *Just Bats.* Toronto: University of Toronto Press. 165 pp.

Free, J. B. 1970. *Insect Pollinations of Crops.* New York: Academic Press. 183 pp.

Fronk, W. D. 1971. Vegetable crop insects. In *Fundamentals of Applied Entomology,* ed. R. E. Pfadt, 375–406. New York: Macmillan Co.

Galé, F. 1991. *Tailless Tale.* Olalla, Wash.: B2 Streamlines. 258 pp.

Garner, J. P., G. K. Taylor, and A. L. R. Thomas. 1999. On the origins of birds: The sequence of character acquisition in the evolution of avian flight. *Proceedings of the Royal Society of London,* B, 266: 1259–66.

Gauthier, J. A. 1986. Saurischian monophyly and the origin of birds. In *The Origin of Birds and the Evolution of Flight,* ed. K. Padian, 1–55. San Francisco: California Academy of Science.

Gewecke, M. 1974. The antennae of insects as air-current sense organs and their relationship to the control of flight. In *Experimental Analysis of Insect Behavior,* ed. L. Barton-Browne, 100–113. Berlin: Springer-Verlag.

Gewecke, M., and M. Niehaus. 1981. Flight and flight control by the antennae in the small tortoiseshell (*Aglais urticae* L., Lepidoptera). I. flight balance experiments. *Journal of Comparative Physiology* 145: 249–56.

Gibbs-Smith, C. H. 1966. *The Invention of the Aeroplane (1799–1909).* London: Faber & Faber. 360 pp.

Gibo, D. L. 1981. Some observations on slope soaring in *Pantala flavescens* (Odonata: Libellulidae). *Journal of the New York Entomological Society* 89: 184–87.

Goodman, L. J. 1965. The role of certain optomotor reactions in regulating stability in the rolling plane during flight in the desert locust, *Schistocerca gregaria. Journal of Experimental Biology* 42: 358–407.

Götz, K. G. 1968. Flight control in *Drosophila* by visual control of motion. *Kybernetic* 4: 199–208.

Gould, S. J. 1985. Not necessarily a wing. *Natural History* 94: 12–25.

Graham, J. B., R. Dudley, N. M. Aguilar, and C. Gans. 1995. Implications of the late Palaeozoic oxygen pulse for physiology and evolution. *Nature* 375: 117–20.

Green, D. S. 1980. The terminal velocity and dispersal of spinning samaras. *American Journal of Botany* 67: 1218–24.

Grosser, M. 1981. *Gossamer Odyssey: The Triumph of Human-Powered Flight.* Boston: Houghton Mifflin. 298 pp.

Haussling, H. J. 1979. Boundary-fitted coordinates for accurate numerical solution of multibody flow problems. *Journal of Comparative Physiology* 30: 107–24.

Hazlehurst, G. A., and J. M. V. Rayner. 1992. Flight characteristics of Triassic and Jurassic Pterosauria: An appraisal based on wing shape. *Paleobiology* 18: 447–63.

Hecht, M. K., J. H. Ostrom, G. Viohl, and P. Wellnhofer, eds. 1985. *The Beginnings of Birds.* Eichstätt, Germany: Freunde des Jura-Museums Eichstätt. 382 pp.

Hedrick, P. W. 1984. *Population Biology: The Evolution and Ecology of Populations.* Boston: Jones & Bartlett. 445 pp.

Heilmann, G. 1927. *The Origin of Birds.* English translation of the 1916 Danish edition. New York: D. Appleton. 209 pp. Reprint. New York: Dover, 1972.

Helbig, A. J. 1991. Experimental and analytical techniques used in bird orientation research. In *Orientation in Birds,* ed. P. Berthold, 270–306. Basel: Birkhäuser Verlag.

———. 1996. Genetic basis, mode of inheritance and evolutionary changes of migratory directions in Palearctic warblers (Aves: Sylviidae). *Journal of Experimental Biology* 199: 49–55.

Hill, D. S. 1997. *The Economic Importance of Insects.* London: Chapman & Hall. 395 pp.

Hirashima, Y., M. Inokuchi, and K. Yamagishi. 1999. Do you believe a "swimming wasp"? *Esakia* 39: 9–11.

Hoerner, S. F., and H. V. Borst. 1975. *Fluid-Dynamic Lift.* Bricktown, N.J.: Hoerner Fluid Dynamics. 491 pp.

Howes, C. 1997. *The Spice of Life: Biodiversity and the Extinction Crisis.* London: Blandford. 192 pp.

Huxley, A. F. 1974. Muscular contraction. *Journal of Physiology* 243: 1–43.

Iddotson, M. R. 1991. A motion-sensitive visual descending neuron in *Apis mellifera* monitoring translatory flow fields in the horizontal plane. *Journal of Experimental Biology* 157: 573–79.

Jander, R., and U. Jander. 1998. The light and magnetic compass of the weaver ant, *Oecophylla smaragdina* (Hymenoptera: Formicidae). *Ethology* 104: 743–58.

Jensen, M. 1956. Biology and physics of locust flight. III. Aerodynamics of locust flight. *Philosophical Transactions of the Royal Society of London,* B, 239: 511–52.

Jepsen, G. L. 1970. Bat origins and evolution. In *Biology of Bats,* ed. W. A. Wimsatt, 1: 1–64. New York: Academic Press.

Ji, Q., P. J. Currie, M. A. Norell, and S.-A. Ji. 1998. Two feathered dinosaurs from northeastern China. *Nature* 393: 753–61.

Kingsolver, J. G., and M. A. R. Koehl. 1985. Aerodynamics, thermoregulation, and the evolution of insect wings: Differential scaling and evolutionary change. *Evolution* 39: 488–504.

———. 1994. Selective factors in the evolution of insect wings. *Annual Review of Entomology* 39: 425–51.

Kirsch, J. A. W., T. F. Flannery, M. S. Springer, and F. J. Lapointe. 1995. Phylogeny of the Pteropodidae (Mammalia: Chiroptera) based on DNA hybridization, with evidence for bat monophyly. *Australian Journal of Zoology* 43: 395–427.

Kirschvink, J. L., D. S. Jones, and B. J. MacFadden, eds. 1985. *Magnetite Biomineralization and Magnetoreception in Organisms: A New Biomagnetism.* New York: Plenum Press. 682 pp.

Kono, T., and C. S. Papp. 1977. *Handbook of Agricultural Pests: Aphids, Thrips, Mites, Snails and Slugs.* Sacramento: State of California Department of Food and Agriculture. 205 pp.

Kukalova, J. 1970. Revisional study of the order Palaeodictyoptera in the Upper Carboniferous shales of Commentry, France, Part III. *Psyche* 77: 1–44.

Kuyt, E. 1992. Aerial radio-tracking of whooping cranes migrating between Wood Buffalo National Park and Aransas National Wildlife Refuge, 1981–84. *Canadian Wildlife Service Occasional Papers* 74: 1–53.

Kvist, A., M. Klaasen, and Å. Lindström. 1998. Energy expenditure in relation to flight speed: What is the power of mass loss rate estimates? *Journal of Avian Biology* 29: 485–98.

Langston, W., Jr. 1981. Pterosaurs. *Scientific American* 224: 122–36.

Lawson, D. A. 1975. Pterosaur from the late Cretaceous of West Texas: Discovery of the largest flying creature. *Science* 187: 947–48.

Lighthill, M. J. 1973. On the Weis-Fogh mechanism of lift generation. *Journal of Fluid Mechanics* 60: 1–17.

———. 1977. Introduction to the scaling of aerial locomotion. In *Scale Effects in Animal Locomotion,* ed. T. J. Pedley, 365–404. New York: Academic Press.

Marden, J. H., and M. G. Kramer. 1994. Surface-skimming stoneflies: A possible intermediate stage in insect flight evolution. *Science* 266: 427–30.

Martin, L. D. 1991. Mesozoic birds and the origin of birds. In *Origins of the Higher Groups of Tetrapods: Controversy and Consensus,* ed. H.-P. Schultze and L. Trueb, 485–540. Ithaca, N.Y.: Comstock Pub. Associates.

Martin, L. D., Z. Zhou, L. Hou, and A. Feduccia. 1998. *Confuciusornis sanctus* compared to *Archaeopteryx lithographica. Naturwissenschaften* 85: 286–89.

Maxworthy, T. 1979. Experiments on the Weis-Fogh mechanism of lift generation by insects in hovering flight. Part 1. Dynamics of the "fling." *Journal of Fluid Mechanics* 93: 47–63.

Maynard Smith, J. 1952. The importance of the nervous system in the evolution of animal flight. *Evolution* 6: 127–29.

McCutchen, C. W. 1977. The spinning rotation of ash and tulip tree samaras. *Science* 197: 691–92.

McGahan, J. 1973. Flapping flight of the Andean condor in nature. *Journal of Experimental Biology* 58: 239–53.

McMahon, T. A. 1984. *Muscles, Reflexes, and Locomotion.* Princeton, N.J.: Princeton University Press. 331 pp.

McMasters, J. H. 1976. Aerodynamics of the long pterosaur wing. *Science* 191: 899.

———— 1986. Reflections of a paleoaerodynamicist. *Perspectives in Biology and Medicine* 29: 331–84.

————. 1989. The flight of the bumblebee and related myths of entomological engineering. *American Scientist* 77: 164–69.

Meffe, G. K., and C. R. Carroll. 1997. *Principles of Conservation Biology.* 2d ed. Sunderland, Mass.: Sinauer Associates. 729 pp.

Mittelstaedt, H. 1950. Physiologie des Glechgewichtssinnes bei fliegenden Libellen. *Zeitschrift für Vergleichende Physiologie* 32: 422–63.

Monath, T. P. 1984. Yellow fever. In *Tropical and Geographical Medicine,* ed. K. S. Warren and A. A. F. Mahmoud, 636–51. New York: McGraw-Hill.

Morse, R. A. 1992. Africanized honey bees in North America. In *Insect Potpourri: Adventures in Entomology,* ed. J. Adams, 151–60. Gainesville, Fla.: Sandhill Crane Press.

Nachtigall, W. 1966. Die Kinematik der Schlagflügelbewegungen von Dipteren: Methodische und analytische Grundlagen zur Biophysik des Insektenflugs. *Zeitschrift für Vergleichende Physiologie* 52: 155–211.

————. 1974. *Insects in Flight: A Glimpse behind the Scenes in Biophysical Research.* New York: McGraw-Hill. 150 pp.

————. 1977. Die aerodynamische Polare des *Tipula*-Flügels und eine Einrichtung zur halbautomatischen Polarenaufnahme. *Fortschritte der Zoologie* 24, 2/3: 347–52.

————. 1980. Some aspects of the kinematics of wingbeat movements in insects. In *Aspects of Animal Movements,* ed. H. Y. Elder and E. R. Trueman, 169–75. Cambridge: Cambridge University Press.

Nachtigall, W., and D. M. Wilson. 1967. The neuromuscular control of dipteran flight. *Journal of Experimental Biology* 47: 77–97.

Nettleship, D. N., and P. G. H. Evans. 1985. Distribution and status of the Atlantic Alcidae. In *The Atlantic Alcidae: The Evolution, Distribution and Biology of the Auks Inhabiting the Atlantic Ocean and Adjacent Water Areas,* ed. D. N. Nettleship and T. R. Birkhead, 53–154. New York: Academic Press.

Newman, B. G., S. B. Savage, and D. Schouella. 1977. Model tests on a wing section of an *Aeschna* dragonfly. In *Scale Effects in Animal Locomotion,* ed. T. J. Pedley, 445–77. New York: Academic Press.

Norberg, R. Å. 1973. Autorotation, self-stability, and structure of single-winged fruits and seeds (samaras) with comparative remarks on animal flight. *Biological Reviews* 48: 561–96.

————. 1975. Hovering flight of the dragonfly *Aeschna juncea* L., kinematics and aerodynamics. In *Swimming and Flying in Nature,* ed. T. Y. Wu, C. J. Brokaw, and C. Brennan, 2: 763–81. New York: Plenum Press.

————. 1985. Function of vane asymmetry and shaft curvature in bird flight feathers; inferences on flight ability of *Archaeopteryx.* In *The Beginnings of Birds,* ed. M. K. Hecht, J. H. Ostrom, G. Viohl, and P. Wellnhofer, 303–18. Eichstätt, Germany: Freunde des Jura-Museums Eichstätt.

Norberg, U. M. 1975. Hovering flight in the pied flycatcher (*Ficedula hypoleuca*). In

Swimming and Flying in Nature, ed. T. Y. Wu, C. J. Brokaw, and C. Brennen, 2: 763–81. New York: Plenum Press.

———. 1976. Aerodynamics, kinematics, and energetics of horizontal flapping flight in the long-eared bat *Plecotus auritus. Journal of Experimental Biology* 65: 179–212.

———. 1985. Evolution of flight in birds: Aerodynamic, mechanical and ecological aspects. In *The Beginnings of Birds,* ed. M. K. Hecht, J. H. Ostrom, G. Viohl, and P. Wellnhofer, 293–303. Eichstätt, Germany: Freunde des Jura-Museums Eichstätt.

———. 1986. On the evolution of flight and wing form in bats. In *Bat Flight—Fledermausflug,* ed. W. Nachtigall, 13–26. Stuttgart: Gustav Fischer.

———. 1990. *Vertebrate Flight: Mechanics, Physiology, Morphology, Ecology and Evolution.* Zoophysiology, vol. 27. Berlin: Springer-Verlag. 291 pp.

———. 1994. Wing design, flight performance and habitat use in bats. In *Ecological Morphology,* ed. P. C. Wainwright and S. M. Reilly, 204–39. Chicago: University of Chicago Press.

Norberg, U. M., and J. M. V. Rayner. 1987. Ecological morphology and flight in bats (Mammalia: Chiroptera): Wing adaptations, flight performance, foraging strategy and echolocation. *Philosophical Transactions of the Royal Society of London,* B, 316: 335–427.

Novas, F. E., and P. F. Puerta. 1997. New evidence concerning avian origins from the Late Cretaceous of Patagonia. *Nature* 387, 6631: 390–92.

Oliver, J. A. 1951. "Gliding" in amphibians and reptiles, with a remark on an arboreal adaptation in the lizard, *Anolis carolinensis* Voigt. *American Naturalist* 85: 171–76.

Olson, S. L., and A. Feduccia. 1979. Flight capability and the pectoral girdle of *Archaeopteryx. Nature* 278: 247–48.

Orr, R. T. 1970. *Animals in Migration.* New York: Macmillan Co. 303 pp.

Osborne, M. F. M. 1951. Aerodynamics of flapping flight with application to insects. *Journal of Experimental Biology* 28: 221–45.

Ostrom, J. H. 1973. The ancestry of birds. *Nature* 242: 136.

———. 1974. *Archaeopteryx* and the origin of flight. *Quarterly Review of Biology* 49: 27–47.

———. 1975. The origin of birds. *Annual Review of Earth and Planetary Sciences* 3: 55–77.

———. 1979. Bird flight: How did it begin? *American Scientist* 67: 46–56.

———. 1985. Introduction to *Archaeopteryx.* In *The Beginnings of Birds,* ed. M. K. Hecht, J. H. Ostrom, G. Viohl and P. Wellnhofer, 9–20. Eichstätt, Germany: Freunde des Jura-Museums Eichstätt.

Owen, O. S., D. D. Chiras, and J. P. Reganold. 1998. *Natural Resource Conservation.* Upper Saddle River, N.J.: Prentice-Hall. 594 pp.

Padian, K. 1983. A functional analysis of flying and walking in pterosaurs. *Paleobiology* 9: 218–39.

———. 1985. The origins and aerodynamics of flight in extinct vertebrates. *Palaeontology* 28: 413–33.

———. 1991. Pterosaurs: Were they functional birds or functional bats. In *Biomechan-*

ics in Evolution, ed. J. M. V. Rayner and R. J. Wootton, 146–60. Cambridge: Cambridge University Press.

Padian, K., and L. M. Chiappe. 1988. The origin of birds and their flight. *Scientific American* 278, 2: 38–47.

————. 1998. The origin and early evolution of birds. *Biological Reviews* 73: 1–42.

Padian, K., and J. M. V. Rayner. 1993. The wings of pterosaurs. *American Journal of Science* 293A: 91–166.

Page, J. 1990. Pushy and brassy, the starling was an ill-advised import. *Smithsonian* 21, 6: 77–85.

Papi, F., and P. Luschi. 1996. Pinpointing "Isla Meta": The case of sea turtles and albatrosses. *Journal of Experimental Biology* 199: 65–71.

Pennycuick, C. J. 1971a. Control of gliding angle in Rüppell's griffon vulture *Gyps rueppellii. Journal of Experimental Biology* 55: 39–46.

————. 1971b. Gliding flight of the dog-faced bat *Rousettus aegyptiacus* observed in a wind tunnel. *Journal of Experimental Biology* 55: 833–45.

————. 1972a. *Animal Flight.* London: Edward Arnold. 68 pp.

————. 1972b. Soaring behaviour and performance of some East African birds, observed from a motor glider. *Ibis* 114: 178–218.

————. 1975. Mechanics of flight. In *Avian Biology,* ed. D. S. Farner and J. R. King, 5: 1–75. New York: Academic Press.

————. 1982. The flight of petrels and albatrosses (Procellariiformes), observed in South Georgia and its vicinity. *Philosophical Transactions of the Royal Society of London,* B, 300: 75–106.

————. 1988. On the reconstruction of pterosaurs and their manner of flight, with notes on vortex wakes. *Biological Reviews* 63: 299–331.

Pennycuick, C. J., T. Alerstam, and B. Larsson. 1979. Soaring migration of the common crane *Grus grus* observed by radar and from an aircraft. *Ornis Scandinavica* 10: 241–51.

Perdeck, A. C. 1958. Two types of orientation in migrating starlings, *Sturnus vulgaris* L. and chaffinches, *Fringilla coelebs* L., as revealed by displacement experiments. *Ardea* 46: 1–37.

Perkins, C. P., and R. E. Hage. 1949. *Airplane Performance Stability and Control.* New York: John Wiley & Sons. 493 pp.

Pettigrew, J. D. 1986. Flying primates? Megabats have the advanced pathway from eye to midbrain. *Science* 231: 1304–6.

Pfadt, R. E. 1971a. Insect pests of cotton. In *Fundamentals of Applied Entomology,* ed. id., 343–73. New York: Macmillan Co.

————, ed. 1971b. *Fundamentals of Applied Entomology.* New York: Macmillan Co. 693 pp.

Pierson, E. D., T. Elmqvist, W. E. Rainey, and P. A. Cox. 1996. Effects of tropical cyclonic storms on flying fox populations on the South Pacific island of Samoa. *Conservation Biology* 10: 438–51.

Pirlot, P. 1977. Wing design and the origin of bats. In *Major Patterns in Vertebrate Evo-*

lution, ed. M. K. Hecht, P. C. Goody, and B. M. Hecht, 375–410. New York: Plenum Press.

Pomeroy, D. 1990. Why fly? The possible benefits for lower mortality. *Biological Journal of the Linnean Society* 40: 53–65.

Pough, H. F., C. M. Janis, and J. B. Heiser. 1999. *Vertebrate Life.* Upper Saddle River, N.J.: Prentice-Hall. 733 pp.

Pratt, H. D. 1992. Malaria control and eradication in the United States. In *Insect Potpourri: Adventures in Entomology,* ed. J. Adams, 56–60. Gainesville, Fla.: Sandhill Crane Press.

Pringle, J. W. S. 1948. The gyroscopic mechanism of the halteres of Diptera. *Philosophical Transactions of the Royal Society of London,* B, 233: 347–84.

———. 1957. *Insect Flight.* London: Cambridge University Press. 133 pp.

Rainey, R. C. 1951. Weather and the movements of locust swarms: A new hypothesis. *Nature* 168: 1057–60.

Rayner, J. M. V. 1979a. A new approach to animal flight mechanics. *Journal of Experimental Biology* 80: 17–54.

———. 1979b. A vortex theory of animal flight. II. The forward flight of birds. *Journal of Fluid Mechanics* 91: 731–63.

———. 1980. Vorticity and animal flight. In *Aspects of Animal Movement,* ed. H. Y. Elder and E. R. Trueman, 177–99. Cambridge: Cambridge University Press.

———. 1981. Flight adaptations in vertebrates. In *Vertebrate Locomotion,* ed. M. H. Day, 137–72. Symposia of the Zoological Society of London, 48. New York: Academic Press.

———. 1985. Mechanical and ecological constraints on flight evolution. In *The Beginnings of Birds,* ed. M. K. Hecht, J. H. Ostrom, G. Viohl, and P. Wellnhofer, 279–88. Eichstätt, Germany: Freunde des Jura-Museums Eichstätt.

———. 1986. Vertebrate flapping flight mechanics and aerodynamics, and the evolution of flight in bats. In *Bat Flight—Fledermausflug,* ed. W. Nachtigall, 27–74. Stuttgart: Gustav Fischer.

———. 1987. The mechanics of flapping flight in bats. In *Recent Advances in the Study of Bats,* ed. M. B. Fenton, P. Racey, and J. M. V. Rayner, 23–42. Cambridge: Cambridge University Press.

———. 1988a. The evolution of vertebrate flight. *Biological Journal of the Linnean Society* 34: 269–87.

———. 1988b. Form and function of avian flight. *Current Ornithology* 5: 1–66.

———. 1990. The mechanics of flight and bird migration performance. In *Bird Migration: Physiology and Ecophysiology,* ed. E. Gwinner, 283–99. Berlin: Springer-Verlag.

———. 1991a. Avian flight evolution and the problem of *Archaeopteryx.* In *Biomechanics in Evolution,* ed. id. and R. J. Wootton, 183–212. Cambridge: Cambridge University Press.

———. 1991b. Wake structure and force generation in avian flapping flight. In *Acta XX Congressus Internationalis Ornithologici,* vol. 2, ed. B. D. Bell, R. O. Cossee, J. E. C.

Flux, B. D. Heather, R. A. Hitchmough, C. J. R. Robertson, and M. J. Williams, 702–15. Wellington, N.Z.: Ornithological Congress Trust Board.

———. 1994. Aerodynamic corrections for the flight of birds and bats in wind tunnels. *Journal of Zoology* 234: 537–63.

———. 1995a. Dynamics of the vortex wake of flying and swimming vertebrates. *Symposium of the Society for Experimental Biology* 49: 131–55.

———. 1995b. Flight mechanics and contraints on flight performance. *Israel Journal of Zoology* 41: 321–42.

———. 1999. Estimating power curves of flying vertebrates. *Journal of Experimental Biology* 202: 3449–61.

Rayner, J. M. V., G. Jones, and A. Thomas. 1986. Vortex flow visualizations reveal change in upstroke function with flight speed in bats. *Nature* 321: 162–64.

Reavis, M. A., and M. W. Luttges. 1988. Aerodynamic forces produced by a dragonfly. *AIAA* 88–0330. 13 pp.

Rees, C. J. C. 1975a. Aerodynamic properties of an insect wing section and a smooth aerofoil compared. *Nature* 258: 141–42.

———. 1975b. Form and function in corrugated insect wings. *Nature* 256: 200–203.

Richardson, W. J. 1991. Wind and orientation of migrating birds: A review. In *Orientation in Birds,* ed. P. Berthold, 226–49. Basel: Birkhäuser Verlag.

Rietschel, S. 1985. False forgery. In *The Beginnings of Birds,* ed. M. K. Hecht, J. H. Ostrom, G. Viohl, and P. Wellnhofer, 371–76. Eichstätt, Germany: Freunde des Jura-Museums Eichstätt.

Rosen, A., and D. Seter. 1991. Vertical autorotation of a single-winged samara. *Journal of Applied Mechanics* 58: 1064–71.

Rosenfeld, C., and R. Cooke. 1982. *Earthfire: The Eruption of Mount St. Helens.* Cambridge, Mass.: MIT Press. 155 pp.

Roskam, J. 1979. *Airplane Flight Dynamics and Automatic Flight Controls.* Part 1. Ottawa, Kansas: Roskam Aviation and Engineering Corp. 643 pp.

Ruben, J. A. 1991. Reptilian physiology and the flight of *Archaeopteryx. Evolution* 45: 1–17.

———. 1993. Powered flight in *Archaeopteryx:* Response to Speakman. *Evolution* 47: 935–38.

———. 1996. Evolution of endothermy in mammals, birds and their ancestors. *Society for Experimental Biology Seminar Series* 59: 347–74.

Ruben, J. A., W. J. Hillenius, N. R. Geist, A. Leitch, T. D. Jones, P. J. Currie, J. R. Horner, and G. Espe III. 1996. The metabolic status of some late Cretaceous dinosaurs. *Science* 273, 5279: 1204–7.

Ruben, J. A., T. D. Jones, N. R. Geist, and W. J. Hillenius. 1997. Lung structure and ventilation in theropod dinosaurs and early birds. *Science* 278, 5341: 1267–70.

Rüppell, G. 1977. *Bird Flight.* New York: Van Nostrand Reinhold. 191 pp.

Rydell, J., and J. R. Speakman. 1995. Evolution of nocturnality in bats: Potential competitors and predators during their early history. *Biological Journal of the Linnean Society* 54: 183–91.

Sater, J. M., ed. 1997. *Smart Structures and Materials, 1997: Industrial and Commercial Applications of Smart Structures Technologies.* Bellingham, Wash.: SPIE—The International Society for Optical Engineering. 490 pp.

Scheffer, R. P. 1997. *The Nature of Disease in Plants.* Cambridge: Cambridge University Press. 325 pp.

Schmidt-Koenig, K., J. U. Ganzhorn, and R. Ranvand. 1991. The sun compass. In *Orientation in Birds,* ed. P. Berthold, 1–15. Basel: Birkhäuser Verlag.

Schmidt-Nielsen, K. 1972. Locomotion: Energy cost of swimming, flying and running. *Science* 177: 222–27.

———. 1990. *Animal Physiology: Adaptation and Environment,* 4th ed. Cambridge: Cambridge University Press. 602 pp.

Schneider, P., and B. Krämer. 1974. Die Steuerung des Fluges beim Sandlaukäfer (*Cicindela*) und beim Maikäfer (*Melolontha*). *Journal of Comparative Physiology* 91: 377–86.

Scholey, K. 1986. The evolution of flight in bats. In *Bat Flight—Fledermausflug,* ed. W. Nachtigall, 1–12. Stuttgart: Gustav Fischer.

Seter, D., and A. Rosen. 1992a. Stability of the vertical autorotation of a single-winged samara. *Journal of Applied Mechanics* 59: 1000–1008.

———. 1992b. Study of the vertical autorotation of a single-winged samara. *Biological Reviews* 67: 175–97.

Shattuck, G. C. 1951. *Diseases of the Tropics.* New York: Appleton-Century-Crofts. 803 pp.

Simmons, N. B. 1995. Bat relationships and the origin of flight. *Symposium of the Zoological Society of London* 67: 27–43.

Simmons, N. B., and J. H. Geisler. 1998. Phylogenetic relationships of *Icaronycteris, Archaeonycteris, Hassianycteris,* and *Palaeochiropteryx* to extant bat lineages, with comments on the evolution of echolocation and foraging strategies in Microchiroptera. *Bulletin of the American Museum of Natural History* 235: 1–182.

Smith, D. S. 1972. *Muscle.* New York: Academic Press. 60 pp.

Smith, J. D. 1977. Comments on flight and the evolution of bats. In *Major Patterns in Vertebrate Evolution,* ed. M. K. Hecht, P. C. Goody, and B. M. Hecht, 427–37. New York: Plenum Press.

Smith, K. G. V., ed. 1973. *Insects and Other Arthropods of Medical Importance.* London: British Museum (Natural History). 561 pp.

Smith, M. J. C. 1996. Simulating moth wing aerodynamics: Towards the development of flapping-wing technology. *AIAA Journal* 34: 1348–55.

Smith, M. J. C., P. J. Wilkin, and M. H. Williams. 1996. The advantages of an unsteady panel method in modeling the aerodynamic forces on rigid flapping wings. *Journal of Experimental Biology* 199: 1073–83.

Solbreck, C. 1985. Insect migration strategies and population dynamics. *Contributions in Marine Science* 27: 641–62.

Somps, C., and M. W. Luttges. 1985. Dragonfly flight: Novel uses of unsteady flow. *Science* 228: 1326–29.

Spedding, G. R. 1986. The wake of a jackdaw (*Corvus monedula*) in slow flight. *Journal of Experimental Biology* 125: 287–307.

———. 1987. The wake of a kestrel (*Falco tinnunculus*) in flapping flight. *Journal of Experimental Biology* 127: 59–78.

———. 1992. The aerodynamics of flight. In *Mechanics of Animal Locomotion,* ed. R. McN. Alexander, 51–111. Berlin: Springer-Verlag.

Spedding, G. R., and T. Maxworthy. 1986. The generation of circulation and lift in a rigid two-dimensional fling. *Journal of Fluid Mechanics* 165: 247–72.

Spedding, G. R., J. M. V. Rayner, and C. J. Pennycuick. 1984. Momentum and energy in the wake of a pigeon (*Columba livia*) in slow flight. *Journal of Experimental Biology* 111: 81–102.

Stange, G., and J. Howard. 1979. An ocellar dorsal light response in a dragonfly. *Journal of Experimental Biology* 83: 351–55.

Swisher, C. C., Y.-Q. Wang, X.-L. Wang, X. Xu, and Y. Wang. 1999. Cretaceous age for the feathered dinosaurs of Liaoning, China. *Nature* 400: 58–61.

Taylor, C. P. 1981. Contribution of compound eyes and ocelli to steering of locusts in flight. I. Behavioural analysis. *Journal of Experimental Biology* 93: 1–18.

Tennekes, H. 1996. *The Simple Science of Flight: From Insects to Jumbo Jets.* Cambridge, Mass.: MIT Press. 137 pp.

Thomas, A. L. R. 1996a. The flight of birds that have wings and a tail: Variable geometry expands the envelope of flight performance. *Journal of Theoretical Biology* 183: 237–46.

———. 1996b. Why do birds have tails? The tail as drag reducing flap, and trim control. *Journal of Theoretical Biology* 183: 247–53.

Thomas, D. W. 1983. The annual migration of three species of West African fruit bats (Chiroptera: Pteropodidae). *Canadian Journal of Zoology* 61: 2266–72.

Thorington, R. W. J., and L. R. Heaney. 1981. Body proportions and gliding adaptations of flying squirrels (Petauristinae). *Journal of Mammology* 62: 101–14.

Thornton, I. W. B. 1996. The origins and development of island biotas illustrated by Krakatau. In *The Origin and Evolution of Pacific Island Biotas, New Guinea to Eastern Polynesia: Patterns and Processes,* ed. A. Keast and S. E. Miller, 67–90. Amsterdam: SPB Academic Publishing.

Torre-Bueno, J. R., and J. Larochelle. 1978. The metabolic cost of flight in unrestrained birds. *Journal of Experimental Biology* 75: 223–29.

Tucker, V. A. 1968. Respiratory exchange and evaporative water loss in the flying budgerigar. *Journal of Experimental Biology* 48: 67–87.

———. 1972. Metabolism during flight in the laughing gull, *Larus atricilla. American Journal of Physiology* 222: 237–45.

———. 1987. Gliding birds: The effects of variable wingspan. *Journal of Experimental Biology* 133: 33–58.

———. 1993. Gliding birds: Reduction of induced drag by wing tip slots between the primary feathers. *Journal of Experimental Biology* 180: 285–310.

———. 1995. Drag reduction by wing tip slots in a gliding Harris' hawk, *Parabuteo unicinctus*. *Journal of Experimental Biology* 198: 775–81.

Tucker, V. A., and C. Heine. 1990. Aerodynamics of gliding flight in a Harris' hawk, *Parabuteo unicinctus*. *Journal of Experimental Biology* 149: 469–89.

Tucker, V. A., and G. C. Parrott. 1970. Aerodynamics of gliding flight in a falcon and other birds. *Journal of Experimental Biology* 52: 345–67.

Unwin, D. M. 1997. Pterosaur tracks and the terrestrial ability of pterosaurs. *Lethaia* 29: 373–86.

Unwin, D. M., and N. N. Bakhurina. 1994. *Sordes pilosus* and the nature of the pterosaur flight apparatus. *Nature* 371: 62–64.

Van den Berg, C., and C. P. Ellington. 1997a. The three-dimensional leading edge vortex of a "hovering" model hawkmoth. *Philosophical Transactions of the Royal Society of London*, B, 352: 329–40.

———. 1997b. The vortex wake of a "hovering" model hawkmoth. *Philosophical Transactions of the Royal Society of London*, B, 352: 317–28.

Vest, M. S., and J. Katz. 1995. Unsteady aerodynamic model of flapping wings. *AIAA Journal* 34: 1435–40.

Villard, H. S. 1968. *Contact! The Story of the Early Birds*. New York: Thomas Y. Crowell Co. 264 pp.

Vogel, S. 1967. Flight in *Drosophila*. III. Aerodynamic characteristics of fly wings and wing models. *Journal of Experimental Biology* 46: 431–43.

———. 1981. *Life in Moving Fluids: The Physical Biology of Flow*. 1st ed. Boston: Willard Grant Press. 352 pp. 2d rev. ed. Princeton, N.J.: Princeton University Press, 1994. 467 pp.

———. 1988. *Life's Devices: The Physical World of Animals and Plants*. Princeton, N.J.: Princeton University Press. 367 pp.

———. 1998. *Cats' Paws and Catapults: Mechanical Worlds of Nature and People*. New York: W. W. Norton & Co. 382 pp.

Wagner, G. P., and J. A. Gauthier. 1999. 1, 2, 3 = 2, 3, 4: A solution to the problem of homology of the digits in the avian hand. *Proceedings of the National Academy of Sciences* 96: 5111–16.

Walker, A. D. 1972. New light on the origins of birds and crocodiles. *Nature* 237: 257–63.

Walsh, J. A. 1984. Estimating the burden of illness in the tropics. In *Tropical and Geographical Medicine*, ed. K. S. Warren and A. A. F. Mahmoud, 1073–85. New York: McGraw-Hill.

Warren, W. H., and D. J. Hannon. 1988. Direction of self-motion is perceived from the optical flow. *Nature* 336: 162–63.

Warrick, D. R., and K. P. Dial. 1998. Kinematic, aerodynamic and anatomical mechanisms in the slow, maneuvering flight of pigeons. *Journal of Experimental Biology* 201: 655–72.

Wehner, R. 1996. Middle-scale navigation: The insect case. *Journal of Experimental Biology* 199: 125–27.

Wehner, R., B. Michel, and P. Antonsen. 1996. Visual navigation in insects: Coupling of egocentric and geocentric information. *Journal of Experimental Biology* 199: 129–40.

Weis-Fogh, T. 1956. Biology and physics of locust flight. II. Flight performance of the desert locust (*Schistocerca gregaria*). *Philosophical Transactions of the Royal Society of London,* B, 239: 459–510.

———. 1973. Quick estimates of flight fitness in hovering animals, including novel mechanisms for lift production. *Journal of Experimental Biology* 59: 169–230.

———. 1975. Flapping flight and power in birds and insects, conventional and novel mechanisms. In *Swimming and Flying in Nature,* ed. T. Y. Wu, C. J. Brokaw, and C. Brennan, 2: 729–62. New York: Plenum Press.

Weis-Fogh, T., and R. McN. Alexander. 1977. The sustained power output obtainable from striated muscle. In *Scale Effects in Animal Locomotion,* ed. T. J. Pedley, 511–25. New York: Academic Press.

Weis-Fogh, T., and M. Jensen. 1956. Biology and physics of locust flight. I. Basic principles of insect flight: A critical review. *Philosophical Transactions of the Royal Society of London,* B, 239: 415–57.

Wellnhofer, P. 1975. Die Rhamphorhynchoidea der Oberjura Plattenkalke Suddeutschlands. *Palaeontographica* A148: 1–33; A149: 1–30.

Wells-Gosling, N. 1985. *Flying Squirrels: Gliders in the Dark.* Washington, D.C.: Smithsonian Institution Press. 128 pp.

Welman, J. 1995. *Euparkeria* and the origin of birds. *South African Journal of Science* 91, 10: 533–37.

Wiest, F. C. 1995. The specialized locomotory apparatus of the freshwater hatchetfish family Gasteropelecidae. *Journal of Zoology* 236, 4: 571–92.

Wigg, D. 1993. *And Then Forgot to Tell Us Why . . . : A Look at the Campaign against River Blindness in West Africa.* Washington, D.C.: World Bank. 44 pp.

Williams, C. B. 1958. *Insect Migration.* New York: Macmillan Co. 235 pp.

Williams, J. D., and R. M. Nowak. 1993. Vanishing species in our own backyard: Extinct fish and wildlife of the United States and Canada. In *The Last Extinction,* ed. L. Kaufman and K. Mallory, 115–48. 2d ed. Cambridge, Mass.: MIT Press.

Willmot, A. P., C. P. Ellington, and A. L. R. Thomas. 1997. Flow visualization and unsteady aerodynamics in the flight of the hawkmoth, *Manduca sexta. Philosophical Transactions of the Royal Society of London,* B, 352: 303–16.

Wiltschko, R. 1991. The role of experience in avian navigation and homing. In *Orientation in Birds,* ed. P. Berthold, 250–69. Basel: Birkhäuser Verlag.

Wiltschko, R., and W. Wiltschko. 1981. The development of sun compass orientation in young homing pigeons. *Behavioral Ecology and Sociobiology* 9: 135–41.

Wiltschko, W., P. Daum, A. Fergenbauer-Kimmel, and R. Wiltschko. 1987. The development of the star compass in garden warblers, *Sylvia borin. Ethology* 74: 285–92.

Wiltschko, W., and F. W. Merkel. 1966. Orientierung zugenruhiger Rotkehlchen im statichen magnetfeld. *Deutsche Zoologische Gesellschaft Verhandlungen* 59: 362–67.

Wiltschko, W., and R. Wiltschko. 1972. Magnetic compass of European robins. *Science* 176: 62–64.

Winston, M. L. 1987. *The Biology of the Honey Bee.* Cambridge, Mass.: Harvard University Press. 281 pp.

Withers, P. C. 1992. *Comparative Animal Physiology.* Fort Worth, Tex.: Saunders College Publishing. 949 pp.

Wootton, R. J. 1990. Major insect radiations. In *Major Evolutionary Radiations,* ed. P. D. Taylor and G. P. Larwood, 187–208. Oxford: Clarendon Press.

Wootton, R. J., and C. P. Ellington. 1991. Biomechanics and the origin of insect flight. In *Biomechanics in Evolution,* ed. J. M. V. Rayner and R. J. Wootton, 99–112. Cambridge: Cambridge University Press.

Zanker, J. M. 1990. The wingbeat of *Drosophila melanogaster.* I. Kinematics. *Philosophical Transactions of the Royal Society of London,* B, 237: 1–18.

Zimmerman, C. H. 1932. Characteristics of Clark Y airfoils of low aspect ratios. *National Advisory Committee on Aeronautics* 431: 1–12.

drag (*cont.*)
153; Reynolds number and, 17; size and, 14; speed and power, 154; stall, *27;* swimming, 150; thrust to overcome, 78; total, 13, 30; viscous, and small body size, 157; viscous or friction, 13, 30, 41
drag coefficient, 15; parachuting seeds, 48; polar diagram, *31;* reference areas, 15, 16
dragonflies: Carboniferous giants, 176; head of, as gravity detector, 134, *135;* hovering, *101,* 103; slope soaring, 68; wing cross section, *34*
Dutch elm disease, 275

eagles, 159, 282; aspect ratio, 62; migration, *222;* reintroductions, 284; wing loading, 64
earwigs, wing folding, 5
echolocation, 212
ectothermic (cold-blooded) animals, 148, 149; standard metabolic rate, 146
Edgerton, Harold, 43
efficiency, walking, 151
Ellington, Charles, 96, 98, 173
Elmqvist, Thomas, 286
elms, 275
Emlen cage (Emlen funnel), *251*
endothermic (warm-blooded) animals, 148; advantages, 149; basal metabolic rate, 146
endothermy: *Archaeopteryx,* 202, 203; flight and, 149; pterosaurs, 189
energetics: locomotion, 148, 150, 151, 162; migration, 223–25; running *vs.* flying, *163*
energy: carbohydrates, 160, 161; in fat, 161; fat for migration, 224, 226; in food, 141; heat from, conversion, 141; high-power, low-cost paradox of flight, 162, *163;* hovering, 157; metabolic rate, 141; migration, *224;* minimum-power and maximum-range speeds, 154, *156;* and muscle, 142, 144; transferred in spermatophore, 162; work and, 140
equilibrium sense, and stability, 132

Eulerian viewpoint, 9
evolution, 2; of complex structures, 167; convergent, 294, 295; extinct birds, 209; insects, *179;* limitations of fossil record, 168; natural selection, 2; pterosaurs, 190–93. *See also* flight evolution
Exocoetidae (flying fish), 43
extinction: Great Auk, 280; passenger pigeon, 281

falcons: gliding, *65;* migration, *222;* peregrine, 281, 282, 284; stoop (high-speed dive), 129
Falealupo Preserve (Samoa), 285–87
fat: energy content, 161; for migration, 226
feathers, 102, *201; Archaeopteryx,* 195, 199, *201,* 202; separated primaries, 62, *63;* wing flexion and, 83; wing structure, 4
fibrils (of muscle cells), 143
figs: dispersed by flying animals, 262; pollination, 273
fig wasps, 273
fish: dispersal to ponds, 264; flying, *43,* 44; hatchetfish, 107–8; lift-based swimming, 104
flapping: active upstroke, 83; amplitude, *125,* 128; *Archaeopteryx,* 201; asymmetrical downstroke and upstroke, 74, 75, 81, *82, 84, 85;* clap-fling mechanism, *94;* downstroke forces, *78;* as evolutionary innovation, 215; forces along span, *80;* frequency, 89; *vs.* helicopter rotor, 71; high-speed photography of, 72; large body size, 159; muscles and linear motion, 108; pigeon wing movements, *84;* pterosaurs, 186–90, 192; relative wind along span, 79; *vs.* rotary motion of engines, 307; *vs.* swimming, *104, 105;* symmetrical stroke, *81;* thrust, *82;* tip path, 72–74, *73;* turns, 124–26; upstroke *vs.* downstroke duration, 74. *See also* gaits
flapping machine: ornithopter, 108–10; stationary, for research, 98
Flettner rotor, 22; samaras (gliding seeds), 54

viscous drag, 13, 30
vision, and stability, 130–31
volcanoes (recolonization after eruption), 262
vortex: bound, 23–25, *23, 24;* starting, *29,* 94; system of, on wing, 28–29, *29;* theory, 96–98; tip or trailing, *29,* 30, 32, 62
vortex wake, *29;* gaits, *76,* 96; gaits, insects, *77*
vultures: aspect ratio, 62; lift-to-drag ratio, 34, 62; *vs.* sailplanes, 41; sinking speed, *57;* soaring, 36, 56–57, 59, 66, 159; wing loading, 64

Wagner effect, 94
wake: stall, *27;* streamlining and, *14;* turbulent, 13; vortex, *29,* 96
walking: cost of locomotion, 151; flying *vs.,* migration, 225; pterosaurs, *185*
wasps: fig pollination, 273; fossil record, 181
weight, *10,* 12; body fat index, 227; flying, 151; gliding, *38, 39;* lift in turns and, *123;* metabolic rate and, 146; migration, and fat storage, 224; samara (gliding seed), 53; shifting (turns), 127; swimming, 150; terminal velocity, 128; walking, 151
wetted area, 15
whales, swimming, 103
wheatears, migration routes of, 256
whooping cranes, 284; migration route, 257, *258*
wind: penetration (soaring), 64; samaras (gliding seeds) and, 55–56; sensing, 132–34
wind tunnels: desert locusts, 91; to study wing movements, 72
wing, 19–20; airfoil of, 19; area, 82; leading and trailing edges, 19; terminology, 20. *See also* airfoil: camber
wing articulation: hovering, 101; insects, 5
wing-beat amplitude, *125*
wing-beat frequency, 89

wing-beat pattern: delayed stall, 96; normal hovering, 101; tip path, 72–74, *73*
wing-flapping device, mechanical, 98
winglets (insects). *See* flight evolution, insect; insect wings
wing loading: birds, 47; *Draco* and flying squirrels, 47; gliding, 39; maneuvering, 139; pterosaurs, 67; soaring, 63–64; speed, 87
wings: amphibian, 105; animal, 3–5, *4,* 290; animal and airplane, 71, 290, 308; bats and birds, 3; insect, cross sections, *34;* insect, veins, *4, 5;* mission-adaptive airplane wing, 301; pterosaurs, 3, *4;* scaling and materials, 307
wing shape: airfoil, 33; albatrosses, 62; control of, 5; crescent- or scimitar-shaped, 35; and lift, 32; planform, 34
wing-stroke angle, 72
wing structure: *Alsomitra* (flying cucumber), 49; animal, *4;* auks, 107; axillary sclerites, 5; bats, 3, *4;* birds, 3, *4;* camber of *Archaeopteryx* wing, 201; *Draco,* 44; hawks and vultures, 62; insect, *4, 5, 34, 166;* pterosaur, 3, *4, 188;* samaras (gliding seeds), 53–55; separated primaries, *63;* stabilizing, in simple gliders, 50; vulture *vs.* sailplane, 41
wing-tip path, 72–74, *73, 74;* hovering, *100, 101*
wing warping: Gossamer Condor, 304; Wrights' control system, 291
Wootton, Robin, 173
work: force and energy, 140; locomotion, 141. *See also* energy
Wright brothers, 290; banked turns, 121, *291*

yaw, 113; hovering turns and, 127
yellow fever, 277

Zanonia. See *Alsomitra macrocarpa* (flying cucumber)
Zimmerman, C. H., 47